AF705232

The Clay Lamps from Ancient Sepphoris

Duke Sepphoris Excavation Reports

Series Editors: Eric M. Meyers
Carol L. Meyers
Benjamin D. Gordon

Volume I	Edited by Eric M. Meyers and Carol L. Meyers, with contributions by Marva Balouka, Anna de Vincenz, and the editors	*The Pottery from Ancient Sepphoris*
Volume II	Eric C. Lapp	*The Clay Lamps from Ancient Sepphoris: Light Use and Regional Interactions*

The Clay Lamps from Ancient Sepphoris

Light Use and Regional Interactions

Eric C. Lapp

Winona Lake, Indiana
Eisenbrauns
2016

© 2016 by Eisenbrauns Inc.
All rights reserved
Printed in the United States of America

www.eisenbrauns.com

Library of Congress Cataloging-in-Publication Data

Names: Lapp, Eric C.
Title: The clay lamps from ancient Sepphoris : light use and regional interactions / Eric C. Lapp.
Description: Winona Lake, Indiana : Eisenbrauns, 2016. | Series: Sepphoris excavation reports ; volume 2 | Includes bibliographical references. | Description based on print version record and CIP data provided by publisher; resource not viewed.
Identifiers: LCCN 2015044274 (print) | LCCN 2015035627 (ebook) | ISBN 9781575064055 (PDF) | ISBN 9781575064048 (hardback : alkaline paper)
Subjects: LCSH: Sepphoris (Extinct city)–Antiquities. | Ceramic lamps–Israel–Sepphoris (Extinct city) | Lamps, Ancient–Israel–Sepphoris (Extinct city) | Ceramic lamps–Israel–Sepphoris (Extinct city)–Catalogs. | Lamps, Ancient–Israel–Sepphoris (Extinct city)–Catalogs. | Pottery, Ancient–Israel–Sepphoris (Extinct city) | Excavations (Archaeology)–Israel–Sepphoris (Extinct city) | Lighting–Social aspects–Israel–Sepphoris (Extinct city) | Sepphoris (Extinct city)–Social life and customs. | Sepphoris (Extinct city)–Commerce–History.
Classification: LCC DS110.S43 (print) | LCC DS110.S43 L36 2016 (ebook) | DDC 933/.45–dc23
LC record available at http://lccn.loc.gov/2015044274

The paper used in this publication meets the minimum requirements of the American National Standard for Information Sciences–Permanence of Paper for Printed Library Materials, ANSI Z39.48-1984.⊗™

No benediction may be said over a lamp
until one can enjoy its light. – m. ber. 8.6

In dedication to my parents,

Nancy and Robert

Contents

List of Illustrations

Figures

Plates

Seventeen plates of lamp drawings, interspersed with photographs, showing 155 of 219 lamps of the Sepphoris corpus. [Note: the catalog lists 217 items, but two of them have an "a" and a "b," for a total of 219 items.]

Tables

Photographs

Preface

In classical and late antiquity, Sepphoris was a cosmopolitan city situated at the crossroads of major trade routes leading from what Josephus describes as the "foreign nations" along the coastal area, Golan, and Decapolis cities to the East. There, Rabbi Judah ha-Nasi ("the Prince") and his rabbinic colleagues codified the Mishnah, one of the great literary works of the ancient world. They resided among a predominantly Jewish population immersed in a Greco-Roman culture that included a variety of performances and games in the city's theater, a zodiac in which the sun-god Helios is portrayed in a synagogue mosaic, and clay oil lamps decorated with mythological and erotic themes. A "culture of light" existed in Sepphoris, and mishnaic texts illuminate that use of lamps in ancient daily life as do the lamps unearthed at the site. Although it is not possible to identify the religious identity of the persons who purchased them, the lamps do reflect the mass clientele and market for which the lighting vessels were intended. The lamps excavated at Sepphoris provide a unique opportunity to learn about Roman, Jewish, Christian, and Islamic art in this Galilean city.

The contents of chap. 2 are a revised and expanded version of my Duke University M.A. thesis, *A Chronological-Typological Study of the Terracotta Oil Lamps from Sepphoris in the Lower Galilee* (1991). Additional lamp finds are included as a result of further excavations at Sepphoris that postdate the completion of my thesis. The contents and data of chap. 4 first appeared in my Duke doctoral dissertation as "Chapter Four: A Regional Survey of the Petrographic Characteristics and Chemical Compositions of Roman Palestinian Clay Lamp Pastes" (1997), supervised by Professor Eric Meyers.

The focus of the work is the typological identification and dating of the clay lamps recovered by the archaeological excavations of the western summit of Sepphoris by the Joint Sepphoris Project (Duke University, Hebrew University; Eric Meyers, Carol Meyers, Ehud Netzer, directors) and the Sepphoris Regional Project (Duke University; Eric Meyers and Carol Meyers, directors). The lamps were unearthed in the residential areas of the western summit. The study further concentrates on the relationship between the lamps found at Sepphoris and those recovered from other cities located in Roman Galilee and the Decapolis. The archaeometric component explores and compares the petrographic and micromorphological features of the clay fabrics of lamp samples belonging to the Palestinian discus, Beit Nattif, bilanceolate, Gilead, and Tarsus types from archaeological sites in both Israel and Jordan.

Prior to the results of the petrographic and trace-element analysis that appeared in my dissertation in 1997, few characterization studies conducted in the region included clay lamps among the samples analyzed. Rather, they focused on household wares and other pottery types. Because petrographic analysis of lamp fabrics at the time comprised less than a handful of samples, I realized the necessity of petrographically analyzing and chemically sourcing lamp samples collected at multiple sites in multiple regions.

This book would not have been possible without the gracious support and help by the following people and institutions. I am most appreciative of the financial funding awarded me by the trustees of the W. F. Albright Institute for Archaeological Research (AIAR) in Jerusalem, Israel (1992–93 Samuel H. Kress Fellowship; 1994–95 United States Information Agency Junior Fellowship); the American Center of Oriental Research (ACOR) in Amman, Jordan (1993 United States Information Agency Junior Fellowship); and the American School of Classical Studies at Athens (1996–97 Samuel H. Kress Joint Athens–Jerusalem Fellowship). I further extend my gratitude to the Dorot Foundation, the Samuel H. Kress Foundation, and the United States Information Agency for sponsoring my research.

The direct current plasma-optical emission spectrometry and petrography of this study were conducted by me in the laboratory of Emily Klein at Duke University in Durham, North Carolina. All DCP sample powders, petrographic thin-sections, and accompanying tables were further prepared by me. The geochemical component of this study was sponsored in part by a grant from the Duke Research Council held under the joint auspices of Eric Meyers and Emily Klein during the 1994–95 academic year.

The archaeometric analysis of this study would not have been possible without the kind permission of Ruth Peleg of the Israel Antiquities Authority and of Safwan Tell and Ghawzi Bisheh of the Department of Antiquities of Jordan. I am further indebted to the following archaeologists in Israel and Jordan who graciously allowed me to study the finds from their excavations and to analyze samples as required: Lawrence Stager and Barbara Johnson (Ashkelon), Rachel Bar-Nathan and Marc Balouka (Beth Shean/Scythopolis), the late Basil Hennessey and Kate da Costa (Pella), the late Harold Mare (Abila), Suzanne Kerner (Umm Qeis/Gadara), Ina Kehrberg and the late Anthony Ostraz (Gerasa/Jerash), and Cherie Lenzen (Beit Ras/Capitolias).

In Israel, I acknowledge the kindness of the late Ehud Netzer and Zeev Weiss of the Hebrew University of Jerusalem. Marva Balouka was most gracious and helpful, and I appreciated our archaeological discussions at the Mt. Scopus campus. Archaeometrists David Adan-Bayewitz, Yuval Goren, Naomi Porat, and Joseph Yellin were very supportive in providing me perspectives on my study, and Joe Zias was always there to help. While a fellow at the W. F. Albright Institute, I further appreciated the insights of Director Seymour Gitin.

In Jordan, I appreciate greatly Abdalla Kasem Abu Aloush and Ahmed Omri for kindly showing me the local clay sources at Umm Qeis/Gadara, and Sultan Shreidah of the local Department of Antiquities in Irbid who permitted me to examine the oil lamps from the Beit Ras/Capitolias excavations. I am most grateful to my fellow lychnologists, Kate da Costa and Ina Kehrberg, for our many "lamp chats" so long ago. I thank Khairieh ʿAmr for her useful points about ceramic sourcing. I also wish to thank Stefanie Lang for her support in Jordan and for her generous hospitality and assistance during my return to Cologne to reexamine the lamp collection of the Römisch-Germanisches Museum. Thomas Parker of the Department of History at North Carolina State University has especially provided me with unconditional support of this study and my work concerning the archaeology of Roman Jordan.

My research on the topics covered in this book were enriched through conversations and travels with my friends and colleagues Jeffrey Blakely, Aaron Brody, Nicoló Bucaria, Shimon Gibson, Ann Killebrew, Carsten Körber, Keith Norman, Hani Nur al-Din, Nicola Stacey, and Leonard Rutgers. I owe John and Kris Jorgensen and Catherine Severns a special thank you. Benjamin Gordon's assistance throughout the preparation of this volume was most appreciated. Jamie Brummitt and Kirstin Rose worked diligently on the lamp database and illustrations archive. A grand thank you is extended to Ravit Nenner-Soriano for her splendid work in arranging the plates with the drawings and interspersed photographs.

I gratefully acknowledge the guidance of Emily Klein of the Nicholas School of the Environment at Duke University. Her geochemical instruction with respect to my preparation of samples and application of dcp-oes in her laboratory at Duke was most appreciated. Allan Boudreau was also quite gracious in permitting me to use his petrographic microscope. I further express my gratitude to Bill and Ellen Meuer and Jack Eggleston, who were graduate students in Duke's geology department at the time of my study. I warmly thank Bruce Marsh of the Department of Earth and Planetary Sciences at the Johns Hopkins University for his support of my current petrographic analyses. Tushar Mittel, formerly of the Department, was also most helpful.

I am most grateful to Profs. Eric Meyers and Carol Meyers of the Department of Religious Studies at Duke University for their encouragement and support of this work. Their devotion to teaching and scholarship strengthened my resolve.

Finally, I extend my warmest gratitude to my wife, Catherine, and to my sons, Louis and Thomas, whose *joie de vivre* guided me through the final preparation of this volume.

Eric C. Lapp
Baltimore, Maryland
October 2015

CHAPTER 1

Introduction

1. Volume Description

The present volume presents a substantial and representative corpus of the clay lamps (photo 1; plates) unearthed on the western summit (residential area) of Sepphoris by the Joint Sepphoris Project (JSP) and the Sepphoris Regional Project (SRP) in the 1980s and 1990s (fig. 1, photo 2).[1] Several other teams have excavated areas of the ancient city: the University of Michigan (1931), the Hebrew University of Jerusalem, Tel Aviv University, and the University of South Florida. This book is the second volume of the Duke Sepphoris Excavation Reports series. It is intended for use in conjunction with vol. I of the series, *The Pottery from Ancient Sepphoris* (Meyers and Meyers 2013b). Chronological designations and dating used in the lamp volume follow those discussed in vol. I.

This chapter provides introductory information. Chapter 2 reports on the various local and regional lamp types represented in the corpus. Chapter 3 presents the imported types recovered from the western summit, including forms from Greece and western Asia Minor. Chapter 4 details the petrographic findings and trace-element results regarding samples collected at Sepphoris and other cities of the Galilee, coastal plain, and Decapolis. Chapter 5 examines the trade, transportation process, and marketing of lamps in Sepphoris. The majority of lamp finds suggest local and regional manufacture. Among these are lighting vessels carrying important implications for Jewish art in the Late Roman period (fig. 2; see discussion in chap. 6). That most of the lamps were recovered from domestic units should further contribute to research on artificial lighting used in the daily lives of the residents of this important Galilean urban center and their interactions with communities both near and far.

---

1. The Joint Sepphoris Project (JSP) excavated the site in 1985–89, the Sepphoris Regional Project in 1993–97. The Sepphoris Acropolis Project (Jonathan Reed, director; Eric Meyers and Carol Meyers, senior advisers) took place in 2000, yet no lamps from that project happen to be included in this volume. Moreover, no lamps from JSP excavation areas that have been or will be published under the auspices of the Hebrew University are published in this volume. For a history of the excavations on the western summit, see Meyers, Meyers, and Gordon forthcoming.

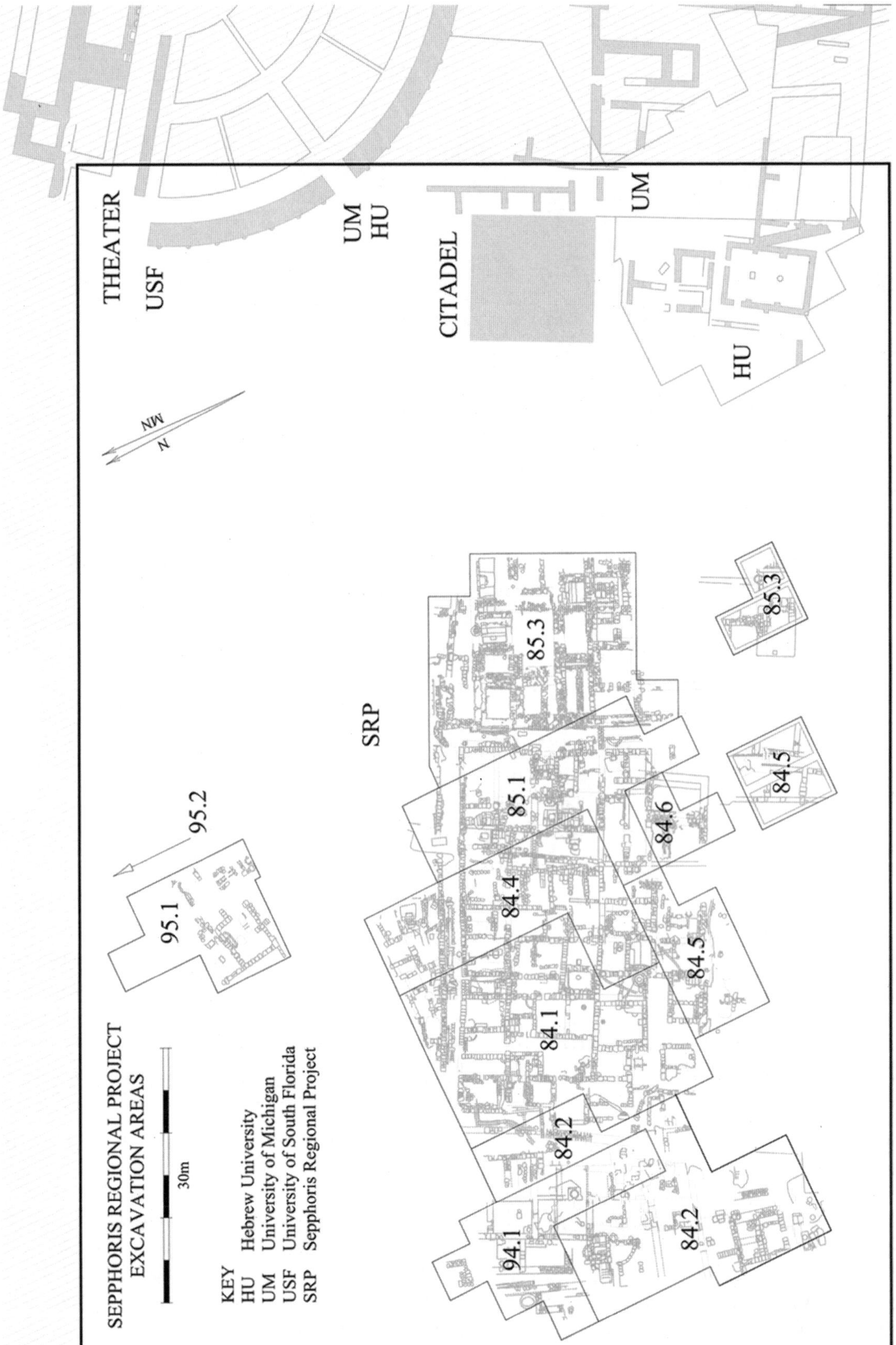

Fig. 1. Plan of western summit.

Photo 1. Clay lamps of various periods recovered from the western summit. Courtesy of Eric and Carol Meyers.

2. *Lamp Profile*

Archaeological sites where clay lamps are excavated have a "lamp profile." The profile refers to the shape-types of local, regional, and imported lamps. By "reading" the lamp profile at any given site (Hodder and Hutson 2003: 166–70), lychnologists–artificial lighting and lamp specialists–and archaeologists alike can piece together patterns of demographic shifts, regional trade connections, and, with respect to the symbols found on lamps, the market-clientele (Lapp 2007). As the findings of this volume show, the Sepphoris lamp profile in many ways resembles the profiles of its urban counterparts at Dor, Caesarea Maritima, and Beth Shean. Shape-type diversity is most pronounced among the regional forms found at Sepphoris (table 1).

3. *Count*

Lychnologists routinely determine the quantities of complete lamps, nearly complete lamps, and lamp fragments of various types unearthed at archaeological sites (see especially da Costa 2003: 2.689–744, appendix 4A; also Hadad 2002: 152, table 1; Magness 2008: 132, table 5.1). By knowing the quantities of various lamp types found at a site, we can construct more accurate distribution maps and gain a greater understanding of regional trade links and demographic shifts (Lapp 1997a:

Photo 2. Western summit of Sepphoris, looking west. Courtesy of Eric and Carol Meyers.

302, fig. 6; 426, fig. 160; da Costa 2010: 84–86, figs. 27–29). A count was conducted on the lamp evidence from Sepphoris, and the results are reported in table 2 under the headings "complete," "nearly complete," and "fragments."

Although lamp counts are useful, they do pose special challenges. First, unless the entire geographical extent of an archaeological site has been excavated, lamp counts do not fully and accurately represent the quantity of lamps and fragments that actually existed at the site, for many still remain to be discovered. For example, the recovery of a few sherds belonging to a specific type at a site may lead to a misleading interpretation as to the actual quantity yet to be found. An unexcavated area of a site where one or more cisterns are located may contain numerous quantities of various lamp types. A relatively new procedure uses mean fragment weight to determine the estimated quantity of preexistent complete lamps at any given site (Barrett 2008: 52, 62–63). Barrett's analysis has shown that the numerous lamp sherds recovered at Petra constitute considerably fewer estimated, preexistent complete lamps by weight which, in turn, can lead to overinterpretation of lamp counts (Barrett 2008: 52, 62–63).

4. *Findspots*

Complete, nearly complete, and fragments of clay lamps (e.g., photo 1) were recovered from nine areas of the western summit (fig. 1) excavated by the Joint Sepphoris Project and the Sepphoris Regional Project, which are now being published

Table 1. Lamp Types from the Western Summit

Type	*Name*	*Other Nomenclature and Date*
Local and Regional		
1	Mold-Made Delphiniform with Kite-Shaped Body and Double Side Lugs (cat. nos. 1–4)	*Date Range:* Second half of second and beginning of first centuries B.C.E. (Rosenthal and Sivan 1978: 13); end of third to second centuries B.C.E. (Hayes 1980: 22–23, pls. 9.84, 85; 16.161); probably late second or first century B.C.E. (Bailey 1975: 272–73, Q592.EA, pl. 112); third or fourth quarter of third century B.C.E. to fourth quarter of second century B.C.E. (Lyon-Caen and Hoff 1986: 53, nos. 115–16)
2	Mold-Made Delphiniform / A-Prime Levantine (cat. nos. 5–6, 8–12)	*Date Range:* Second century B.C.E. to first half of first century B.C.E.; second and first centuries B.C.E. (Lyon-Caen and Hoff 1986: 57, nos. 128–31); 200 to 100 B.C.E., extending perhaps as late as 75 B.C.E. (Młynarczyk 1997: 110, table 2); mid-second to mid-first centuries B.C.E., and possibly extending to end of first century B.C.E. (Rosenthal and Sivan 1978: 13)
	Variant 1 (cat. no. 7)	*Date Range:* Second century B.C.E. to first half of first century B.C.E. (Hadad 2002: 13–14, Type 2, no. 3)
	Conical-Shaped Nozzles (cat. nos. 13–17)	*Date Range:* Probably Hellenistic, second–first centuries B.C.E.
3	Northern Collar Neck/Galilean Collar Neck (cat. no. 18)	*Other Nomenclature:* Ginnosar Type; Kinneret Type; Fernandez Type L1; Gärtner Early Roman Type 2.2
		Date Range: Mid- to late-first century B.C.E. to early first century C.E., about 50 B.C.E.–50 C.E. (Sussman 1990: 97–98)
4	Herodian (cat. nos. 19–43)	*Other Nomenclature:* Kennedy Type 3; Gärtner Early Roman Type 2.3
		Date Range: Late first century B.C.E. to mid-second century C.E. (Rosenthal and Sivan 1978: 80 nn. 63–64; Barag and Hershkovitz 1994: 44–47, 50–51, and 53–56; Lapp 1997: 17 n. 4, 21–22; late first century B.C.E. to first century C.E., to 135 C.E. (Sussman 2008: 219–21, nos. 19–23)
5	Sepphorean Spatulate (cat. nos. 44–52)	*Other Nomenclature:* Huqoq Type (d) (Kahane 1961: 138, "Herodian" mold-made); Gärtner Early Roman Type 2; Sussman RH4 (Sussman 2012: 92–94, fig. 66.1, 3)
		Date Range: Mid-first to early second centuries C.E. (Kahane 1961: 168); around 70 C.E., or slightly earlier (Sussman 2012: 94)
6	Darom (cat. nos. 53–70)	*Other Nomenclature:* Kennedy Type 7; Molded Judean; Gärtner Early Roman Type 2.4
		Date Range: Second half of first century C.E. to first half of second century C.E., or from about 70 C.E. to 135 C.E. (Rosenthal and Sivan 1978: 82; Sussman 1982: 16; Lapp 1997a: 34–39, figs. 19–20); 70–135 C.E. (Sussman 2008: 222, nos. 28–29, Samaria type); 70–135 C.E. (Hadad 2002: 16, nos. 13–18)

Table 1. Lamp Types from the Western Summit (cont.)

Type	*Name*	*Other Nomenclature and Date*
	Darom, Square-Bodied (cat. nos. 60–61)	*Other Nomenclature:* Kennedy Type 7
		Date Range: Second half of first century C.E. to first half of second century C.E., or from about 70 C.E. to 135 C.E. (Rosenthal and Sivan 1978: 82; Sussman 1982: 16; 2008: 222; Lapp 1997b: 34–39, figs. 19–20)
7	Molded Judean Floral (cat. nos. 71a, 72–73)	*Other Nomenclature:* Barag and Hershkovitz Type D I–III, Molded Lamps with Floral Decorations; Fernandez Type L 6
		Date Range: Second half of first century C.E., about 50–66 C.E. (Barag and Hershkovitz 1994: 69)
8	Monolychnoi, Ring-Lamps with Elongated Nozzles with Circular Heads (cat. nos. 74–76)	*Date Range:* First century C.E.
9	Polylychnoi, Box-Like Rectangular Lampstand (cat. no. 77)	*Date Range:* First century C.E.
10	Palestinian Discus (cat. nos. 78–142)	*Other Nomenclature:* Kennedy Type 5; Palestinian round with decorated discus; Broneer Type XXV, version; Elgavish Type 20, Fernandez Types L 9.1 a–c; Lapp Type 85; Rosenthal-Heginbottom 1995: Type 26 A–H; Sellers-Baramki 1953: Type II; Gärtner Early Roman Type 2.5
		Date Range: Late first–third centuries C.E. (Lapp 1997: 39; da Costa 2003: 1.50; last third of first to third centuries C.E. (Sussman 2008; 229–331, nos. 71–86); Hadad 16–20, Type 7, Variant 1, nos. 19–29, late first [after 70] through second century C.E.)
11	Syrian Discus (cat. nos. 143–44)	*Other Nomenclature:* Loeschke Type VIII
		Date Range: Third–fourth centuries C.E. (Sussman 2008: 234–36, nos. 104–17)
12	Jebel Jofeh (cat. nos. 145–54)	*Other Nomenclature:* Kennedy Type 8; Gärtner Early Middle–Late Roman Type 3.11
		Date Range: Mid-third–mid-fourth centuries C.E. (Hadad 2002: 22–24, Type 13, nos. 38–45); third–fourth centuries C.E. (Sussman 2008: 238, no. 124)
13	Northern Stamped (cat. nos. 155–61)	*Other Nomenclature:* Gärtner Early Middle–Late Roman Type 3.5
		Date Range: Late third–fifth centuries C.E. (Sussman 1989: 42); Fourth century C.E. (Bailey 1988: 281, Q2313–2320)

Table 1. Lamp Types from the Western Summit (cont.)

Type	*Name*	*Other Nomenclature and Date*
14	Beit Nattif, Northern Versions (cat. nos. 162–66)	*Other Nomenclature:* Kennedy Type 11; Hadad Type 17; Elgavish Types 34, 37; Gärtner Early Middle–Late Roman Type 3.6
		Date Range: Second half of third century C.E. to fifth century C.E.; fourth–fifth centuries C.E. (Hadad 2002: 26, 30–35, Type 17, nos. 74–113); third/fourth–fifth centuries C.E. (Sussman 2008: 238–39, nos. 125–26); end of third–late fifth/early sixth century C.E. (da Costa 2003: 1.68; correspond to da Costa's bow-shaped nozzle type); second half of third to fifth century C.E. (Magness 2008: 130)
15	Gilead (cat. no. 167)	*Other Nomenclature:* None
		Date Range: Fourth–fifth centuries C.E. (da Costa 2003: 1.97, 100)
16	Bilanceolate (cat. nos. 168–72)	*Other Nomenclature:* Kennedy Type 10; Hadad Type 16; Fernandez Type L 12; Gärtner Early Middle–Late Roman Type 3.14
		Date Range: Third to early fifth centuries C.E. (da Costa 2003: 1.57, 66–67); third–fourth centuries C.E. (Rosenthal and Sivan 1978: 110–11); fourth–fifth centuries C.E. (Hadad 2002: 26, Type 16, nos. 53–73)
17	Galilean (cat. nos. 173–76)	*Other Nomenclature:* Hadad Type 22; da Costa "Galilean"; Loffreda Type l6; Gärtner Byzantine Type 4.5
		Date Range: Fifth to early sixth centuries C.E. (Hadad 2002: 56–61, Type 22, nos. 255–70)
18	Caesarean Discus (cat. nos. 177–81)	*Other Nomenclature:* Caesarea 1 (Sussman 2008: 245–47, nos. 170–87); Hadad Type 23
		Date Range: Fourth–fifth centuries C.E. (Sussman 1980: 76); mid-fourth and fifth centuries C.E. (Sussman 2008: 245–47, nos. 170–87); fourth–sixth centuries C.E. (Israeli and Avida 1988: 112); sixth–seventh centuries C.E. (Vine and Hartelius 2000: T26A); fifth–sixth centuries C.E. (da Costa 2003: 1.177, 180–81); fifth to early sixth centuries C.E. (Hadad 2002: 61, Type 23, nos. 271–72)
19	Galilean Discus (cat. nos. 182–88)	*Date Range:* Fourth century C.E. (MacDonnell 1988: 119–20, no. 101, fig. 6-7, pl. 6-4)
20	"Sepphorean" Discus (cat. nos. 189–208)	*Other Nomenclature:* Gärtner Middle–Late Roman Type 3.8
		Date Range: Fourth century C.E. (MacDonnell 1988: 135–36, no. 98 = Sepphorean Round-bodied with Large Filling Hole, fig. 6-7, pl. 6-4)

Table 1. Lamp Types from the Western Summit (cont.)

Type	*Name*	*Other Nomenclature and Date*
21	Jerash with Zoomorphic or Tongue Handle (cat. no. 209)	*Other Nomenclature:* With zoomorphic handle: Scholl Group III; da Costa Jerash; Group 4, Lamp with High Curved Handle (Rosenthal and Sivan 1978: 139, no. 574); Hadad Type 29; Gärtner Early Islamic Type 5.1
		Date Range: With zoomorphic handle: end of first half of seventh to mid-eighth centuries C.E. (Scholl 1991: 66, 69–72; end of sixth and first half of eighth centuries C.E. (Kehrberg 1989: 88f); mid-(?) to late sixth to ninth centuries C.E. (da Costa 2003: 1.192, 196); Hadad 2002: 71.
22	Slipper (Hadad Type 30) (cat. nos. 210–12)	*Other Nomenclature:* Elgavish Type 58
		Date Range: Umayyad, sixth to mid-seventh centuries C.E. (Hadad 2002: 70–72, Type 30, nos. 313–16)
23	Slipper (Hadad Type 36) (cat. no. 213)	*Other Nomenclature:* Dobbin's Type 27
		Date Range: Early eighth to early Abbasid period (Hadad 2002: 82–95, Type 36, nos. 356–419)
24	Islamic with Large Filling-Hole and Tongue Handle (cat. no. 214)	*Other Nomenclature:* Resembles Hadad Type 37
		Date Range: End of eighth or ninth century C.E.
Imported		
25	Ephesus, Broneer Type XIX (cat. nos. 215–16)	*Other Nomenclature:* Howland Type 49A
		Date Range: Late second century B.C.E. to early first century C.E. (Hayes 1980: 15); first century B.C.E. to early first century C.E. (Bailey 1994: 86); mid- to second half of second century B.C.E. to first century B.C.E. (125–75 B.C.E.; Sussman 2008: 215, nos. 5–7)
26	Corinthian, Broneer Type XXVIIC (cat. no. 217)	*Date Range:* Second half of second to third centuries C.E. (Bailey 1988: 402)
27	Athenian, Broneer Type XXVIII, Later Version (cat. nos. 218–19)	*Date Range:* Late fourth to early fifth centuries C.E. (Karivieri 1996: 51–52)

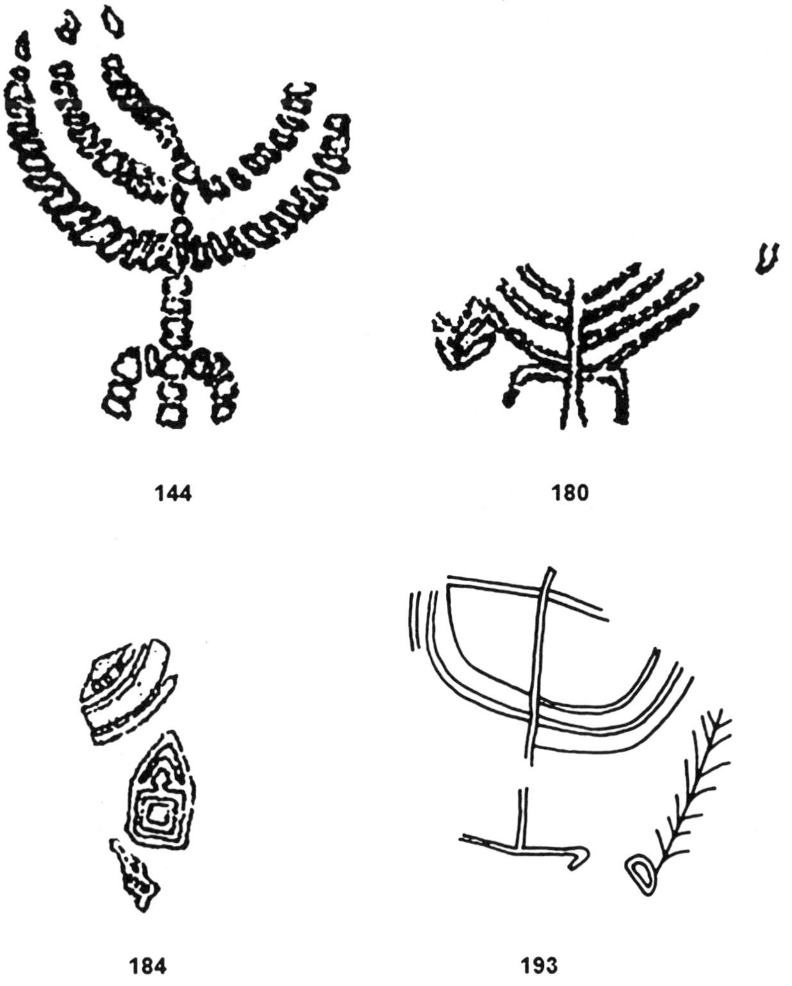

Fig. 2. Jewish symbols: images of the menorah on discus lamps.

by the Duke team: 84.1, 84.2, 84.4, 84.5, 84.6, 85.1, 85.3, 94.1, and 95.1; see appendix A for the context of many of the lamps in this catalog. The lamp evidence was recovered from secondary depositional contexts, primarily fill found in domestic rooms, cisterns, and *miqva'ot*. The most significant deposit of clay lamps and lamp fragments was in L84.1068, a ceramically rich fill in a large cistern that was used as a dump by the end of the Middle Roman period. The lamps and other pottery date to the Early and Middle Roman periods with a few fourth-century wares. The numismatic evidence recovered from the cistern does not extend beyond the fourth century. The latest is a Constantinopolis issue, reg. no. 84.1068.03C (Ariel and Bijovsky forthcoming: IAA 113714), dated 383–395 C.E. That the cistern yielded numerous Palestinian discus lamps and fragments decorated with non-Jewish imagery, including mythological themes and erotic scenes, has implications regarding Roman and Jewish interactions. Bronze figurines of Pan and Prometheus were found in an adjoining cistern, further underscoring the cosmopolitan nature of the city in the Middle to Late Roman period (Cormack 1996: 171–72; forthcoming).

Table 2. Lamp Quantities from the Western Summit and Other Areas of Sepphoris

Type	*Name*	*Western Summit* Complete	*Western Summit* Nearly Complete	*Western Summit* Fragmentary	*Other Areas*[a]
Local and Regional					
1	Mold-Made Delphiniform with Kite-Shaped Body and Double Side Lugs		1	3	
2	Mold-Made Delphiniform/A-Prime Levantine	1		6	
	Variant 1			1	
3	Northern Collar Neck/Galilean Collar Neck			1	10
4	Herodian	1	3	21+	143
5	Sepphorean Spatulate		2	8+	1
6	Darom	1		22	80
	Darom, Square-Bodied			2	
7	Molded Judean Floral		4		
8	Monolychnoi, Ring-Lamps with Elongated Nozzles with Circular Heads			3	
9	Polylychnoi, Box-Like Rectangular Lampstand			1	
10	Palestinian Discus	3	15	53+	586
11	Syrian Discus			2	
12	Jebel Jofeh	1	1	8	18
13	Northern Stamped	6	1	7+	59
14	Beit Nattif, Northern Versions			6	28
15	Gilead		1		
16	Bilanceolate	5	2	9+	53
17	Galilean	3		1	106
18	Caesarean Discus			5	
19	Galilean Discus			7	
20	"Sepphorean" Discus		1	19	35
21	Jerash with Zoomorphic or Tongue Handle			1	6
22	Slipper (Hadad Type 30)	2		1	6
23	Slipper (Hadad Type 36)	1			
24	Islamic with Large Filling-Hole and Tongue Handle		1		
Imported					
25	Ephesus, Broneer Type XIX			2	
26	Corinthian, Broneer Type XXVIIC			1	
27	Athenian, Broneer Type XXVIII, Later Version			2	

a. Lamps recovered by Hebrew University excavations as reported in Gärtner 1999.

5. Methodology

Complete, nearly complete, and fragments of ceramic oil lamps excavated at Sepphoris during the 1985–89 field seasons of the Joint Sepphoris Project were selected for macroscopic examination. A mix of conventional macroscopic examination procedures, as used by other lychnologists and as reported by Bailey (1988), was adopted in this study. A special effort was made to select representative lamp examples of each typological group from every chronological period represented at the site. Physical properties and features such as the clay, core, and slip colors, hardness, inclusion types, existence of carbon, and measurements, in addition to the typological and decorative elements of each example were studied in order to achieve sound chronological readings. A catalog of the lamps and lamp fragments belonging to a specific type follows each section in order to facilitate the reader's study of the type and to avoid irksome thumbing to the end of the chapter. Illustrations and photographs picturing representative samples from each lamp type are provided in the plates section.

The theory behind the macroscopic analysis of the Sepphoris corpus is that each lamp's typological, decorative, and physical fabric properties can contribute to the dating of strata and, on a broader scale, to a better understanding of the ancient culture that produced it. Further, because lychnology is a descriptive science, its methodology demands the recording of as many visible and physical properties of the artifact as possible, in essence representing a three-dimensional artifact by textual description supplemented by drawings and photographs. Also, because clay lamps are comparable to metamorphosed sedimentary rocks as are other types of pottery (Rice 1987: 376), the Sepphoris lamp fabrics in this study were analyzed accordingly. Physical features such as clay, core, and slip colors, hardness, clay composition (specifically the inclusion types present and their grain sizes and frequency of occurrence), and existence of burnish were noted. Each surface of the lamp, exterior and interior, was examined for traces of carbon (its black residue is an indicator of burning and use). Dimensions, such as the length, width, and height of the example, were recorded. A general assessment of the lamp's physical condition—namely, to what degree the object has weathered—is also included, as is the typology and ornamentation of the example. As many data as possible have been collected on each example in order to understand from what type of clay it was made, the method of its manufacture, where it was produced, and in what chronological period it was fashioned.

Dimensions

The dimensions of each sample were measured using an aerospace stainless steel, 15-cm caliper to determine the length (L), width (W), and height (H). All dimensions are provided in centimeters (cm). For incomplete examples, particularly body fragments, the greatest dimension (GD) was measured, specifically, the longest diagonal length. A dimension that could not be measured is marked "ud" (undetermined).

Completeness/Preservation

The preservation of a lamp refers to the completeness of the lamp find. Obviously, if a lamp is fully preserved, it is described as "complete." If only a fragment is preserved, it is rated "incomplete," and a percentage of preservation is provided. A fragment typically originates from either the upper or lower half of a complete lamp, so one is already beginning at 50%. From this starting point, one must determine how much of the half is preserved. The completeness of a lamp sample refers to the percentage of the lamp body preserved (e.g., "Incomplete, 35% preserved"). Thus, for example, if only a portion of a lamp's upper half is preserved, one begins at 50% of the lamp and subtracts what is missing. The majority of lamp finds from the western summit are fragmentary, reflecting their secondary depositional history.

Shape-Type

Each complete lamp or fragment is identified as belonging to a specific shape-type, which comprises all the macroscopic features that define a specific lamp group: morphology, design, patterns, symbols, inscriptions, lamp makers' marks, decoration, slip, fabric, inclusions, firing, and any other visible feature shared by members of that group. When these elements are compiled as whole shape-types, each type tends to be unique to a specific workshop and to have been manufactured in a specific chronological period as determined by its recovery among coins and other pottery types from stratigraphic sequences at widespread geographical sites. In order to avoid any confusion, I use conventional nomenclature that most accurately describes the lamp group in terms of (1) its place of manufacture, as determined by the discovery of similar molds or actual workshops, and (2) the region of greatest distribution and concentration, which points to the epicenter of production.

Date

The date assigned to a lamp follows the chronological designations and dating described by Meyers and Meyers (2013a: 6–10). Lamp dates are further based on information about comparative lamp finds from other sites.

Description

Each catalog entry includes a description of the shape, decorative patterns, symbols, inscriptions, and lamp makers' marks of a lamp. These features are followed by a description of the physical properties of the clay fabric of the lamp, such as color, inclusions, hardness, voids, and condition (of the fabric itself). The technique of manufacture, whether wheel-thrown or mold-made, is also identified.

Color

The color of a lamp sample's clay fabric (and if present, of the core and slip) was measured using the international standard, *Munsell Color Soil Charts* (Munsell

1971). The color characteristics for all samples were measured on dry examples, particularly from a broken edge as opposed to a weathered, discolored outer surface, to obtain the most accurate reading. If the slip color varied in hue, several readings were performed and the most prevalent noted. If present, the color of slip stains on the interior surface of the lamp was recorded.

Inclusions

A broken edge of each lamp sample was examined using a 10× Rupex hand-magnifying lens in order to identify any inclusions and voids in the fabric. The inclusion's color was reported, and whenever possible, the inclusion's angularity, roundness, and sorting were noted. The most common inclusion colors are red, white, orange, black, and gray. In some fabrics, it was possible to identify the inclusion as an actual mineral or rock fragment. For example, white is typically calcium carbonate, and red, hematite. Quartz sand is often identifiable as orange. Black is generally magnetite, and gray, chert. The actual mineral identifications are included in the catalog when absolutely certain; otherwise, only the inclusion color is given.

Categories of roundness/sphericity and angularity of inclusion grains are based on Pettijohn, Potter, and Siever 1973, as is visual estimation of sorting. The relative frequency of occurrence for each inclusion type was determined: high frequency (HF), medium frequency (MF), and low frequency (LF). The grain size for each inclusion type was based on the Wentworth scale and was recorded (see Tucker 1981: 11, table 2.1); clay, silt, sand, granules, pebbles, etc., are recorded as coarse sand (CS), fine sand (FS), very fine sand (VFS), and silt (Si). In some fabrics, it was possible to record the degree of sorting exhibited by the inclusion types: well sorted, moderately sorted, poorly sorted. Although in some fabrics no inclusions were visible, this in itself does not mean none exists in the lamp's fabric; perhaps heavy encrustation covers the inclusions that would otherwise be visible on the surface. The recording of inclusions—and their particle sizes, color, and degree of sorting—occurring in lamp fabrics is an important first step leading to eventual petrographic and trace-element analysis. It aids the lychnologist in identifying any variations among lamp fabrics and in selecting fabrics for thin section and chemical analyses.

Estimated Porosity

The presence of cracks, voids, and vapor veins in the Sepphoris lamp fabrics was recorded as a first step in identifying lamp samples for future petrographic and water absorption analyses. Such analyses, in turn, will help determine estimated porosity (P_E) and better understand the effects of thermal stress on lamp fabrics resulting from firing and use (Lapp 2012: 77–78).

Mohs Hardness

The hardness of a fired ceramic object, such as a lamp, is difficult to determine. To date, no satisfactory means of estimating the hardness of baked clays has been devised. The term *hardness* itself is problematic but usually means resistance to scratching or abrasion (Searle and Grimshaw 1959: 834). It is often measured by observing the resistance of the materials being tested to abrasion by various harder substances. To measure this resistance, a "scratch test" has been developed whereby minerals of known and varying hardness are drawn across the surface of the material tested in order to determine on a relative and comparative basis which is the softer of the two (Searle and Grimshaw 1959: 835). By far the most extensively used and convenient collection of minerals applied to the scratch test can be found arranged in the Mohs scale. For this reason, I used the Mohs scale to determine the hardness of the Sepphoris lamps. Because the hardness of a ceramic sample may vary under similar firing conditions, many scholars have chosen to disregard this physical property, considering it too controversial. I recognize the problems inherent to this property (see discussion in Rice 1987: 356) but, nonetheless, consider it important, for it provides us an additional piece of information that can be used to compare lamps and to determine whether variation occurs among lamp groups, thus suggesting different firing practices.

Condition

The physical condition of the clay fabric of each lamp sherd was assigned a general rating: excellent, good, fair, or poor. The condition of a lamp and lamp sherd has much to do with the artifact's depositional history at any given site. In this study, *condition* refers only to the state of the clay fabric of the lamp and how severely it has been weathered by wind abrasion, groundwater, leaching, or salt erosion as indicated by the pock-marked or cratered surface of the lamp sherd (on identifying salt erosion and related issues concerning clay lamp fabrics, see Lapp 2012: 71). For example, a complete lamp with a friable or flaking fabric would receive a "poor" rating. A broken or incomplete lamp sample, however, would receive a "good" rating if its fabric is in good shape and exhibits no signs of wear, weathering, or cratering caused by salt erosion.

Burning

The presence of soot on a lamp's nozzle is usually indicative of primary burning and is one of the most important clues as to whether the lighting vessel had been used. A lack of evidence of burning would suggest that a lamp was intended for sale, for storage for later use, or for use as a light symbol or photoamulet in a ritual or burial context (Lapp in press b). For example, a large quantity of lamps showing no signs of burning found in a room may be evidence of a lamp shop or a storehouse, such as at Capernaum (Loffreda 1974: 131). Or, a number of unused lamps recovered from a tomb or shrine, such as at Pella, Beth Guvrin, and Banias, suggests that the lamps were used as symbols of light. At Pella, for example, da Costa reports that seven of the lamps recovered from Tomb 39A were unlit (da Costa 2003: 2.703; Lapp 1997a: 210 n. 71). At Banias/Caesarea Philippi, Berlin (1999: 37, 40) notes that a high percentage of lamps deposited at the shrine of Pan were unused. Da Costa

(2003: 1.356) observes that "in funerary contexts lamps were increasingly used symbolically, and not actually lit." On the symbolic meaning of lamps, see also Smith 1966: 9–11.

Origin

A lamp type's manufacturing workshop or production center is noted when there exists sufficient petrographic and trace-element analyses and/or evidence of lamp workshops (e.g., molds, wasters, and kilns). For example, the hippodrome lamp workshop discovered at Jerash establishes the manufacture of a number of fourth- to early fifth-century C.E. lamp types there (Kehrberg 2001: 231–39). Neutron Activation Analysis (NAA) results of Herodian lamp fabrics excavated at Sepphoris suggest a pronounced preference among Jewish Sepphoreans for Herodian lamps manufactured in the Jerusalem area (Adan-Bayewitz et al. 2008: 55, 77).

Finger Impressions

The Sepphoris lamp sherds were examined for finger impressions. With the application of 2D scanning technology, fingerprints can be extracted from the surfaces of lamp fragments with relative ease (Lapp and Nicoli 2014: 8–10, figs. 12–13). In the Sepphoris corpus, imprints were identified on northern stamped (no. 155) and "Sepphorean" discus (cat. nos. 192, 194–95, 197–98, 202, 208) sherds. They were made when the lamp maker pressed the clay in its prefiring leather stage into the concave cavities of the mold. Fingerprinting lamps recovered from archaeological sites such as Sepphoris will help identify various lamp makers of some of the lamp types and determine the geographical distribution and trade of their products (Lapp and Nicoli 2014: 10).

Comparative

Comparative lamp examples for various shape-types found at other sites are listed. A "similar," "close," or "exact" match to a parallel lamp is noted. Finding an exact match implies that the lamp or lamp fragment was made from the same mold as the Sepphoris example.

CHAPTER 2

Local and Regional Types

1. Introduction

Twenty-seven types of clay oil lamps were identified among the finds of the Duke excavations at Sepphoris (tables 1–2, pp. 5ff., 10 above). Lamps dating to the Hellenistic, Roman, Byzantine, and Umayyad periods are represented. This section reports on a representative sampling of those lamps.

Hellenistic Period

The earliest surviving evidence for the use of clay lamps from the western summit dates to the second–first centuries B.C.E. The corpus of lamps from this period is relatively small, comprising eight body fragments (cat. nos. 1–5) and nine cone-shaped nozzles (cat. nos. 6–13). No complete Hellenistic oil lamps were found in stratigraphic contexts at the site. Three types of Hellenistic-period lamps and versions thereof are represented at Sepphoris: (1) mold-made kite/A-prime Levantine type, (2) mold-made delphiniform type with kite-shaped body and double side lugs, and (3) Judean coastal plain type with circular nodules. Several inclusion types are found in the Sepphoris examples: chert, hematite, white, and a black mineral, possibly magnetite. That magnetite tends to be uniquely associated with coastal beach sand may corroborate NAA findings suggesting Phoenician coastal towns as the origin of the A-prime lamps. The rough execution of the patterning exhibited by some examples suggests local and rural Galilean manufacture or simply the poor skill of the lamp-mold carver. The Sepphoris examples are fashioned in an orange, yellow, or dark reddish brown clay. Most examples are coated in a black or dark reddish-brown slip. The lamps' walls are thick, and vestiges of strap handles are found on two fragments (cat. nos. 1, 3). The lamps are massive and rough in appearance.

Roman Period

The lamp types representing the Roman period on the western summit include: northern collar neck (cat. no. 18); Herodian (cat. nos. 19–43); Sepphorean spatulate (cat. nos. 44–52); Darom (cat. nos. 53–70); Judean floral fine ware or molded with floral pattern (cat. nos. 71–73); monolychnoi, ring-lamps with elongated nozzles with circular heads (cat. nos. 74–76); polylychnoi, box-like rectangular lampstand (cat. no. 77); Palestinian discus (cat. nos. 78–142); Syrian discus (cat. nos. 143–44); Jebel Jofeh (cat. nos. 145–54); northern stamped (cat. nos. 155–61); and Beit Nattif, northern version (cat. nos. 162–66). They are local and regional forms. A Roman-period lamp fragment originating from a Corinthian workshop outside Palestine was also recovered.

Byzantine Period

Eight main lamp types dating to the Byzantine period are represented in the lychnological record of the western summit: Gilead (cat. no. 167); bilanceolate (cat. nos. 168–72); Galilean (cat. nos. 173–76); Caesarean discus (cat. nos. 177–81); Galilean discus (cat. nos. 182–88); "Sepphorean" discus (cat. nos. 189–208); and Jerash with zoomorphic or tongue handle (no. 209). Several likely originate from workshops in Jerash of the Decapolis, suggesting connections between the two cities. A few indicate manufacture in Galilee, perhaps even in Sepphoris or in nearby Shikhin. Bilanceolate lamps were an especially popular lamp type in the Decapolis, particularly at Pella and Gadara, and their discovery at Sepphoris further suggests contact with these prosperous mercantile cities. That Caesarean discus lamps were found in Sepphoris indicates links with the port of Caesarea Maritima, where such lighting devices were produced.

Early Islamic Period

Three main lamp groups dating to the Early Islamic period are found in the Sepphoris corpus: slipper, Hadad Type 30 (cat. nos. 210–12); slipper, Hadad Type 36 (cat. no. 213); Islamic with large filling-hole and tongue handle (cat. no. 214).

2. *Typology*

The typology presented in this chapter and the next is the result of sorting the 219 complete clay oil lamps and fragments discovered in excavations on the western summit of Sepphoris into types: 24 local and regional types, described in this chapter; and 3 imported types, described in chap. 3. These items have been assigned catalog numbers from 1 to 219. Drawings of 153 of them are presented in the plates. (Many of the very small fragments were not drawn.)

Type 1. Mold-Made Delphiniform with Kite-Shaped Body and Double Side Lugs (cat. nos. 1–4)

A nearly complete double convex lamp with an elongated nozzle was found at Sepphoris (cat. no. 1). T. Levine refers to this mold-made lamp type as "kite-shaped with pointed lugs" because of two prominent lugs on each side of the body which create a kite-like impression (2003: 115). Similarly, the kite-shaped body of the Sepphoris lamp in reconstruction resembles examples of this group found at Maresha (T. Levine 2003: 115–16, nos. 155–59, fig. 6.15). Typically, the nozzle-bridge of such lamps is decorated with floral patterns or is plain. An inverted and concentric *V* design with volutes in high relief decorates the nozzle-bridge of the Sepphoris example, creating an elegant impression. A row of chevrons between grooves decorates the nozzle of a kite-shaped lamp excavated at Maresha (Ambar-Armon 2014: 36, no. 25, fig. 5.9), and two winged erotes supporting palmettes between their heads flank the lamp's filling-hole (T. Levine 2003: 116, no. 156). A rounded ridge surrounds a central filling-hole that is small- to medium-sized in diameter. These lamps have a round flat base.

Three further examples of the kite-shaped lamp type were excavated at Sepphoris (cat. nos. 2–4). However, they differ from the above lamp in the stylistic treatment of the patterning; it is rather rough, perhaps reflecting a less-skilled carver of the lamp mold. One of them (cat. no. 2) consists of multiple mended fragments. The wall of this lamp is thick, and the fabric highly friable. A geometric pattern of broad lines and semicircles decorates both the upper and lower halves of the lamp's body. Two double, perpendicular lines in high relief intersect inside the ring base, creating a cross. This may represent a lamp maker's mark, or it was intended as decoration or had apotropaic symbolic significance. The other two examples of the kite-shaped type (cat. nos. 3–4) have thick walls and geometric patterning suggestive of local Galilean folk art with possible cultic or ritual meaning.

Hayes (1980: 22–23, pls. 9.84–85, 16.161) dates examples of this type to the end of the third to second centuries B.C.E. Bailey (1975: 272–73, Q592.EA, pl. 112) dates an example in the British Museum to probably the late second or first century B.C.E. Rosenthal and Sivan (1978: 13, photo 24) identify examples in the Schloessinger Collection and date them to the second half of the second and the beginning of the first centuries B.C.E. Lyon-Caen and Hoff (1986: 53, nos. 115–16) date two examples in the Musée du Louvre to the third or fourth quarter of the third century B.C.E. to the fourth quarter of the second century B.C.E.

1. Reg. no. 85.3096.3L.2. GD = 8.1 cm. Incomplete, 20% preserved. Mold-made delphiniform with kite-shaped body with double lugs. Second half of the second and the beginning of the first centuries B.C.E. (Rosenthal and Sivan 1978: 13, photo 24). Upper body: complete round nozzle-head, elongated nozzle-bridge, and shoulder fragment. Part of rim around filling-hole and nozzle-underside preserved. Rim around filling-hole: 7.0 mm wide with a slight depression in center. Wick-hole, 1.1 cm in diameter. Below wick-hole: 1.5 mm wide line across nozzle-bridge. On nozzle-bridge: an inverted *V* motif with curls on each point, resembling a gable. To left: two parallel lines delineate where nozzle-bridge meets shoulder. On shoulder: radials in high relief. Clay color: 10YR 7/3 very pale brown. Slight traces of slip. Slip color: 5YR 6/6 reddish yellow. Inclusions: rounded, well-sorted gray (VFS, VHF); sub-angular, poorly sorted gray (FS, MF and CS, LF); gray (VCS, VLF); red hematite (CS, LF). Hardness not measured. Fair condition. Heavy burning on nozzle. Mold-made.

2. Reg. no. 85.3115.2L.1. GD = 8.9 cm. Incomplete, 80% preserved. Mold-made delphiniform with kite-shaped body with double lugs. Second half of the second and the beginning of the first centuries B.C.E. (Rosenthal and Sivan 1978: 13, photo 24). Nozzle, shoulder, nozzle-underside, and base fragment. Multiple mended fragments. On upper and lower body: roughly executed geometric pattern of wide lines and semicircles in high relief. Ring base. On base, a lamp maker's mark: two double, intersecting perpendicular lines in high relief, creating a cross. Degenerate side lug, right shoulder. Filling-hole, 1.5 cm diameter. Clay color: 7.5YR 7/4 dull orange. Slip color: 10R 5/6 red. Inclusions: gray (CS, LF and VCS, VLF). Hardness not measured. Very poor condition. Highly friable fabric. Burning on nozzle. Mold-made.
3. Reg. no. 85.3056.1L. GD = ud. Nearly complete, 95% preserved. Mold-made delphiniform with kite-shaped body with double lugs. Second half of the second and the beginning of the first centuries B.C.E. (Rosenthal and Sivan 1978: 13, photo 24). Nozzle missing. On body, upper half: roughly executed geometric pattern of wide, parallel radial and straight lines in high relief. Ring base. Degenerate side lug, left shoulder. Wide rim around filling-hole. Clay color: 7.5YR 7/4 dull orange. Slip color: 10R 5/6 red. Inclusions: gray (CS, LF and VCS, VLF). Hardness not measured. Poor condition. Highly friable fabric. Mended two fragments. Burning on nozzle. Mold-made.
4. Reg. no. 85.3037b.3L. GD = 6.0 cm. Incomplete, 20% preserved. Mold-made delphiniform with kite-shaped body with double lugs. Second half of the second and the beginning of the first centuries B.C.E. (Rosenthal and Sivan 1978: 13, photo 24). Part of shoulder, rim of filling-hole, and nozzle. On upper shoulder: one set of three parallel lines; on left side, repeated zigzag pattern and two curved parallel lines and nodule. Filling-hole, diameter 1.6 cm. Rim width around filling-hole, 8.0–9.0 mm. Below wick-hole ridge: a single line in relief, 1.5 mm width. At base of nozzle-bridge: band of ladder delineating shoulder. Elongated nozzle with slight bulge on upper surface. Clay color: 10YR 7/2 light gray. Core color: in section, 7.5YR 7/4 pink. Slip color: 7.5YR 6/4 light brown. Hardness not measured. Medium fired. Thick walled: 5.0–8.0 mm wide. Inclusions: sub-angular gray (LF, FS). Elongated voids (MF). No burning. Fair (+) condition. Pock-marked surface, evidence of salt erosion. Mold-made.

Type 2. Mold-Made Delphiniform/A-Prime Levantine (cat. nos. 5–17)

The delphiniform lamp type was very common throughout the eastern Mediterranean.[1] In fact, one of the largest groups of delphiniform lamps—728 in all—was found on the Cycladic island of Delos (Bruneau 1965: 81–88); these differ from the Sepphoris examples in that an "S-coil" decorates the lug. This type dates to the second and first centuries B.C.E. (Rosenthal and Sivan 1978: 13; Lyon-Caen and Hoff 1986: 57). Its popularity may be explained by its decorative geometric and floral patterning, which would have appealed to most any market-clientele. Also, its thick walls guaranteed durability and long-term use, and the large filling-hole facilitated the introduction of oil into the fuel chamber.

1. For the origin of the nomenclature, see Walters 1914: 42–44.

The delphiniform form with an S-shaped side lug and flat base corresponds to Młynarczyk's Type A-Prime, which was perhaps modeled on wheel-made Rhodian lamps (so Młynarczyk 1997: 22–25). At Athens, the delphiniform group corresponds to Howland's Type 45A, which is dated to the mid- to third quarter of the third century B.C.E. (Howland 1958: 143–45) and which, in turn, relates to lamps categorized as Group IV at Tarsus and dated to the late third and early second centuries B.C.E. (Goldman and Jones 1950: 88, no. 39). In Hellenistic Palestine, this type relates to Type 12A at Dor and is dated to 200–50 B.C.E. (Rosenthal-Heginbottom 1995: 238, nos. 1–6, figs. 5.16:11–13, 5.17:1–3).

Three shoulder fragments of the Hellenistic delphiniform type were found at Sepphoris (cat. nos. 5–7). They are characterized by a rounded, sharply curving shoulder (see Lapp 1991: nos. 1–2; figs. 70–71, respectively). Two of the fragments (cat. nos. 5–6) have a prominent, well-rounded ridge surrounding a medium- to small-sized filling-hole. Catalog no. 5 exhibits the remnants of a strap handle. An irregular radial pattern with deep grooves reminiscent of the Early-Hellenistic-period sunburst form decorates the shoulder of cat. no. 9. A similar lamp fragment exhibiting patterning and the remnants of a strap handle were found at Meiron (Meyers, Strange, and Meyers 1981: pl. 9.14:10). A deeply impressed, stamped leaf pattern decorates the shoulder of cat. no. 6. Lamps of this type have a characteristic gray fabric. The clay of the Sepphoris fragments is yellowish gray (2.5Y 5/1). Whereas cat. no. 5 is coated in a black slip (10YR 1.7/1), cat. no. 6 is slipped in brownish gray (10YR 4/1). White calcium carbonate represents the most common inclusion type of the two fragments.

A face with a two-part beard, probably Serapis, decorates the lower part of the nozzle-bridge of a delphiniform lamp fragment (cat. no. 8). A double rope pattern of two parallel lines delineates where the elongated nozzle meets the body of the lamp. The fabric of the lamp suggests a regional Palestinian origin.

Similar lamps were found at Tel Anafa and were dated to after the mid-second century B.C.E. (Israeli and Avida 1988: 16–17). At Maresha, excavations of Subterranean Complexes 21 and 70 yielded similar delphiniform/A-Prime lamps (T. Levine 2003: 117–19, nos. 16–166; Ambar-Armon 2014: 33, nos. 26–35, fig. 5.6). Several delphiniform lamps with red fabrics were also found there (T. Levine 2003: 120–21, nos. 167–71, fig. 6.17). Excavations in the upper town of Tel Sandahannah/Marisa in the early twentieth century recovered several examples (Bliss and Macalister 1902: pls. 62.8, 20; 62.9–10; 63.6), as did later excavations in the lower town (Kloner and Hess 1985: 130, fig. 4.10–12). Tel Anafa yielded a significant quantity of delphiniform lamps postdating the mid-second century B.C.E. (Weinberg 1971: 103–5, pl. 18.b). At Beth Shean, a lamp with an S-coil side lug was found (Hadad 2002: 13–14, Type 2, no. 2). A Sepphoris lamp fragment (cat. no. 1) resembles an example in the Schloessinger Collection dated to the mid-second to mid-first centuries B.C.E. and possibly extending to the end of the first century B.C.E. (Rosenthal and Sivan 1978: 13, photo 22). For this reason, a mid-second to early first-century B.C.E. date for the Sepphoris specimens is plausible.

Delphiniform lamps were also found at ʿAin Shems/Beth Shemesh (Zoitopoulou and Fossey 1992: 54–55, no. 14, fig. 15), Beth Zur (Sellers 1933: 51, fig. 42),

Caesarea Maritima (Holum et al. 1988: 44–45), Gezer (Macalister 1912: pl. 183.1, 7, 12, 16–20; Gitin 1990: pls. 35.26, 41.26), Masada (Barag and Hershkovitz 1994: 13f, Class B), Tel Michal (Fischer 1989: fig. 13.1:13–14; fig. 13.2:24), and Tirat Yehuda (Yeivin and Edelstein 1970: 61, fig. 10.1–3). In Hellenistic Arabia, an A-Prime lamp was found in a context dating to the first half of the second century B.C.E. at Pella of the Decapolis (McNicoll et al. 1992: 113, 116, pl. 81.1). Four examples in the Musée du Louvre date to the second and first centuries B.C.E. (Lyon-Caen and Hoff 1986: 57, nos. 128–31). Młynarczyk suggests a chronological range for the A-Prime Type from 200 to 100 or perhaps as late as 75 B.C.E. (Młynarczyk 1997: 110, table 2).

5. Reg. no. 95.1016.10L. L = 6.4 cm. W = 2.3 cm. Incomplete, 30%. Delphiniform. Hadad Type 2. Rosenthal-Heginbottom Type 12A. Second century B.C.E. to the first half of the first century B.C.E.; mid-second to mid-first centuries B.C.E., and possibly extending to the end of the first century B.C.E. (Rosenthal and Sivan 1978: 13, photo 22). Body fragment with broad, steeply sloping rounded shoulder. Large central circular filling-hole surrounded by broad, prominent well-rounded ridge. Irregularly spaced groups of deeply impressed vertical ridges tapering outward from discus decorate shoulder. Broken handle. Thick walls. Clay color: 2.5Y 5/1 yellowish gray. Slip color: 10YR 1.7/1 black. Hardness: 2.0. Inclusion(s): gray (VFS, very high frequency); white (FS-MS, low frequency). Fair (-) condition. Mold-made. Parallels: Amman (Harding 1946: pl. 20.1), Dor (Stern 1981: 107) (exact). Tel Mevorakh (Stern 1978: pl. 22.1).
6. Reg. no. 84.2144.2L.1. L = 4.5 cm. W = 2.4 cm. Incomplete, 15% preserved. Delphiniform. Hadad Type 2. Rosenthal-Heginbottom Type 12A. Second century B.C.E. to the first half of the first century B.C.E. Broad, steeply sloping, and rounded shoulder fragment. Large central circular filling-hole encircled by wide, prominent well-rounded ridge. Deeply impressed, repeated pattern of tapering leaf enclosed between the discus ridge and outer shoulder ridge by an impressed vertical line of connected rounded triangles. Thick walls. Clay color: 10YR 6/2 grayish-yellow brown. Slip color: 10YR 4/1 brownish gray. Burnish, slight traces on exterior surface. Hardness: 2.0. Inclusion(s): white (FS, very low frequency); black (FS, very low frequency). Burning, interior surface. Good condition. Mold-made.
7. See p. 23 for cat. no. 7.
8. Reg. no. 85.3152.1L.1. GD = 3.9 cm. Incomplete, 5% preserved. Delphiniform. Hadad Type 2. Rosenthal-Heginbottom Type 12A. Second century B.C.E. to the first half of the first century B.C.E. Nozzle-bridge and shoulder fragment. On nozzle-bridge: a face with a two-part beard, probably Serapis. Double rope pattern of two parallel lines delineates where elongated nozzle meets body. Two deep parallel grooves delineate channel from face to wick-hole. Clay color: 7.5YR 7/4 pink. Slip color: 2.5YR 4/6 red. Hardness: not measured. Inclusion(s): red, hematite, rounded, poorly sorted (FS, low frequency); gray, sub-rounded, poorly sorted (FS, medium frequency). Heavy burning traces, exterior and interior surfaces. Fair condition. Fabric indicates local manufacture. Mold-made.
9. Reg. no. 85.3037b.3L.2. L = 10.7 cm. W = 6.1 cm. H = 3.4 cm. Complete. Delphiniform radial/sunburst. Hadad Type 2. Rosenthal-Heginbottom Type 12A. Second century B.C.E. to the first half of the first century B.C.E. Steeply sloping shoulder. On shoulder: relatively evenly spaced, wide, radial linear grooves. Part of left shoulder, plain. Filling-hole, diameter 1.8 cm. Rim width around filling-hole, 4.0–8.0 mm.

Elongated nozzle with bulbous long top, perhaps in order to lend support to nozzle that would otherwise break under its own weight and length. Wick-hole, 1.5 cm in diameter. Round pedestal base. Clay color: 10YR 7/2 light gray. Slip color: 2.5YR 4/4 reddish brown. Slip traces on nozzle and shoulder. Hardness: undetermined. Medium fired. Inclusions: white, sub-angular (VCS, MF); gray, sub-rounded (FS, HF); red (VFS, LF). Burning around nozzle. Poor condition. Heavy weathering, underside exterior surface. Highly friable. "Dirty" fabric. Mold-made.

10. Reg. no. 85.3022b.1L. GD (preserved L) = 8.8 cm. H = 2.3 cm. Incomplete, 80% preserved. Delphiniform radial/sunburst. Hadad Type 2. Rosenthal-Heginbottom Type 12A. Second century B.C.E. to the first half of the first century B.C.E. Part of shoulder and underside missing. Curved, sloping shoulder. On shoulder: wide, evenly spaced, deeply impressed grooves. Filling-hole (diameter 1.8 cm) encircled by rounded flat ridge (4.0 mm wide). Elongated nozzle with slight bulge on upper surface. On nozzle-bridge: a single vertical groove; another groove across nozzle-bridge delineates nozzle. Round nozzle-head with wick-hole, 1.0 cm in diameter. Rough groove delineates round flat base, approximately 3.0 cm in diameter. Clay color: 10YR 7/2 light gray. Slip color: 2.5Y 4/0 dark gray. Hardness: undetermined. Medium fired. No visible inclusions. Burning around nozzle. Fair condition. Pockmarked surface, evidence of salt erosion. Mold-made. Comparisons, shape-type: Adler Collection (Adler 2004: 11, no. 55).
11. Reg. no. 85.3230.4L. Not shown. GD = 3.1 cm. Incomplete, 5% preserved. Delphiniform radial/sunburst. Hadad Type 2. Rosenthal-Heginbottom Type 12A. Second century B.C.E. to the first half of the first century B.C.E. Part of shoulder and filling-hole. Steep shoulder. On shoulder: wide radials in high relief. Prominent rounded ridge (width, 3.0 mm) around filling-hole, then a groove, and next, a second very prominent ridge (width, 6 mm). Clay color: 7.5YR 4/0 dark gray. Slip color: 7.5YR 2/0 black. Slight traces of slip. Hardness: undetermined. No visible inclusions. No burning. Fair (–) condition. Heavy salts on exterior surface. Mold-made.
12. Reg. no. 85.3313.2L. Not shown. GD = 5.2 cm. Incomplete, 10% preserved. Delphiniform radial/sunburst Hadad Type 2. Rosenthal-Heginbottom Type 12A. Second century B.C.E. to the first half of the first century B.C.E. Part of shoulder and filling-hole. Very steep shoulder. On shoulder: deep radial grooves. Very prominent rounded ridge (width, 7 mm) around filling-hole (preserved arc, 1.7 cm). Clay color: 2.5YR 8/2 white. No slip. Hardness undetermined. Thickness: approximately 0.5 cm. Inclusions: gray, angular, fair to well sorted (VFS, VHF). Yellow marl clay. Soft fired. No burning. Fair condition. Mold-made.

Variant 1 (cat. no. 7)

A version of the delphiniform lamp was found at Sepphoris and is characterized by a wide, flat rim surrounding the central filling-hole and a steeply sloping shoulder (cat. no. 3). A radial pattern consisting of raised, closely spaced vertical ridges decorates the shoulder. A slight nodule on the shoulder indicates a pseudo-handle. The base of this mold-made lamp is flat and undecorated. Its fabric differs from the fabrics of cat. nos. 1–2 in that it is a dull orange (7.5YR 7/4). A close parallel of this lamp was recovered at Beth Sheʿarim (Avigad 1976: no. 33, pl. 71). A similar one was found in the foundation level under the Early Roman basilica at Beth Shean and is dated to the second century B.C.E. to first half of the first century B.C.E. (Hadad 2002: 13–14, Type 2, no. 3). Note, however, that Hadad identifies this Beth Shean lamp as a Type 2 form; but unlike a Type 2 lamp found at the site with an S-coil lug, this lamp has no lug and therefore may represent a later version (see Hadad 2002: 13–14; cf. no. 2 with an S-coil lug to no. 3 with no lug). Based on this example, the Sepphoris lamp dates from the second century B.C.E. to the first half of the first century B.C.E.

Młynarczyk's observation (1997: 25) that lamps belonging to the A-prime type were manufactured in more than one area is highly plausible. She supports her hypothesis on the variation of fabrics by citing the numerous lamps of this type widely diffused in Palestine (Negev, Shephelah, Judea, Samaria), along the Syro-Palestinian coast, Cyprus, Delos, Alexandria, Carthage, and Morocco. As she points out, some of these fabrics can be attributed to specific regions: Cyprus (pinkish-buff fabric); the Jerusalem area (buff/light-brown fabric); the Maresha region of the Shephelah (a chalky fabric), and several Phoenician coastal towns that presumably had manufactured lamps of this sort out of porous gray ware and a hard, fine ashy gray ware (Młynarczyk 1997: 25). Trace-element analyses of lamp fabrics further substantiate her claim. Neutron Activation Analysis (NAA) conducted on a lamp found at Masada indicates that it was manufactured in the Jerusalem area (Yellin 1994: 13, no. 2, nn. 15–18). NAA analysis has also established the Phoenician coast (perhaps the vicinity of Tyre) as the manufacturing origin of the gray-ware Tel Anafa lamps of the A-Prime type (Dobbins 1995: 29–30 n. 26). Taking this into consideration, Młynarczyk suggests that the provenance of the gray-ware lamps found at Alexandria, Pella of the Decapolis, and Delos was also Phoenicia. Similarly, the Sepphoris examples of this lamp type may have originated in Phoenicia, perhaps in the vicinity of Tyre. However, until the Sepphoris lamp fabrics of this type are analyzed by means of petrography and trace-element analysis, the verdict is still out as to whether they come from workshops in Phoenician coastal towns or were locally manufactured in Galilee, perhaps even at or in the vicinity of Sepphoris. Late Roman lamp molds found at Sepphoris and rabbinic texts alluding to pottery manufacture associated with the city suggest the possibility of lamp manufacture there.

7. Reg. no. 94.1005.4L.1. L = 5.9 cm. W = 3.9 cm. H = 3.2 cm. Incomplete, 40% preserved. Delphiniform, Variant 1. Hadad Type 2. Rosenthal-Heginbottom Type 12A. Second century B.C.E. to the first half of the first century B.C.E. Body fragment with flat, wide filling-hole rim, then steeply sloping shoulder. Small circular nodule on shoulder. Raised, consecutive, closely spaced vertical grooves form radiating pattern that decorates shoulder. Medium-sized central filling-hole without encircling ridge. Thick wall. Flat undecorated base. Clay color: 7.5YR 7/4 dull orange. Hardness: 3.0. Inclusion(s): white (CS, low frequency). Rock fragments (?) (VCS, low frequency). Burning traces, exterior and interior surfaces. Poor condition. Mold-made. Parallels: Beth Sheʿarim (Avigad 1976: pl. 71.33) (exact); similar wide flat rim around filling-hole and consecutive radials on shoulder: Hagoshrim (Sussman 2009: 142–43, no. 263, dated to the second–first centuries B.C.E.).

Variant 2: Cone-shaped Nozzles (cat. nos. 13–17)

The cone-shaped nozzles from Sepphoris belong to one of two groups: elongated and pointed (cat. nos. 13–14, 16) and comparatively short and square-ended (cat. no. 17). The ends of the nozzles slope steeply downward between a 45° and

30° angle. Some examples are coated in either a black (10YR 1.7/1), reddish-brown (2.5YR 4/6), or dark red (10R 3/4) slip. White calcium carbonate and gray chert are the most common inclusion types found among the examples. Hematite and possible magnetite also occur in some of the clay fabrics. The elongated nozzles with thick walls and slightly pedestaled bases suggest that they had initially belonged to Hellenistic forms, such as the delphiniform type. They are included here, although their fragmentary preservation makes it difficult to obtain an accurate typological identification. They generally resemble the locally mold-made lamps of the second–first centuries B.C.E. and thus can be roughly dated to those centuries.

13. Reg. no. 94.1058.1L.1. L = 7.3 cm. W= 3.7 cm. Incomplete, 20% preserved. Hellenistic nozzle. Second–first centuries B.C.E. A long, narrow, cone-shaped nozzle with lip around circular, slightly sloping wick-hole. Thick wall. Flat undecorated base. Clay color: 2.5Y 6/1 yellowish gray. Slip color: 10YR 1.7/1 black (exterior only). Hardness, 3.5. Burning on nozzle. Poor condition. Mold-made.
14. Reg. no. 84.2244.1L.1. L = 6.5 cm. W = 5.4 cm. Incomplete, 20% preserved. Hellenistic nozzle. Second–first centuries B.C.E. Bottom portion of a long cone-shaped nozzle with raised ring base. Thin wall. Clay color: 5YR 6/6 orange. Slip color, exterior and interior surfaces: 2.5YR 4/6 reddish brown. Hardness: 2.8. Inclusion(s): gray (FS-CS, very low frequency); black (?) (FS, very low frequency). Burning on nozzle. Fair condition. Mold-made.
15. Reg. no. 84.2201.8L.1. L = 3.5 cm. W = 3.2 cm. Incomplete, 10% preserved. Hellenistic nozzle. Second–first centuries B.C.E. A broad, short cone-shaped nozzle with round, sharply sloping wick-hole. Thick wall. Clay color: 2.5YR 3/6 dark reddish brown. Hardness: 3.0. Inclusion(s): white (VFS-FS, low frequency); black (?) (FS, very low frequency). Burning on nozzle. Fair condition. Mold-made.
16. Reg. no. 84.2066.3L.1. Not shown. L = 4.4 cm. W = 3.0 cm. Incomplete, 25% preserved. Hellenistic nozzle. Second–first centuries B.C.E. A long and narrow nozzle with oval upright wick-hole. Nozzle-head with squared corners. Rounded shoulder. Wide rim around filling-hole. Thin wall. Clay color: 10YR 7/2 dull yellow-orange. Hardness: 2.5. Inclusion(s): gray (VFS, low frequency). Fair condition. Mold-made. Parallels: Capernaum (Loffreda 1974, fig. 28.2–3); Meiron (Meyers, Strange, and Meyers 1981: pl. 9.14:8).
17. Reg. no. 85.3407.7L.1. Not shown. GD = 4.8 cm. Incomplete, 5% preserved. Hellenistic with elongated square nozzle. Second–first centuries B.C.E. Complete nozzle with square head. Wick-hole, 1.2 cm × 1.1 cm diameter. Clay color: 10YR 6/1 light gray to gray. Slip color: 2.5Y 4/0 dark gray. Meticulously and evenly buffed surface. Nonvisible inclusions. Voids: round, poorly sorted (MF, VFS). Hardness: undetermined. Good condition. No burning. Possible Attic import or local imitation thereof. Mold-made.

Type 3. Northern Collar Neck (cat. no. 18)

Northern collar-neck lamps are distinguished by a tall neck with a curved and rounded collar. Sussman (1990: 97–98) points out that this important feature was fashioned in the Ephesus style from the last quarter of the second century B.C.E. to the early part of the first century C.E. The collar neck encircles a large filling-hole with a plain rim that greatly facilitates the pouring of oil into the fuel chamber. The collar neck represents nearly a third of the lamp in height. The collar tapers inward where it is joined to the upper body of the lamp. These lamps are further characterized by a globular body with a steeply sloping, rounded shoulder. The upper and lower surfaces of the nozzle bulge slightly. The wick-hole is large and rectangular in shape, and it slopes downward at approximately a 45° angle. The base of the lamp is round and slightly concave. A loop handle was attached to the collar and upper shoulder at the back of the lamp. Similar to the knife-pared Herodian lamps, the collar-neck lamps were thrown on a wheel, the evidence of which is visible as spiraling marks on the interior surface of their bases. Akin to Herodian lamp nozzles, the northern collar-neck nozzles were fashioned separately and attached to the body.

This type of lamp is not decorated, but the brick-red (5YR 5/6) fabrics caused by the iron-oxide-rich clay used for the manufacture of some examples are appealing to the eye. They are typically fired hard. Macroscopically visible, the fabric inclusions of this lamp type belong to a type associated with sedimentary rocks: angular quartz and moderate to high frequency of coarse, sand-size white (calcium carbonate) inclusions.

The geographic distribution of the northern collar-neck type is relatively restricted to northern sites (thus the "northern" component of this type's nomenclature). The probable origin of this lamp group is Galilee, where the most significant concentration of this type occurs. For this reason, it is surprising that only one, nearly complete lamp of this local Galilean type was unearthed in the western summit at Sepphoris (cat. no. 18; for additional examples of the northern collar-neck lamp found at the site, see Gärtner 1999: 27–28). The occurrence of northern collar-neck lamps is relatively rare, suggesting a small local Galilean manufacturing operation. Because it is wheel-made, any potter working in rural estates, villages, towns, and cities could have manufactured lamps of this type.

Northern collar-neck lamps are commonly recovered from tombs. In 1987, however, a complete example was found in association with an Early Roman boat from the Sea of Galilee (Sussman 1990: 97–98, fig. 12.1). It was recovered as part of the boat's fishing equipment, indicating that Galilean fishermen used lamps probably in the early morning as they set out to fish. The tall neck characteristic of this lamp type prevented oil spillage, making it all the more appealing to a fisherman in a swaying boat; it may even have been the lamp of choice among fishermen. Northern collar-neck lamps have also been excavated at the northern sites of Afula, Abu-Shusha/Geva, Capernaum, Karm er-Ras, and Nazareth (Fernandez 1983: 32; Feig 1990: 73–75, fig. 9.10–11).

Sussman (1990) reports two lamp fragments from this class that were recovered from Jerusalem, suggesting that Jewish pilgrims from Galilee may have used these lamps on their arduous journey to Jerusalem, perhaps during the three annual pilgrimage festivals. Papyrological evidence indicates that travelers took oil lamps with them on their journeys (Roberts and Turner 1952: 104–5, 118, no. 627). Perhaps they were also given to friends and relatives while people were visiting Jerusalem, as souvenirs from the Galilee. The only occurrence of this lamp type outside Roman Palestine is a complete example found in a tomb at Abila in Transjordan (Fuller

1987: 490, fig. 116.A–B). Their occurrence in Judea, the Jezreel Valley, and Transjordan indicates contact between the regions. Because of similarities to Early Roman wares, Sussman (1990) suggests that this lamp type was probably made from the mid- to late first century B.C.E. to the early first century C.E. (about 50 B.C.E.–50 C.E.).

18. Reg. no. 85.3300.4L.1. L, preserved = 9.0 cm. W, preserved = 6.2 cm. Incomplete, 40% preserved. Northern collar neck. Mid- to late first century B.C.E. to the early first century C.E. (about 50 B.C.E.–50 C.E.). Most of base and part of nozzle-bridge preserved. Conical nozzle. Raised round pedestal base (diameter, 3.4 cm). Wick-hole, preserved diameter: 1.1 cm. Clay color: 5YR 5/6 yellowish red. No slip. Inclusions: dark gray, sub-rounded (CS, VLF); gray (Csi, VHF). Hardness: undetermined. Fair condition. Hard fired. Burning around nozzle wick-hole. Wheel impressions on interior surface. Wheel-made. Shape-type: Caesarea Maritima (Sussman 2008: 218–19, no. 17, identified as Teapot 3 type and dated to first century C.E.).

Type 4. Herodian (cat. nos. 19–43)

Herodian lamps represent one of the most distinctive types produced in Roman Palestine. With its characteristic hand-fashioned splayed nozzle and large round oil chamber, the Herodian lamp was widely popular among the inhabitants of the region. This popularity can be explained in part by several of the Herodian lamp's features: its comparatively large oil reservoir provided fuel for prolonged lighting; a large wick-hole ensured a large flame, providing more lighting capability; and its body was rugged and durable. These attributes made the Herodian lamp the workhorse of the everyday lighting devices of Roman Palestine. Its widespread distribution in Roman Palestine and Arabia can also be explained by the ease with which this form could be manufactured; it could be made by any potter with a wheel, with no need for a mold, which generally required a skilled craftsperson to carve.

The term *Herodian* refers to the Herodian Dynasty, beginning in 37 B.C.E. and ending with the death of Herod Agrippa II, presumably in 100 C.E. This designation for the lamp is misleading, given that this lamp type was not manufactured until the later years of Herod the Great's reign, if not afterward, according to the archaeological evidence (Kahane 1961: 126–43; Avigad 1970: 129–40). The term further implies that this type was intimately linked to Herodian-controlled lands, when in fact it was also manufactured in neighboring Roman provinces, such as Nabatea, under different rulership (Barag and Hershkovitz 1994: 24). This lamp type also corresponds to Kennedy's Type 3 (1963: Type 3, 71–72, no. 487, pl. 20).

The spatulate-shaped nozzle and round wheel-made body of the Herodian lamp have typological similarities to Greco-Roman prototypes, as does the technique of paring the sides of the nozzle and oil reservoir with a knife. Traces of knife-paring are still visible on numerous examples. Reminiscent of the Cnidian type oil lamps of the second–first centuries B.C.E., the wide and sharply splayed nozzle of this lamp type was fashioned separately by the lamp maker and attached to a wheel-thrown body (Rosenthal and Sivan 1978: 80 n. 66). The lamp maker used a knife or scraper to pare the lamp and to trim any excess clay from its body; he or she finished the lamp by paring its base, sides, and nozzle. Evidence for such lamp makers' tools made from broken Roman common wares (and even sherds of glass vessels) have been uncovered in the hippodrome workshop at Gerasa/Jerash (Kehrberg 1992: 451–64). Paring took place when the clay had reached the leather stage just before firing. In addition to the Herodian type, lamps belonging to the Sepphorean spatulate, Darom, and Beit Nattif with Beit Nattif types also have spatulate-shaped nozzles.

The characteristic spatulated nozzles were fashioned separately and attached to the relatively large, globular wheel-made body before firing. That the nozzles are shaped from a thick clump of clay explains why this component of the lamp is often the only part that survives, as is the case with the Sepphoris examples. Traces of knife-paring are still visible on the outer surfaces of the bodies and nozzles of most lamps. The clay fabrics of Herodian lamps tend to be well levigated and smooth to the touch. A low-to-medium frequency of coarse, sand-size, white calcareous inclusions are typically visible in low-to-medium frequency on the surfaces of many of the lamps. Whether these inclusions represent naturally occurring temper or temper added by the lamp maker has not yet been determined by means of thin-section analysis.

The wick-hole and channel were created by the insertion of a wooden instrument held at an angle through the top of the nozzle and down its neck. A hole was punched through one of the side walls of the lamp's body either before or after the attachment of the nozzle, thereby enabling the wick channel of the nozzle to meet the hole made through the wall of the oil reservoir. One or more ridges, often of varying widths, surround the large filling-hole; some examples lack a rim altogether. The base of the body is usually flat and undecorated, and no lamp makers' marks occur on the base. In contrast to the Palestinian discus lamp type whose small filling-holes made it quite difficult to fill the lamp's chamber and required the breaking of the discus itself, the medium to large filling-hole of the knife-pared Herodian lamp greatly facilitated the introduction of oil into the reservoir. Knife-pared ring lamps also occur with multiple nozzles. The clay fabric of the knife-pared lamp is a buff color, but other samples are fashioned in yellow, red, or orange. Special examples have gray fabrics with a glossy black slip.

The decoration of these lamps is limited to simple geometric patterns of impressed concentric circles, lines, and roulettes. Decoration typically occurs on the neck or angular tips of the nozzle, though isolated specimens with decorated shoulders have also been found. This kind of ornamentation is continued in the Darom, Beit Nattif, and Samaritan traditions but in molded rather than in incised relief. The upper surface of the nozzles is often decorated with either two horizontally positioned concentric circles or a varied pattern of up to five circles. The circles are typically enclosed within two impressed simple or wavy lines. The lines are incised across the neck of the nozzle. In some instances, the line is impressed with a comb-like instrument. The association of combs with respect to Roman-period lamps is further attested in the Mishnah, which mentions the use of their broken billets presumably as wick adjusters (*m. Kelim* 13:8).

Herodian lamps inspired lamp-mold carvers of the Darom type: three Herodian lamps decorate the shoulder of a Darom lamp (Sussman 1982: 119, no. 231). Sussman (1982: 119) observes that the lamp motifs may also represent undecorated Darom lamps. Inscriptions on Herodian lamps are rare. A graffito ("Yeḥohanan") written in ink on the shoulder of a Herodian fragment unearthed at Masada may refer to the Jewish owner of the lamp (Barag and Hershkovitz 1994: 71, no. 125).

Barag and Hershkovitz (1994: 24–58) describe the conventional Herodian lamp as Type C and have distinguished 13 subgroups found at Masada. Six of these types are based on the size or shape of the body and/or rim (the diagnostic feature of each subgroup in square brackets): (1) Type CI [a broad rim]; (2) Type CII [a medium rim]; (3) Type CIII [a narrow rim]; (4) Type CIV [a large body and narrow rim]; (5) Type CV [a body with straight sides]; (6) Type CVI [a convex top]. Five of these types are characterized by knife-pared versions manufactured in gray clay fabrics: (1) Type CVII (the gray ware, broad and medium rim); (2) Type CVIII (the gray ware with handle); (3) Type CIX (the gray ware with various nozzle decorations); (4) Type CX (the gray ware with two wick-holes); (5) Type CXI (the gray ware three-nozzled lamp). And two of these types are characterized by the number of nozzles or ring bodies: (1) Type CXII (the four-nozzled lamps); (2) Type CXIII (the ring-lamps). The features that define each of these types among the Herodian group may be a sort of trademark: an attempt by their respective lamp makers to differentiate their product from those made by other potters.

Although several attempts have been made to establish a chronological order of the various types of Herodian lamps (Smith 1961: 53–65; Kahane 1961: 126–43; see discussions in Rosenthal and Sivan 1978: 80; MacDonnell 1988: 116–17; Barag and Hershkovitz 1994: 45–58; Fernandez 1983: 32–35), Rosenthal and Sivan's observation (1978: 82) that the different types may merely reflect differences of style between contemporaneous workshops throughout Palestine is certainly plausible. The knife-pared wheel-made lamps were probably introduced near the end of Herod's reign or soon afterward. Excavations have demonstrated their widespread distribution throughout Roman Palestine and in neighboring provinces during the first century C.E. (Rosenthal and Sivan 1978: 81; Barag and Hershkovitz 1994: 45–58).

Lychnologists favor restricting the chronological range of the Herodian lamp type to the Early Roman period—namely, from the last quarter of the first century B.C.E. to approximately 70 C.E., with some examples continuing as late as the mid-second century C.E. (Rosenthal and Sivan 1978: 80 nn. 63–64; Barag and Hershkovitz 1994: 44–47, 50–51, 53–56; Lapp 1997a: 17 n. 4, 21–22). Israeli and Avida (1988: 38) support a somewhat later range: from the first to the second century C.E. Considering the dates of Herodian lamps at other sites and the chronology of this lamp type at Sepphoris in relation to numismatic evidence, I support a chronological range at Sepphoris from the late first century B.C.E. to the second century C.E.

The Herodian lamp enjoyed a wide distribution; it has been found at many sites in Roman Galilee, Palestine, and Transjordan (Lapp 1997a: 19, 302–3, fig. 6). However, these lamps are rare outside Roman Palestine and Arabia; only two such lamp fragments have been found outside the region, both belonging to the version made of gray fabric with black slip. A splayed Herodian nozzle was recovered at Sidi Krebish Benghazi (modern Berenice) in Libya (Bailey 1985: 90, C628bis), and another was excavated at Palaepaphos, Cyprus (Bezzola 2004: 51–52, no. 228, pls. 6, 22). One complete Herodian lamp of unknown provenance is in the collection of the *Akademisches Kunstmuseum* at the University of Bonn (Hübinger 1993: 40, pl. 7.59).

At Sepphoris, a significant number of Herodian lamp fragments and nozzles (cat. nos. 19–43) were found in a variety of archaeological deposits in the western summit. They exhibit the characteristic splayed nozzle, large wick-hole, and, with respect to cat. nos. 27–30, simple geometric decoration (circles and lines) impressed on the neck and tips of their nozzles. One group (cat. nos. 19–21) is characterized by a rounded shoulder and a ledge (approximately 3 mm wide) followed by a prominent ridge around the central filling-hole. The shoulder of cat. no. 20 is wider than that of cat. no. 21. The most dominant inclusion type in the Herodian examples at Sepphoris is white calcium carbonate. Red inclusions of hematite are found in several lamps (cat. nos. 19, 20). These examples are fashioned in varying hues of orange and yellow clay. Catalog no. 32 is fashioned in a light gray-gray clay (10YR 6/1) and is covered in a black slip (2.5Y 2/0), suggesting manufacture in the Jerusalem area. Catalog nos. 35–38 and 41–42 also have gray fabrics and black slips indicative of a Jerusalem origin. Nozzles are either plain (cat. no. 19–22) or bear an incised or comb-impressed line across the neck (cat. nos. 27–30), sometimes with two opposing circles (cat. nos. 27–28).

One nozzle (cat. no. 29) is decorated with two parallel incised lines, between which are three concentric circles. This design, however, does not represent the special oddity of this example but, rather, the fact that it is made of a light-gray clay and has a black slip. In many respects, therefore, this example is paralled by other relatively rare occurrences of black-slipped Herodian lamps (Barag and Hershkovitz 1994: 53–58, nos. 78–101).

Several studies have examined specimens of the Herodian lamp type in an effort to determine its provenance. The earliest scientific study of a Herodian lamp was conducted by Asaro (1981: 141–43, "Meir-9," table 4), who analyzed a single fragment from Meiron. The analysis of a single sherd cannot in itself determine the provenance of its manufacture, for, as Gunneweg and Perlman (1984–85: 80) point out, it is necessary to find a compositional match with reference material. The importance of Asaro's analysis lies in the fact that, by conducting the first chemical analysis of any lamp fabric recovered from a stratigraphic context in Roman Palestine, he was the first to recognize the importance and the potential of trace-element analyses of lamp fabrics.

Gunneweg and Perlman (1984–85: 80–82) conducted further neutron activation analysis on six Herodian lamp specimens, three from Jerusalem and three from Avdat (Oboda); their work revealed that the six fragments share similar compositions and also match the lamp from Meiron. Because they also match the chemical compositions of ceramic reference material from Jerusalem, comprising a group of 20 vessels dating to the Early Roman period, in addition to clay collected from the Jerusalem vicinity, Gunneweg and Perlman (1984–85: 80–81, table 1) have shown that the Herodian lamp specimens from Meiron, Avdat, and Jerusalem reflect the composition of pottery manufactured in Jerusalem. These chemical matches suggest cultural contact between the regions of the Negev, Judea, and Galilee. Petrographic thin-section and trace-element analyses of select Herodian lamp samples from Sepphoris confirm that this lamp type was manufactured in different regions and presumably various workshops (Adan-Bayewitz et al. 2008).[2] As further thin-section analyses will most likely show, numerous workshops located in rural and urban centers of widespread geographic range were manufacturing Herodian lamps using clay gathered at a number of different sources. The Jerusalem area represents just one of the many sources for the manufacture of knife-pared lamps.

Yellin's NAA (Yellin 1994: 109, table 2b) of several fragments of the Herodian lamp type from Masada suggests that 12 of the samples originated from the Jerusalem area (Masada samples MSDL 7–11, 13–19) and one from the north coast (Masada sample MSDL 12). His findings corroborate those of Gunneweg and Perlman (1984–85: 80), who argue for a Jerusalem origin for several Herodian lamps found at Avdat, Jerusalem, and Meiron. These analyses may mean that the Herodian lamps from Sepphoris also have originated from workshops in the Jerusalem area and/or the north coast. If so, the Jewish community of Sepphoris likely had contact with their counterparts in those regions. But in order to determine whether the Herodian lamps from the western summit at Sepphoris did indeed originate from either the Jerusalem area, the north coast, or both localities, they first must be analyzed using a combination of trace-element (e.g., dcp-oes [direct coupled plasma optical emission spectrometry], icp-oes [inductively coupled plasma atomic emission spectrometry], or NAA) and petrographic thin-section analyses, the results of which must be calibrated to and compared with the lamp fabrics and clay samples of the studies carried out by Yellin, Gunneweg, and Perlman. It is also possible that future analyses of this kind may indicate that all or several of the Herodian lamps from the western summit at Sepphoris were manufactured at other workshops in different locations. Because Herodian lamps are wheel-made, any pottery workshop could have manufactured them (Lapp 1997a: 23; 2001c: 296; 2006b: 315). This may explain in part their wide distribution in Roman Palestine, Galilee, and Transjordan (Lapp 1997a: 19, 302–3, fig. 6).

The most comprehensive, regional analytical study on the origins of the Herodian lamp was conducted at the Ernest Orlando Lawrence Berkeley National Laboratory (Adan-Bayewitz et al. 2008). The Berkeley analysis measured 176 Herodian lamps using instrumental neutron activation and high-precision X-ray fluorescence

2. Previous petrographic observations also reported clay variation among the fabrics of Herodian nozzles from Sepphoris, thus suggesting possible different sources (Lapp 1997: 23). The observations were preliminary in nature. Because Herodian lamp fabrics were not part of the archaeometric study of my dissertation, no petrographic descriptions were presented for this lamp type (cf. Adan-Bayewitz et al. 2008). However, detailed macroscopic, petrographic, and chemical compositional descriptions were included for each sample of the three main lamp types selected for analysis in my study: the Palestinian discus; the bilanceolate; and the Beit Nattif, northern version groups (Lapp 1997). I adopted a two-method approach for the archaeometric component of my dissertation, combining petrographic thin-section and dcp-oes chemical analyses of samples belonging to the latter three types and others. The samples were collected from multiple sites from various regions of Israel and Jordan.

analyses, 96 of which were further analyzed petrographically. The lamp samples came from the coastal plain site of Dora, the Galilean sites of Sepphoris and Iotapata, Gamala in the Golan, and the Decapolis city of Scythopolis. Thirteen micromorphological groups were identified among the samples: Jerusalem and Ha-Motza; Iotapata; Dora (Groups 1–4); Sepphoris (Groups 1–2); and Scythopolis (Groups 1–5; see Adan-Bayewitz et al. 2008: 50–69, table 3). The study showed that the large majority of Herodian lamps from the predominantly Jewish sites of Sepphoris, Iotapata, and Gamala came from workshops in the Jerusalem area, while most of those from the predominantly Gentile cities of Dora and Scythopolis were manufactured in the north (Adan-Bayewitz et al. 2008: 68–69, 77). The study concludes that "[t]he pronounced preference shown by the Jews of Gamala, Iotapata, and Sepphoris for Herodian lamps specifically from the near vicinity of Jerusalem apparently demonstrates their strong ties with that area in the late Second Temple period" (Adan-Bayewitz et al. 2008: 77).

19. Reg. no. 84.1120.3L.2. L = 8.6 cm. W = 6.3 cm. H = 2.7 cm. Complete. Herodian. Late first century B.C.E. to mid-second century C.E. Circular wheel-made body with a rounded shoulder. Slightly sloping spatulated nozzle with a wick-hole slightly wider than long. Small filling-hole with a rim surrounded by a relatively wide ledge and a prominent ridge. Flat undecorated base. Clay color: 10YR 8/4 light yellow-orange. Hardness: 2.5. Inclusion(s): gray (FS, low frequency); red (VFS, low frequency); white (FVS, low frequency); mica (CSi, very low frequency). Heavy burning on nozzle. Very good condition. Parallels, shape-type: Caesarea Maritima (Sussman 2008: 219–20, nos. 19–23); Khirbet Shemaᶜ (Meyers, Kraabel, and Strange 1976: pl. 8.11:11–12), Jerusalem (Barag 1973: 103, fig. 3.5) (similar); Meiron (Meyers, Strange, and Meyers 1981: pl. 9.14:14–15; pl. 9.15:1–19); Pella (da Costa 2003: 2.503, nos. 1–5); provenance unknown (Rosenthal and Sivan 1978: 81, nos. 14–15; Israeli and Avida 1988: 40, fig. 59, exact); Adler Collection (Adler 2004: 16–31, nos. 71–167).
20. Reg. no. 84.1068.17L.1. L = 9.3 cm. W = 8.2 cm. Incomplete, 45% preserved. Herodian. Late first century B.C.E. to mid-second century C.E. Wide, circular body fragment with rounded shoulder. Slightly sloping spatulated nozzle with a wick-hole slightly longer than wide. Small filling-hole, the rim of which is surrounded by a relatively wide ledge and a prominent ridge. Flat base. Clay color: 10YR 8/4 light yellow-orange. Hardness: 2.5. Inclusion(s): gray (FS, low frequency); red (VFS, low frequency); white (VFS, low frequency); mica (CSi, very low frequency). Extensive burning on nozzle. Good condition. Wheel-made. Parallels, shape-type: Caesarea Maritima (Sussman 2008: 219–20, nos. 19–23); provenance unknown (Israeli and Avida 1988: 40, figs. 60–62) (exact).
21. Reg. no. 85.3125.2L.3. L = 7.3 cm. W = 5.2 cm (preserved width only). H = 2.5 cm. Nearly complete, 85% preserved. Herodian. Late first century B.C.E. to mid-second century C.E. Complete nozzle and shoulder fragment. Filling-hole missing. Wick-hole diameter, 1.3 cm × 1.2 cm. Pronounced ridge (3.5 cm diameter) around filling-hole. Thin wall, 0.2 cm. Clay color: 5YR 7/4 pink. No slip, but a wash fired to 10YR 8/3 very pale brown. Inclusions: white sub-angular (LF, CS). Voids (HF, VFS). Hardness: 3.0. Fair condition. Lime-popping. Moderate burning on nozzle. Straw impressions on base. Wheel-made. Parallels, shape-type: Caesarea Maritima (Sussman 2008: 219–20, nos. 19–23).

22. Reg. no. 84.1068.18L.1. L = 8.0 cm. W = 4.7 cm. Incomplete, 35% preserved. Herodian. Late first century B.C.E. to mid-second century C.E. Sharply spatulated nozzle with a wick-hole whose width is less than its length. Level, nonsloping nozzle. Significant portion of lamp body (upper and lower) is missing. Remnant of rounded shoulder. Flat undecorated base. Clay color: 5YR 7/6 orange. Core color: 10YR 7/1 light gray. Hardness: 3.0. Inclusion(s): white (MS-CS, medium frequency). Burning on nozzle. Excellent condition. Wheel-made. Parallels, shape-type: Caesarea Maritima (Sussman 2008: 219–20, nos. 19–23).
23. Reg. no. 84.1068.17L.3. L = 7.9 cm. W = 5.8 cm. H = 2.6 cm. Nearly complete, 95%preserved. Herodian. Late first century B.C.E. to mid-second century C.E. Circular wheel-made body with a rounded shoulder. Slightly upward projecting nozzle with a wick-hole whose width is less than its length. Tips of nozzle are sharply angular. Small filling-hole with a rim surrounded by a ledge and a prominent ridge. Two finely incised lines encircle ridge on the shoulder. Walls of medium thickness. Impressed base. Clay color: 5YR 8/4pale orange. Hardness: 2.8. Inclusion(s): white (MS, medium frequency). Extensive burning on nozzle. Fair condition. Wheel-made. Parallels, shape-type: Caesarea Maritima (Sussman 2008: 219–20, nos. 19–23); Capernaum (Loffreda 1974: fig. 28.4) (exact); Jordan Valley, east of Tubas (C. A. Kennedy 1963: pl. 20.487) (exact); provenance unknown (Israeli and Avida 1988: 40, fig. 59) (exact).
24. Reg. no. 84.4134a.3L.1. L = 6.4 cm. W = 5.9 cm. Incomplete, 45% preserved. Herodian. Late first century B.C.E. to mid-second century C.E. Slightly sloping spatulated nozzle with a wick-hole whose width is greater than its length. Circular body with a flat shoulder. Small filling-hole surrounded on the outside by a very slight ledge and a fairly prominent ridge. A single incised line composed of a consecutive series of tightly spaced rectangles, thus suggesting the application of a comb-like instrument, crosses neck of nozzle. Two impressed concentric circles are represented directly beneath the line. Clay color: 7.5YR 7/6 orange. Core color: 10YR 6/3 dull yellow-orange. Hardness: 3.8. Inclusion(s): red (FS-CS, low frequency); white (FS-CS, low frequency). Extensive burning on nozzle. Good condition. Wheel-made. Parallels, shape-type: Amman (Harding 1946, pl. 20.2); Caesarea Maritima (Sussman 2008: 219–20, nos. 19–23); Maresha (Oren and Rappaport 1984: 137, fig. 14) (close); provenance unknown (Israeli and Avida 1988: 39, fig. 55) (close); Qumran (Patrich and Arubas 1989: pl. 5.C, fig. 3.12) (similar).
25. Reg. no. 84.1068.17L.2. L = 8.2 cm. W = 4.9 cm. Incomplete, 55% preserved. Herodian. Late first century B.C.E. to mid-second century C.E. Circular body with flat shoulder. Nozzle-head and most of nozzle-bridge missing. Small- to medium-sized filling-hole surrounded on the outside by a ledge approximately twice the width of the thin outer ridge. A single incised line composed of a series of tightly spaced rectangles (thus suggesting the application of a comb-like instrument) crosses the neck of the nozzle. Two impressed concentric circles are represented directly beneath the line. Flat undecorated base. Clay color: 5YR 8/3 pale orange. Hardness: 3.0. Inclusion(s): white (MS-CS, medium frequency; VCS, low frequency). Burning on nozzle and body. Fair condition. Wheel-made. Parallels, shape-type: Caesarea Maritima (Sussman 2008: 219–20, nos. 19–23).
26. Reg. no. 84.5001a.4L.1. L = 8.7 cm. W = 5.1 cm. Incomplete, 45% preserved. Herodian. Late first century B.C.E. to mid-second century C.E. Wide circular body with flat shoulder. Non-sloping, slightly spatulated nozzle with a filling-hole wider than long.

Significant portion of upper body is missing. Two parallel incised lines composed of a series of indiscriminately spaced rectangles cross the neck of the nozzle. Flat undecorated base. Clay color: 5YR 6/6 orange. Hardness: 3.0. Inclusion(s): white (MS, medium frequency). Burning around wick-hole. Good condition. Wheel-made. Parallels, shape-type: Caesarea Maritima (Sussman 2008: 219–20, nos. 19–23).

27. Reg. no. 84.4134a.1L.1. L = 4.0 cm. W = 1.4 cm. Incomplete, 10% preserved. Herodian. Late first century B.C.E. to mid-second century C.E. Spatulated nozzle with a wick-hole whose width is greater than its length. Sloping nozzle. A single incised line with wavy edge (oriented toward the wick-hole) crosses neck of nozzle. Two impressed concentric circles in juxtaposition rendered directly beneath line. Trace of evenly spaced, impressed squares (produced by a comb-like instrument) located to the lower left of the circle. Clay color: 2.5Y 6/1 yellowish gray. Slip color: 10YR 3/1 brownish black (exterior only). Burnish: Very well burnished. Hardness: 3.0. Inclusion(s): white (MS–CS, low frequency). Good condition. Wheel-made. Parallels, shape-type: Caesarea Maritima (Sussman 2008: 219–20, nos. 19–23); Khirbet Shemaʿ (Meyers, Kraabel, and Strange 1976: pl. 8.11:11–12); Meiron (Meyers, Strange, and Meyers 1981: pl. 9.14:14–15; pl. 9.15:1–19); provenance unknown (Rosenthal and Sivan 1978: 81, nos. 331–34).

28. Reg. no. 84.5001a.6L.1. L = 3.4 cm. W = 3.7 cm. Incomplete, 10% preserved. Herodian. Late first century B.C.E. to mid-second century C.E. Spatulated nozzle with a wick-hole whose width is greater than its length. Sloping nozzle. A single incised line composed of indiscriminately spaced rectangles under which are two impressed concentric circles. Clay color: 2.5YR 7/6 orange. Slip color: 7.5YR 6/6 orange, slip flaking on exterior surface. Hardness: 3.0. Inclusion(s): red (FS, very low frequency); white (FS-MS, medium frequency). Fair (+) condition. Wheel-made. Parallels, shape-type: Caesarea Maritima (Sussman 2008: 219–20, nos. 19–23).

29. Reg. no. 84.1068.17L.4. L = 3.6 cm. W = 1.3 cm. Incomplete, 10% preserved. Herodian. Late first century B.C.E. to mid-second century C.E. Spatulated nozzle with a wick-hole whose width is greater than its length. Sharply sloping nozzle. Two impressed concentric circles with three inner rings are situated on each point of the nozzle-head. Three concentric circles with three internal rings are closely impressed across the neck of the nozzle. These circles are enclosed from above and below by an incised line created with a comb-like instrument with rectangular-shaped teeth. Clay color: 10YR 7/1 light gray. Slip color: 10YR 1.7/1 black (exterior and interior). Hardness: 2.0. Inclusion(s): white (FS, low frequency). Burning on nozzle. Fair condition. Wheel-made. Parallels, shape-type: Caesarea Maritima (Sussman 2008: 219–20, nos. 19–23); Jerusalem (Amiran and Eitan 1970: pl. 7.B [first nozzle, upper left] = *EI* 11, pl. 53, fig. 1; Hamilton 1940: 30, fig. 14.8) (exact); Khirbet el-Mukhayyat (Alliata 1988: 343, fig. 10.17) (exact).

30. Reg. no. 84.5030.3L.1. L = 3.6 cm. W = 3.2 cm. Incomplete, 10% preserved. Herodian. Late first century B.C.E. to mid-second century C.E. A sharply spatulated nozzle with a wick-hole whose width is considerably less than its length. Nozzle slopes slightly. A single incised line of well-articulated, evenly spaced rectangles at base of nozzle neck. Slight trace of similar comb-made line of rectangles situated directly beneath the upper line. Clay color: 5YR 6/6 orange. Hardness: 3.0. Inclusion(s): white (FS, low frequency). Burning on nozzle. Fair condition. Wheel-made. Parallels, shape-type: Caesarea Maritima (Sussman 2008: 219–20, nos. 19–23).

31. Reg. no. 85.4001.5L.1. L = 6.0 cm. W = 3.4 cm. Incomplete, 10% preserved. Herodian. Late first century B.C.E. to mid-second century C.E. Spatulated nozzle with a wick-hole whose width is slightly greater than its length. Nozzle slopes slightly. A single line beneath which two impressed concentric circles with one inner ring are represented on the neck of the nozzle. Clay color: 10YR 7/3 dull yellow-orange. Hardness: 3.0. Inclusion(s): white (VFS, very low frequency). Burning on nozzle. Good condition. Wheel-made. Parallels, shape-type: Caesarea Maritima (Sussman 2008: 219–20, nos. 19–23); Capernaum (Loffreda 1974: fig. 43.8–9) (exact).
32. Reg. no. 85.1208.1L.4. Not shown. GD, splayed nozzle width, tip to tip = 3.1 cm. Incomplete, 10% preserved. Herodian. Late first century B.C.E. to mid-second century C.E. Nozzle fragment. On nozzle: two concentric circles, one on each tip. On nozzle-bridge: two impressed, parallel billet lines between which are two impressed concentric circles. Wick-hole, 1.6 cm × 1.4 cm diameter. Underside: two flat planes. Clay color: 10YR 6/1 light gray-gray. Slip color: 2.5Y 2/0 black. Inclusions: fair sorting, white angular (HF, VFS). Hardness: 3.0. Good condition. Medium firing. No burning. Wheel-made. Parallels, shape-type: Caesarea Maritima (Sussman 2008: 219–20, nos. 19–23).
33. Reg. no. 85.1208.1L.1. Not shown. GD = 4.3 cm. Incomplete, 10% preserved. Herodian. Late first century B.C.E. to mid-second century C.E. Nozzle fragment. Plain nozzle-bridge. Wick-hole, 1.5 cm diameter. Underside: two wide flat planes and one narrower plane. Clay color: 5YR 7/4 pink. Slip color: 7.5YR 7/6 reddish yellow. Inclusions: well sorted, white angular (HF, VFS). Hardness: 3.0. Good condition. Medium firing. Burning near wick-hole. Wheel-made. Parallels, shape-type: Caesarea Maritima (Sussman 2008: 219–20, nos. 19–23).
34. Reg. no. 85.1181.3L.2. Not shown. GD = 3.8 cm. Incomplete, 10% preserved. Herodian. Late first century B.C.E. to mid-second century C.E. Nozzle fragment. Wick-hole, 1.7 cm × 1.3 cm diameter. Underside: two wide splays. Clay color: 5YR 7/6 reddish yellow. No slip. Discoloration from firing on left side of splay and on interior surface (7.5YR 7/6 reddish yellow). Inclusions: white angular (HF, VFS). Hardness: 3.0. Good condition. Medium firing. No burning. Wheel-made. Parallels, shape-type: Caesarea Maritima (Sussman 2008: 219–20, nos. 19–23).
35. Reg. no. 85.3100.5L.1. GD = 5.7 cm. Incomplete, 10% preserved. Herodian. Late first century B.C.E. to mid-second century C.E. Nozzle fragment. On nozzle-bridge: two impressed concentric circles, one on each splay tip; two parallel, impressed comb-billet lines across bridge. Wick-hole, 1.5 cm diameter. Underside: two wide flat planes. Clay color: 10YR 6/1 light gray-gray. Slip color: 2.5Y 3/0 very dark gray. Slip unevenly applied. Inclusions: poorly sorted, white sub-angular (MF, VFS). Hardness: 3.0. Good condition. Medium firing. No burning. Wheel marks. Probable Jerusalem origin. Wheel-made. Parallels, shape-type: Caesarea Maritima (Sussman 2008: 219–20, nos. 19–23).
36. Reg. no. unmarked. Not shown. GD = 4.6 cm. Incomplete, 5% preserved. Herodian. Late first century B.C.E. to mid-second century C.E. Complete nozzle fragment. On nozzle-bridge: asymmetrical positioning of two impressed concentric circles, one on each splay tip; two parallel, impressed comb-billet lines diagonally across bridge. Wick-hole, 1.5 × 1.25 cm diameter. Underside: two wide flat planes. Clay color: 10YR 6/1 light gray-gray. Slip color: 2.5Y 3/0 very dark gray. Inclusions: poorly sorted, white sub-angular (MF, VFS). Hardness: 3.5. Good condition. Medium-hard

firing. No burning. Probable Jerusalem origin. Wheel-made. Parallels, shape-type: Caesarea Maritima (Sussman 2008: 219–20, nos. 19–23).

37. Reg. no. 85.3103.2L.1. GD = 5.4 cm. Incomplete, 20% preserved. Herodian. Late first century B.C.E. to mid-second century C.E. Nozzle, shoulder, underside, and handle fragment. Filling-hole, greatest preserved arc 2.0 cm. Underside: two wide flat planes. Complete loop handle with two prominent grooves. Loop 1.4 cm diameter. Pronounced ridge around filling-hole. Thin wall, 0.2 cm. Clay color: 10YR 6/1 light gray-gray. Slip color: 2.5Y 3/0 very dark gray. Inclusions: white sub-angular (MF, VFS). Hardness: 5.0. Fair condition. Hard fired, clings when tapped. No burning. Probable Jerusalem origin. Wheel-made. Parallels, shape-type: Caesarea Maritima (Sussman 2008: 219–20, nos. 19–23).
38. Reg. no. 85.1000.1L. GD = 6.0 cm. Not shown. Incomplete, 10% preserved. Herodian. Late first century B.C.E. to mid-second century C.E. Complete nozzle, shoulder, and underside fragment. On nozzle-bridge: row of three impressed overlapping concentric circles beneath impressed comb-billet lines across bridge. Wick-hole, 1.4 × 1.2 cm diameter. Prominent ridge at edge of body. On upper shoulder: a single air-hole, 0.1 cm diameter. Underside: three wide flat planes. Clay color: 10YR 6/1 light gray-gray. No slip. Discoloration 10YR 7/2 light gray in places caused by secondary firing. Inclusions: black angular (VLF, VFS). Hardness: 3.5. Fair condition. Chipped underside. Medium fired. Slight burning around wick-hole. Probable Jerusalem origin. Wheel marks on interior surface. Two clumps of excess clay resulting from instrument pushing through oil chamber wall to create wick-hole passage. Wheel-made. Parallels, shape-type: Caesarea Maritima (Sussman 2008: 219–20, nos. 19–23).
39. Reg. no. 85.1145.2L.3. Not shown. GD = 4.2 cm. Incomplete, 20% preserved. Herodian. Late first century B.C.E. to mid-second century C.E. Nozzle, shoulder, handle, and underside fragment. On nozzle: single impressed concentric circle on left splay tip. Wick-hole, 2.0 cm arc. Loop handle. Clay color: 10YR 6/1 light gray-gray. Slip color: 2.5Y 3/0 very dark gray. Inclusions: white sub-angular (MF, VFS). Hardness: 3.0. Poor condition. Medium firing. No burning. Probable Jerusalem origin. Nozzle L85.1145.3L.1 (GD = 3.0 cm) probably belongs to this lamp. Wheel-made. Parallels, shape-type: Caesarea Maritima (Sussman 2008: 219–20, nos. 19–23).
40. Reg. no. 85.3318.3L.1. Not shown. GD = 5.2 cm. Incomplete, 5% preserved. Herodian. Late first century B.C.E. to mid-second century C.E. Very large nozzle-head. Splay width tip-to-tip, 4.1 cm. On nozzle: single impressed concentric circle on left splay tip. Wick-hole, 2.4 cm × 1.6 cm diameter. Nozzle-underside: three wide flat planes. Loop handle. Clay color: 5YR 6/6 reddish yellow. Core color: 7.5YR 7/6 reddish yellow. Inclusions: white angular to sub-angular (LF, pebble); poorly sorted, gray sub-angular (MF, VFS); angular voids (VHF, F-VFS) filled in with groundwater salts creating impression of inclusions. Buffed surface. Hardness: 5.0. Hard fired. Poor condition. Conchoidal fracture. Burning near wick-hole. Probable Jerusalem origin. Wheel-made. Parallels, shape-type: Caesarea Maritima (Sussman 2008: 219–20, nos. 19–23).
41. Reg. no. 85.3301.2L.1. Not shown. GD = 5.0 cm. Incomplete, 5% preserved. Herodian. Late first century B.C.E. to mid-second century C.E. Complete nozzle and shoulder underside fragment. Splay width tip-to-tip, 3.2 cm. On nozzle: single impressed concentric circle on left splay tip. Wick-hole, 1.4 cm × 1.4 cm. Nozzle-underside: two flat wide planes. Clay color: 10YR 6/1 light gray-gray. Slip color: 2.5Y 3/0 very dark

gray. Inclusions: poorly sorted, white angular to sub-angular (MF, FS and VLF, CS); black (LF, FS). Hardness: Mohs 3.0. Medium firing. Poor condition. Lime-popping. No burning. Probable Jerusalem origin. Wheel-made. Parallels, shape-type: Caesarea Maritima (Sussman 2008: 219–20, nos. 19–23).

42. Reg. no. 85.3313.2L. Not shown. GD = 2.8 cm. Incomplete, 3% preserved. Herodian. Late first century B.C.E. to mid-second century C.E. Nozzle fragment. Splay width tip-to-tip, 3.1 cm. On nozzle: two impressed concentric circles, one on each tip. Wick-hole, width only 1.3 cm. Nozzle-underside: two flat wide planes and tool marks. Clay color: 10YR 6/1 light gray-gray. Slip color: 2.5Y 3/0 very dark gray. Inclusions: poorly sorted, white angular to sub-angular (MF, FS and VLF, CS); black (LF, FS). Hardness: Mohs 3.0. Medium firing. Fair condition. No burning. Probable Jerusalem origin. Wheel-made. Parallels, shape-type: Caesarea Maritima (Sussman 2008: 219–20, nos. 19–23).
43. Reg. no. 85.3216.7L.1. Not shown. L = 9.1 cm. W = 6.8 cm. H = 2.5 cm. Nearly complete, 80% preserved. Herodian. Late first century B.C.E. to mid-second century C.E. Two mended fragments. Complete nozzle and shoulder underside fragment. Splay width tip-to-tip, 3.2 cm. On nozzle: single impressed concentric circle on left splay tip. Wick-hole, 2.0 cm × 1.4 cm. Filling-hole, 2.4 cm diameter. Very prominent ridge around filling-hole. Nozzle-underside: two, flat wide planes and a third, central, narrower plane. On base: many straw impressions and tool marks; definitive matting impressions. Clay color: 2.5YR 5/6 and 4/6 red. No slip. Inclusions: fair-to-good sorting, round to sub-angular voids (VHF, VFS) filled in with groundwater salts creating impression of inclusions. Hardness: 5.0. Hard fired. Good condition. Heavy burning on nozzle. Wheel impressions on interior surface. Uncharacteristic red brick fabric similar to that of the northern collar-neck type. Wheel-made. Parallels, shape-type: Caesarea Maritima (Sussman 2008: 219–20, nos. 19–23).

Type 5. Sepphorean Spatulate (cat. nos. 44–52)

Mold-made lamps with spatulate-shaped nozzles, circular bodies, and simple linear decoration have been found at Sepphoris (cat. nos. 44–52). A lamp mold used to make this lamp type was found at the site, suggesting the mold may have been carved in a workshop located in the city or nearby. Its discovery further suggests manufacture of the Sepphorean spatulate type in one or more workshops located in the city or surrounding area (see discussion in chap. 5). They belong to the Huqoq Type (d) group, also referred to as the mold-made version of the Herodian lamp because of their splayed nozzles and round bodies (Kahane 1961: 168). This nomenclature is misleading, however, because the contemporary mold-made Darom and Masada D lamp types also have spatulated nozzles and circular bodies. To avoid this confusion and to reflect the discovery of a lamp mold at Sepphoris suggesting its manufacture there, the name *Sepphorean spatulate* is preferable and was adopted for this corpus (Lapp 1991: 61–63). By contrast, the production centers for the Darom type are Judea and Samaria, and for the Herodian, throughout Palestine. The splayed nozzle is one of their chief diagnostic characteristics.

Sepphorean spatulate lamps are a distinct lamp type and should not be confused with the Herodian, Darom, and Masada Type D forms. Several diagnostic characteristics—a pronounced, circular, and slightly convex base; a pronounced throat-shaped bulge on the nozzle-underside—distinguish them from those other types. Additionally, no decorative device such as a leaf, which is characteristic of the Masada D lamps, is found on the underside of the Sepphorean spatulate lamps. And, unlike many Herodian lamps, no circle devices decorate their bodies or nozzles. Their fabric is consistently pale white and finely levigated and has a calcium carbonate rich matrix. In addition, no ridge surrounds the medium-sized filling-hole.

Sepphorean spatulate lamps have similarities to the wheel-made Herodian (Masada CI–III; see Barag and Hershkovitz 1994: 59–78), Darom, and Masada Type D (molded with floral patterns) forms. Like Herodian lamps, the Sepphorean spatulate lamps' shoulder lacks ornamentation, thus differentiating it from the Darom variety, which is typically decorated with Judean folk art: a geometric or floral pattern or daily-life objects. In contrast, the nozzles and shoulders of Darom and Masada Type D lamps tend to be richly ornamented. At most, a single band of two closely juxtaposed parallel ridges in high relief cross the midsection of the nozzle-neck. Another similar band of double lines follows the outer edge of the shoulder, which flares slightly upward in profile view. In section, the elongated flattened body of the Sepphorean spatulate lamp resembles that of Darom lamps. Also like Darom lamps, the Sepphorean spatulate type tends to be unslipped. Simple in appearance, the Sepphorean spatulate lamp's round body, spatulated nozzle, and brief ornamentation resembles the wheel-made Herodian lamp.

The Sepphorean spatulate lamp type is found mainly at Sepphoris and suggests that it was manufactured there or at nearby Shikhin. It appears at few other sites in Galilee or elsewhere in Israel and Jordan. Examples have been found at Jalame (MacDonnell 1988: 116–17, 122, pl. 6-1.4, probably dated to the late first or early second century C.E.) and Beth Shean (Hadad 2002: 16–17, no. 12, Type 6, misidentified as a Darom lamp). Two others were found at Wadi el-Badhan and Silet edh-Dhahr in northern Israel (Kahane 1961: 138). Three were found at Qumran, ʿAin Feshkha, and Murabbaʿat in the Dead Sea area (Kahane 1961: 138). That it occurs in greater abundance at Sepphoris than at any other site suggests possible manufacture there. Its close typological similarities to the Darom and Masada Type D lamps suggest contemporary manufacture and use. The Sepphorean spatulate dates to the mid-first–early second centuries C.E. (Kahane 1961: 168). It is the Galilean answer to the Judean- and Samaritan-produced Darom lamp and is the common ware equivalent of the finer, more aesthetically appealing Darom lamp.

44. Reg. no. 84.1180.1L.1. L = 9.2 cm. W = 5.9 cm. H = 2.5. Nearly complete, 95%. Sepphorean spatulate. Mid-first–early second centuries C.E. Circular body. Slightly sloping shoulder. Large central filling-hole. Spatulated nozzle with a circular wick-hole. Bulging frog-throat nozzle bottom. Two closely juxtaposed raised lines follow the outer ridge of the circular body. Raised circular-shaped flat base. Clay color: 5YR 7/4 dull orange. Hardness: 3.0. Extensive burning on nozzle and on interior and exterior surfaces. Fair condition. Mold-made. Parallels, shape-type: Jalame (MacDonnell 1988: 116–17, 122, pl. 6–1.4); Beth Shean (Hadad 2002: 16, 17, no. 12, Type 6, misidentified as a Darom lamp); IAA Collection (Sussman 2012: 92–94, nos. 907–19).

45. Reg. no. 84.1068.15L.7. L = 9.2 cm. W = 6.0 cm. Nearly complete, 85%. Sepphorean spatulate. Mid-first–early second centuries C.E. Circular body. Slightly sloping shoulder. Large central filling-hole. Spatulated nozzle. Upright, level nozzle. Approximately one-half of nozzle and a portion of the body top and base are missing. A single band of two closely juxtaposed ridges of equal width crosses the midsection of the nozzle. Slightly elevated, wide discus base. Clay color: 10YR 8/2 light gray. Hardness: 2.0. Burning on nozzle. Fair condition. Mold-made. Parallels, shape-type: IAA Collection (Sussman 2012: 92–94, nos. 907–19).
46. Reg. no. 84.2051.3L.1. L = 6.6 cm. W = 5.9 cm. Incomplete, 35%. Sepphorean spatulate. Mid-first–early second centuries C.E. Circular body. Slightly sloping shoulder. Large central filling-hole. Spatulated nozzle. Upright, level nozzle. Two closely juxtaposed raised lines follow outer ridge of circular body. A single band of two closely aligned ridges of equal width crosses the lower portion of the nozzle neck. Part of nozzle top and side, the base, and a considerable portion of the shoulder are missing. Clay color: 10YR 8/4 light yellow-orange. Hardness: 3.0. Inclusion(s): white (VCS, very low frequency). Burning on nozzle. Fair condition. Mold-made. Parallels, shape-type: IAA Collection (Sussman 2012: 92–94, nos. 907–19).
47. Reg. no. 84.1068.16L.2. L = 7.0 cm. W = 4.7 cm. H = 2.5 cm. Incomplete, 30%. Sepphorean spatulate. Mid-first–early second centuries C.E. Circular body. Slightly sloping shoulder. Large central filling-hole. Spatulated nozzle with an almost circular wick-hole, the width of which is slightly greater than the length. Sloping nozzle. Two closely juxtaposed raised lines surround outer ridge of circular body. A single raised line delineates both sides of the nozzle and extends to a single band of two closely juxtaposed ridges of equal width crossing the midsection of the nozzle. A small portion of the nozzle top and a considerable amount of the body (top and base) are missing. Raised flat discus base. Clay color: 10YR 7/3 dull yellow-orange. Hardness: 2.5. Inclusion(s): white (FS, low frequency); mica (CSi, very low frequency). Burning on nozzle. Fair (+) condition. Mold-made. Parallels, shape-type: IAA Collection (Sussman 2012: 92–94, nos. 907–19).
48. Reg. no. 84.1313.2L.1. L = 5.0 cm. W = 1.5 cm. H = 1.4 cm. Incomplete, 30% preserved. Sepphorean spatulate. Mid-first–early second centuries C.E. Circular body. Slightly sloping shoulder. Large central filling-hole. The nozzle, a considerable portion of the body, and the base are missing. Two closely juxtaposed raised lines follow the outer ridge of the circular body. Clay color: 7.5YR 7/6 orange. Hardness: 3.0. Inclusion(s): white (VFS-CS, medium frequency). Poor condition. Mold-made. Parallels, shape-type: IAA Collection (Sussman 2012: 92–94, nos. 907–19).
49. Reg. no. 84.1094.7L.7. L = 6.8 cm. W = 6.1 cm. H = 2.6 cm. Incomplete, 75%. Sepphorean spatulate. Mid-first–early second centuries C.E. Circular body. Slightly sloping shoulder. Large filling-hole. Missing nozzle. Two closely juxtaposed raised lines following outer ridge of circular body. Raised, round flat base. Clay color: 5YR 7/4 dull orange. Hardness: 3.0. Inclusion(s): white (MS-CS, medium frequency); black (FS, very low frequency). Burning on exterior surface. Fair condition. Mold-made. Parallels, shape-type: IAA Collection (Sussman 2012: 92–94, nos. 907–19).
50. Reg. no. 85.1140.2L. GD = 6.9 cm. FH = 2.3 cm. Incomplete, 20% preserved. Sepphorean spatulate. Mid-first–early second centuries C.E. Four mended fragments. Most of upper shoulder and back sidewall. Upper surface slightly convex. Double ridges follow contour of body. Pronounced ridge around filling-hole. Clay color: 5YR 5/4 reddish brown. No slip. Hardness: not measured. Inclusion(s): red, rounded hema-

tite (LF, FS). Magnetite (?) (FS, very low frequency). Burning on exterior surface. Black discoloration of fabric resulting from secondary firing or use. Tool mark on shoulder. Medium fired. Fair condition. Some salts. Minor lime-popping. Mold-made. Parallels, shape-type: IAA Collection (Sussman 2012: 92–94, nos. 907–19).

51. Reg. no. 84.1068.16L.2. Not shown. L = 7.0 cm. W = 4.7 cm. H = 2.5 cm. Incomplete, 55% preserved. Sepphorean spatulate. Mid-first–early second centuries C.E. Circular body. Slightly sloping shoulder. Large central filling-hole. Spatulated nozzle with an almost circular wick-hole (the width is slightly greater than the length). Sloping nozzle. Two closely juxtaposed raised lines surround the outer ridge of the circular body. A single raised line delineates both sides of the nozzle and extends to a single band of two closely juxtaposed ridges of equal width crossing the midsection of the nozzle. A small portion of the nozzle top and a considerable amount of the body (top and base) are missing. Raised flat discus base. Clay color: 10YR 7/3 dull yellow-orange. Hardness: 2.5. Inclusion(s): white (FS, low frequency); mica flakes (CSi, very low frequency). Burning on nozzle. Fair (+) condition. Mold-made. Parallels, shape-type: IAA Collection (Sussman 2012: 92–94, nos. 907–19).
52. Reg. no. 84.1068.15L.9. L = 4.7 cm. W = 2.3 cm. Incomplete, 20% preserved. Sepphorean spatulate. Mid-first–early second centuries C.E. Spatulated nozzle. Evidence suggests a once circular wick-hole. A single band of closely juxtaposed ridges of unequal width, the upper one nearest to the wick-hole being the thinner of the two. Volute defined on the bottom of the nozzle by a raised, curved edge. A considerable portion of the upper and lower nozzle is missing. A single ring base is composed of three ridges of equal width. Clay color: 2.5Y 8/4 pale yellow. Hardness: 2.5. Inclusion(s): gray (FS, low frequency). Burning on nozzle. Good condition. Mold-made. Parallels, shape-type: IAA Collection (Sussman 2012: 92–94, nos. 907–19).

Type 6. Darom (cat. nos. 53–70, 71b)

Darom lamps are arguably the most visually expressive of lamp forms made in Roman Palestine. The rich variety of images of objects from daily life that decorate the shoulders and/or nozzles of these lamps is highly reflective of local Judean folk art. The images include earrings, combs, pitchforks, amphorae, fluted chalices, fruits, bird cages, and baskets. These daily-life images on Darom lamps set this type apart from other Roman-period lamps of the region, especially the Palestinian discus group whose motifs tend to be mythological in theme. The ornamentation on select Darom lamps found on the western summit includes: an almond-shaped device with lily motif (cat. no. 53); a pattern of raised concentric circles (cat. no. 54); "water-drops" (cat. no. 55); and spiraling vines with pointed maple leaves (cat. no. 59). Unlike numerous examples of contemporary Palestinian discus type lamps, no mythological images decorate the Darom lamps. Their absence may reflect stricter observance of Jewish law in the south, where lamps of this type were presumably manufactured, or at least a conservative attitude toward the use of such imagery on lamps.

Regev (2008: 178–79) suggests that the patterns depicted on Darom lamps are reflective of an "assertive" style and connects them with similar patterns found on Jewish ossuaries. Sussman (1981: 17) ascribes the manufacture of Darom lamps to

Jewish potters on the basis of the motifs decorating their shoulders and nozzles. She identifies menorahs and flora associated with Sukkoth and other Jewish festivals on several Darom lamps. If this is ascription is correct, the occurrence of the Darom lamps at Sepphoris may suggest that Sepphoris had one or more Jewish traders, immigrants, or tourists from Judea. Perhaps a production center located in Galilee also manufactured Darom lamps using molds brought by Jewish lamp makers who had immigrated to Galilee from Judea after the destruction of the Second Temple in 70 C.E.

The term *Darom*, which means "south" in Hebrew, refers to the area where the most concentrated distribution of this type occurs, specifically in southern Israel in Judea and the Shephelah. They are also called "southern" lamps. The Darom fragments from Sepphoris possess all the features common to this lamp type found elsewhere. They have thin walls, and their surfaces are not slipped. The delicate execution of the patterns indicates the skill of the lamp mold maker. Like the Herodian and Sepphorean spatulate types, Darom lamps have a splayed nozzle. The mold-made Darom type, however, is lighter in weight than the heavier wheel-made Herodian type. Several of the mold-made body fragments feature an upright pierced handle (cat. nos. 54, 56–59). The Sepphoris examples are fashioned in clays ranging from several hues of orange, brown, and gray. A black mineral, possibly magnetite occurs in several examples (see cat. nos. 53, 55). The average hardness of the lamps is 3.0 on Mohs scale.

Square Body. Two Darom lamp fragments with square-shaped bodies were found at Sepphoris (cat. nos. 60–61). One sherd (cat. no. 60) is characterized by a small pinched handle with a medium-sized filling-hole surrounded by a single prominent ridge and a single raised ring base. Similar square lamps date to the Roman period (Rosenthal and Sivan 1978: 108). Delicately executed, overlapping circles forming rosettes decorate the base of the other square-bodied Darom lamp from Sepphoris (cat. no. 61). The precision and symmetry of the design were achieved with the use of a compass, the points of which were etched into the lamp mold. The compass-made rosette on this example is similar in execution to those found on Jewish ossuaries in Judea, suggesting that ossuary carvers may also have carved the Darom lamp molds.

The distribution of Darom lamps indicates that they originate in Judea, for they are most common at sites in Judea and the Judean Shephelah. For example, in the Shephelah, seven complete Darom lamps were recovered from a sealed, rock-cut burial cave at Ḥorvat Zefiyya (Nahshoni 2002: 55, fig. 7.1–4; 57, fig. 8.1–3). In recent years, more examples have been discovered at Galilean sites such as Sepphoris and ʿIblin (Lapp 1991b: 123–25, figs. 104–9.35–41; 1996: 218, no. 107; 1997a: 36 n. 48). Darom lamps are rarely found outside Palestine; a complete Judean molded lamp was reportedly found at Ambelikou on Cyprus (Oziol 1977: 169, pl. 27.513), and several fragments were recovered at the Nabatean port of Roman Aila on the Red Sea (e.g., reg. nos. 91191, 58501, 17473 in Lapp forthcoming a). Darom lamps were manufactured from the second half of the first century C.E. to the first half of the second century C.E., or from about 70 to 135 C.E. (Rosenthal and Sivan 1978: 82; Sussman 1982: 16; Lapp 1997a: 34–39, figs. 19–20).

Petrographic and neutron activation analyses indicate that the Darom lamps originate in the Jerusalem area and the Shephelah, probably in the region of Beth Guvrin, Hebron, and Maresha, where a large concentration of Darom oil lamps was discovered, suggesting a production center in this area (see examples in Sussman 1982: 17, 31–127, 129–33, figs. 1–246, A–Z, Aa, Bb). Instrumental neutron activation analysis (INAA) of three Darom lamp fragments—from Masada, the Cave of Horror, and Aroer—indicates that two of the sherds were manufactured in the Shephelah (Masada sample MSDL 24 and Aroer sample MSDL 27) and one in the Jerusalem area (Cave of Horror sample MSDL 25; Yellin 1994: 109).

The occurrence of Darom lamp fragments at Sepphoris (cat. nos. 53–70) suggests contact with the Jerusalem area and the Shephelah. However, Darom lamps have been found throughout Israel (Lapp 1997a: 36, 315, fig. 20): in the Negev, the coastal plain, Samaria, the northern Jordan Valley, and Galilee, as well as in Judea and the Shephelah. The significant quantities found at Sepphoris and also in tombs at ʿIblin suggest that either there was a second production center in Galilee or that Darom lamps were sent to northern customers by Judean lamp makers (Lapp 1991b: 63–64, figs. 104–9.35–40; 1997a: 39; and personal communication with Nurit Feig of the Israel Antiquities Authority in Jerusalem in July 1997). In order to determine whether the Darom lamps found at Galilean sites (Sepphoris, ʿIblin, and Nabratein) and at the Decapolis city of Beth Shean/Scythopolis (Hadad 2002) originated from a Judean production center requires that they be analyzed using a combination of trace-element (e.g., dcp-oes, icp-oes, or NAA) and petrographic thin-section analyses. The results then must be calibrated to and compared with the lamp fabrics and clay samples of the Masada characterization study. A chemical and petrographic match with the Masada samples would further establish the origin of the Darom lamps in the Jerusalem area and/or the Shephelah. It is also possible that future analyses may establish that all or several of the Darom lamps from Sepphoris came from other workshops in entirely different localities. Their somewhat widespread occurrence indicates that either the workshops in the Shephelah and Jerusalem area were solely responsible for the manufacture and trade of these forms throughout Roman Palestine, or additional contemporaneous workshops operating elsewhere, such as in the Galilee, were also responsible for their manufacture and distribution.

53. Reg. no. 84.1120.3L.1. L = 5.2 cm. W = 4.2 cm. Incomplete, 15% preserved. Darom. Second half of the first century C.E. to the first half of the second century C.E., or from about 70 C.E. to 135 C.E. Nozzle and base fragment. Spatulated nozzle with two slightly flaring volutes. Large wick-hole whose length is greater than its width. Directly beneath the wick-hole, a single band of two closely juxtaposed ridges crosses the neck of the nozzle. A raised, well-executed, inverted lily motif is centered in the middle of the nozzle neck. From its top, on each side a stylized vine extends downward and near its end curves upward. A round bulb with short and centralized points at its top is rendered on each end of the vine and appears to represent a pomegranate. The remnant of a single band consisting of two juxtaposed ridges had encircled the body of the lamp. Thin wall. Clay color: 7.5YR 6/6 orange. Hardness: 2.0. Inclusion(s): magnetite (?)(FS, very low frequency). Slight burning around wick-hole. Excellent condition. Mold-made. Parallels, shape-type: Caesarea Maritima

(Sussman 2008: 222, nos. 28–29); Pella (da Costa 2003: 2.503–5, nos. 6, 8–9, 11); Umm Qais (da Costa 2003: 2.504–5, nos. 7, 10, 12–13); provenance unknown (Sussman 1982: 109, nos. 204, 115, and no. 221 [close]); Adler Collection (Adler 2004: 34–51, nos. 168–295).

54. Reg. no. 84.1068.17L.5. L = 7.6 cm. W = 3.4 cm. H = 2.7 cm. Incomplete, 45% preserved. Darom. Second half of the first century C.E. to the first half of the second century C.E., or from about 70 C.E. to 135 C.E. Upper portion of circular body fragment with rounded shoulder. A single flaring volute in relief situated on what had been the shoulder of the nozzle neck. Broken loop handle. Nozzle, portion of shoulder top, and base are missing. Randomly positioned concentric circles composed of four inner rings of varying radii decorate the shoulder of the lamp. Large filling-hole surrounded by a narrow ledge and a fairly prominent ridge. Thin walls. Ring base composed of a single band, two closely juxtaposed ridges. Clay color: 10YR 8/3 light yellow-orange. Inclusions: not visible. Hardness: 3.0. Burning around nozzle. Good condition. Mold-made. Parallels, shape-type: Caesarea Maritima (Sussman 2008: 222, nos. 28–29).
55. Reg. no. 84.2169.1L.1. L = 5.6 cm. W = 3.3 cm. Incomplete, 30% preserved. Darom. Second half of the first century C.E. to the first half of the second century C.E., or from about 70 C.E. to 135 C.E. Upper portion of circular body fragment with a rounded shoulder and part of a base side. Trace of two extended, composite sweeping volutes in relief situated at the point where the base of the nozzle neck met the shoulder of the lamp body. Large central filling-hole surrounded by a very narrow ledge and a band consisting of a single ridge. A considerable portion of the shoulder, the entire nozzle, and the base of the lamp are missing. On shoulder: a band of evenly spaced, consecutive "water drops" in positive relief is enclosed between the outermost band surrounding the filling-hole and a band of two closely juxtaposed ridges following the outer edge of the circular lamp body. Small, broken pinched handle. Thin walls. Clay color: 5YR 7/6 orange. Hardness: 3.0. Inclusion(s): white (VFS, low frequency); magnetite (?)(CS, low frequency). Good condition. Mold-made. Parallels, shape-type: Caesarea Maritima (Sussman 2008: 222, nos. 28–29).
56. Reg. no. 85.1100.2L.2. L = 8.0 cm. W = ud. Complete. Darom. Second half of the first century C.E. to the first half of the second century C.E., or from about 70 C.E. to 135 C.E. Round body. On shoulder: three bands of two parallel ridges. Round nozzle. A band of two ridges follows the outer contour of the nozzle. Two pronounced flaring volutes on each side of nozzle-bridge. Large central filling-hole surrounded by a very narrow ledge and a band consisting of a single ridge. Small, pierced, pinched handle. Circular, slightly convex base delineated by two bands of raised double rings. Thin walled. Clay color: 10R 4/8 red. No slip. Hardness: 3.0. Hard fired. Inclusion(s): white, well sorted (FS, high frequency). Good condition. Mended. Heavy burning on nozzle. Mold-made. Published: Lapp 1996: 218, no. 107. Parallels, shape-type: Caesarea Maritima (Sussman 2008: 222, nos. 28–29).
57. Reg. no. 85.3103.1L.2. Balk. GD = 3.8 cm. Incomplete, 10% preserved. Darom. Second half of the first century C.E. to the first half of the second century C.E., or from about 70 C.E. to 135 C.E. Shoulder fragment and complete loop handle (aperture, 1.2 cm in diameter). On shoulder: a single vine with pronounced round nodule at the end of a curl, a fruit. Single groove on handle. Double ridge follows contour of body. Two parallel grooves on handle. A ridge around filling-hole with a center groove. Clay color: 5YR 6/6 reddish yellow. No slip. Inclusions: gray, moderately

sorted, sub-angular (VFS, high frequency). Hardness: 3.5. Hard fired. Good condition. No burning. Mold-made. Parallels, shape-type: Caesarea Maritima (Sussman 2008: 222, nos. 28–29).

58. Reg. no. 99.9999.99L. Balk. GD = 3.5 cm. Incomplete, 5% preserved. Darom. Second half of the first century C.E. to the first half of the second century C.E., or from about 70 C.E. to 135 C.E. Shoulder fragment and complete loop handle (aperture, 1.4 cm in diameter with drill-spin impressions). On shoulder: consecutive elongated petals. Single groove on handle. Double ridge follows contour of body. Double ridge around filling-hole. Clay color: 7.5YR 7/4 pink. No slip. Inclusions: red, hematite (LF, FS). Hardness: not measured. Good condition. No burning. Mold-made. Parallels, shape-type: Caesarea Maritima (Sussman 2008: 222, nos. 28–29).
59. Reg. no. 84.1120.1L.1. L = 5.1 cm. W = 2.5 cm. Incomplete, 20% preserved. Darom. Second half of the first century C.E. to the first half of the second century C.E., or from about 70 C.E. to 135 C.E. A shoulder fragment with thin walls. A portion of the lower base shoulder remains. Complete loop handle. Remnant of a single, fairly prominent ridge surrounding a broad ledge which encircles the central filling-hole. Raised, finely executed representation of a stylized grapevine, one branch of which ends in a representation of a seven-pointed leaf decorates the shoulder of the lamp. Clay color: 2.5YR 6/6 orange. Core color: 10YR 8/3 light yellow. Hardness: 3.0. Inclusion(s): gray (FS-MS, high frequency); red, hematite (FS, medium frequency). Excellent condition. Mold-made. Parallels, shape-type: Caesarea Maritima (Sussman 2008: 222, nos. 28–29); unknown provenance (Sussman 1982: 118, nos. 229–30) (similar).
60. Reg. no. 95.1016.14L.3. L = 6.4 cm. W = 2.3 cm. Incomplete, 45% preserved. Darom. Second half of the first century C.E. to the first half of the second century C.E., or from about 70 C.E. to 135 C.E. Square-shaped body fragment. Small pinched handle. Medium-sized central filling-hole surrounded by a single prominent ridge. A narrow band of closely juxtaposed double ridges surrounds ridge of filling-hole. A thin raised ridge follows outline of outer shoulder. Single raised ring base. Clay color: 5YR 7/4 dull orange. Slip color: 10R 4/6 red (exterior surface only). 10R 3/6 dark red (interior surface only). Hardness: 2.0. Inclusion(s): gray (CS, low frequency). Good condition. Mold-made. Parallels, shape-type: Caesarea Maritima (Sussman 2008: 222, nos. 28–29).
61. Reg. no. 85.3174.7L.2. GD = ud. Incomplete, 20% preserved. Darom. Second half of the first century C.E. to the first half of the second century C.E., or from about 70 C.E. to 135 C.E. Square-shaped base. On base: delicately executed, thin linear rosette with six petals, several of which fade into fabric. Rosette incised into mold using a compass. Single prominent ridge (width, 1.5 cm) follows square outline of base. Slightly tapering wall, presumably to facilitate removal of lamp from mold. Clay color: 2.5Y 6/4 light reddish brown. No slip, as natural color of fabric is quite appealing. Fabric discoloration from firing. Inclusions: white, angular, well sorted (VHF, VFS). Hardness: 4.0. Hard fired. Good condition. No burning. Mold-made. Parallels, shape-type: Caesarea Maritima (Sussman 2008: 222, nos. 28–29).
62. Reg. no. 84.1120.3L.7. Not shown. L = 2.6 cm. W = 2.2 cm. H (including handle) = 2.5 cm. Incomplete, 3% preserved. Darom. Second half of the first century C.E. to the first half of the second century C.E., or from about 70 C.E. to 135 C.E. A small, thin shoulder fragment with a pierced handle. Two parallel incised lines run the length of the handle surface. A single lily motif consisting of a narrow almond-

shaped flower bud surrounded on each side by a stylized leaf that breaks into a curl at approximately the height of the bud. The lily motif is probably a remnant member of a band stretching around the shoulder whose space is defined by a single, narrow ridge running around the filling-hole and the inside of the body on the outside. Thin walls. Clay color: 7.5YR 5/4 dull brown. Discoloration of surface due to intense heat. Hardness: 3.0. Inclusion(s): white (VFS, low frequency). Magnetite (?)(VFS, low frequency). Burning on interior surface. Good condition. Mold-made. Parallels, shape-type: Caesarea Maritima (Sussman 2008: 222, nos. 28–29).

63. Reg. no. 84.2090.1L.1. Not shown. L = 4.9 cm. W = 2.2 cm. Incomplete, 5% preserved. Darom. Second half of the first century C.E. to the first half of the second century C.E., or from about 70 C.E. to 135 C.E. Thin shoulder fragment. A single band of repeated and raised almond-shaped devices whose outline is followed on the outside by a single fine line in positive relief. A band consisting of closely spaced repeating miniscule nodules surround the outer edge of the lamp body. Clay color: 10YR 4/1 brownish gray (exterior). 10R 5/4 reddish brown (interior). Hardness: 4.0. Inclusion(s): white (CS, very low frequency). Fair (+) condition. Mold-made. Parallels, shape-type: Caesarea Maritima (Sussman 2008: 222, nos. 28–29); unknown provenance (Sussman 1982: 115, no. 221) (close).
64. Reg. no. 84.5085.2L. Not shown. Darom. Second half of the first century C.E. to the first half of the second century C.E., or from about 70 C.E. to 135 C.E. GD = 6.5 cm. Incomplete, 25% preserved. Shoulder, handle, filling-hole, and underside fragment. Plain, flat, and slightly curving shoulder. Complete loop handle with single groove in center (aperture 1.1 cm diameter). Rim (0.8 cm wide) around filling-hole. Clay color: 7.5YR 7/6 reddish yellow. Slip color: 2.5YR 5/6 red. Splashes of slip on interior surface. Discoloration on exterior surface: 5YR 5/6 yellowish red and in places 7.5YR 6/6 reddish yellow. Buffed upper surface, particularly on shoulder to hide the join between the upper and lower halves. Splotches on surface caused by secondary firing. Surface exhibits a smooth shiny appearance. Inclusions: gray (MF, VFS). Hardness: not measured. Medium fired. Good condition. No burning. Mold-made. Parallels, shape-type: Caesarea Maritima (Sussman 2008: 222, nos. 28–29).
65. Reg. no. 84.5085.11L. Not shown. GD = 5.5 cm. Incomplete, 25% preserved. Darom. Second half of the first century C.E. to the first half of the second century C.E., or from about 70 C.E. to 135 C.E. Thickness: 0.2 cm. Shoulder, handle, filling-hole, and underside fragment. On shoulder: delicately fashioned spiraling vines with grape leaves, scientific in detail. Three pointed tips on each side of leaf with a central tip, and veins depicted inside each leaf. One complete leaf is preserved and two-thirds of a second, both positioned to the left; on the right, remnant of a third leaf. Complete loop handle with single groove in center (aperture 1.1 cm diameter). Prominent double ridge (0.3 cm wide) around filling-hole. Clay color: 10R 5/6 red. No slip. Discoloration caused by firing, 10R 5/6 red. Most of body: 10YR 6/1 light gray-gray. Inclusions: gray (LF, VFS). Hardness: Not measured. Medium fired. Good condition. No burning. Mold-made. Parallels, shape-type: Caesarea Maritima (Sussman 2008: 222, nos. 28–29).
66. Reg. no. 84.5085.3L. Not shown. GD = 4.0 cm. Incomplete, 10% preserved. Darom. Second half of the first century C.E. to the first half of the second century C.E., or from about 70 C.E. to 135 C.E. Nozzle with nearly complete flaring volute and nozzle-underside fragment. On nozzle-bridge: a long leaf decorated with consecutive *V*s and a remnant of a spiraling vine curl; outer line of leaf follows contour of flaring

volute. Clay color: 5YR 6/4 light reddish brown. No slip. No visible inclusions. Hardness: not measured. Medium fired. Good condition. Burning on underside. Mold-made. Parallels, shape-type: Caesarea Maritima (Sussman 2008: 222, nos. 28–29).

67. Reg. no. 84.6004.2L. Not shown. GD = 3.4 cm. Incomplete, 3% preserved. Darom. Second half of the first century C.E. to the first half of the second century C.E., or from about 70 C.E. to 135 C.E. Nozzle-bridge fragment with part of flaring volute. On nozzle-bridge: a vertical almond motif flanked by a wide curve composed of two parallel lines (0.1 cm wide) in high relief across bridge. Clay color: 7.5YR 7/4 pink. No slip. Inclusions: gray (MF-LF, VFS). Hardness: not measured. Medium fired. Fair condition. Heavy burning on remnant wick-hole edge. Mold-made. Parallels, shape-type: Caesarea Maritima (Sussman 2008: 222, nos. 28–29).
68. Reg. no. 85.3302.3L. Not shown. GD = 2.8 cm. Incomplete, 3% preserved. Darom. Second half of the first century C.E. to the first half of the second century C.E., or from about 70 C.E. to 135 C.E. Two mended fragments. Flaring volute and shoulder fragment. On shoulder: a single spiraling vine with partially preserved grape cluster; part of grape leaf preserved to the lower right of cluster. Clay color: surface fired to 10YR 7/3 very pale brown. No slip. Fabric discoloration from firing. Inclusions: well sorted, angular, white (VHF, VFS). Hardness: 3.0. Medium fired. Fair condition. No burning. Mold-made. Parallels, shape-type: Caesarea Maritima (Sussman 2008: 222, nos. 28–29).
69. Reg. no. 85.3252.2L. GD = 5.8 cm. Incomplete, 15% preserved. Darom. Second half of the first century C.E. to the first half of the second century C.E, or from about 70 C.E. to 135 C.E. Nozzle volute, shoulder, and underside fragment. On shoulder: a single spiraling vine; one major grape cluster, partially preserved; two minor thistle leaves. All decoration in delicate high relief. On underside: straw impressions from matting. Clay color: 7.5YR 7/4 pink. No slip. Inclusions: gray (LF, VFS). Hardness: not measured. Hard fired. Good condition. Some salts on upper shoulder. No burning. Mold-made. Parallels, shape-type: Caesarea Maritima (Sussman 2008: 222, nos. 28–29).
70. Reg. no. 85.3174.2L.2. Not shown. GD = 3.7 cm. Incomplete, 3% preserved. Darom. Second half of the first century C.E. to the first half of the second century C.E., or from about 70 C.E. to 135 C.E. Nozzle-bridge and part of volute. Remnant of double ridge running along outline of body. Remnant of join between upper and lower halves. Clay color: 5YR 6/4 light reddish brown. No slip. Fabric discoloration from firing. Inclusions: white, angular, well sorted (VHF, VFS). Hardness: not measured. Good condition. No burning. Mold-made. Parallels, shape-type: Caesarea Maritima (Sussman 2008: 222, nos. 28–29).

71b. Reg. no. 85.1150.2L.1. GD = 5.9 cm. Incomplete, 25% preserved. Darom. Second half of the first century C.E. to the first half of the second century C.E., or from about 70 C.E. to 135 C.E. Part of shoulder and base (3.7 cm diameter). Complete loop handle. On shoulder, in high relief: 5 pomegranates (to the left, 3; to the right, 2) attached to vines. Prominent ring base ridge. Clay color: 5YR 6/6 reddish yellow. No slip. Inclusions: white, angular (VLF, FS); gray, rounded (LF, FS). Hardness: not measured. Fair condition. Eight mended fragments. No burning. Mold-made.

Type 7. Molded Judean Floral (cat. nos. 71a, 72–73)

An abundance of intricate floral designs decorate the shoulders of the molded Judean floral lamp type, which corresponds to Barag and Hershkovitz Type D I–III (Molded Lamps with Floral Decorations; Barag and Hershkovitz 1994: 59–71). The floral species are so accurately rendered that they can be identified with various plants or trees growing in the Judean Hills and northern Palestine: these include, for example, oak leaves and acorns, capers, plantains, ivy and acanthus leaves, myrtle twigs, and olive twigs with leaves and fruit (Barag and Hershkovitz 1994: 59–71). These well-executed patterns are indicative of a highly skilled lamp-mold maker. Such lamps are commonly fashioned in a finely levigated, grayish-yellowish brown (10YR 6/2) fabric with a characteristic black (10YR 1.7/1) slip. The expert, evenly applied slip and general firing of these lamps reflect the work of a superb lamp maker. This highly aesthetic and well-made lamp was undoubtedly coveted by many lamp purchasers and, like the Beit Nattif and Darom lamp types, was surely considered a fine-ware lamp, a luxury product among the clay lamps manufactured in Palestine.

The molded Judean floral lamps have a round body with usually one and sometimes two spatulate nozzles. The handle belongs to the pierced loop variety. They have a low, single-ring base. Although they rarely have lamp makers' marks, a stamped Greek seal reading "Joseph" occurs on the underside of a molded-with-floral-pattern lamp recovered from a room in the casemate wall at Masada. The lamps of this type have a fine-grained fabric, typically containing a low frequency of fine, sand-size white inclusions. Neutron activation analyses have shown that several specimens from Masada and Aroer were made in the Jerusalem area (Yellin 1994: 109, 118).

This lamp group is similar to and was probably influenced by the knife-pared Herodian type in terms of its rounded body and splayed nozzle. It also resembles the Broneer Type XXI lamps (see, e.g., Rosenthal and Sivan 1978: 19–21, nos. 40–50). In several respects—flaring volutes, concentric circles on each tip of the spatulate nozzle, outstanding workmanship, thin walls, and the fact that it is mold-made—this lamp type is similar to the Darom group.

This type is most often found in Judea: at Antipatris, Jerusalem, Masada, Nahlat-Yehuda, and Qumran (Barag and Hershkovitz 1994: 59–71). The most significant deposits were at Masada. In the Negev, they have been discovered at Aroer and Oboda. They have also been uncovered at Gezer, which is located on the last of the foothills of the central mountains of the northern Shephelah. In Galilee, they have been found at Capernaum and Sepphoris. In Nabatea, examples have been found at Petra. They date to the second half of the first century C.E., about 50–66 C.E. (Barag and Hershkovitz 1994: 69).

Sepphoris cat. no. 71a exhibits all of the features associated with the molded Judean floral type: a finely executed, sharply spatulated nozzle featuring outward flaring volutes that meet at the bottom of the nozzle, where a delicately rendered leaf device was added as decoration; two opposing concentric circles are incised on the points of the spatulate nozzle. Catalog nos. 71a and 72 are fashioned in a gray clay that is coated with a black slip. The slip is well-burnished, producing a smooth, hard surface with a dull luster.

71a. Reg. no. 84.1056.1L.5. L = 3.4 cm. W = 3.4 cm. Incomplete, 15% preserved. Molded Judean floral; Barag and Hershkovitz Type D I–III. Second half of the first century C.E., about 50–66 C.E. Sharply spatulated nozzle characterized by outward flaring volutes that meet at the bottom of the nozzle, where a delicately rendered leaf device was added for decoration. Two opposing concentric circles are incised on the points of the nozzle. Clay color: 10YR 6/2 grayish yellow-brown. Slip color: 10YR 1.7/1 black. Hardness: 2.0. Inclusion(s): white (FS, low frequency). Excellent condition. Mold-made. Parallels, shape-type: Caesarea Maritima (Sussman 2008: 221–22, nos. 26–27); Jerusalem (Amiran and Eitan 1970: pl. 7.B, lower left).

72. Reg. no. L85.3216.2L. GD = 6.3 cm. Thickness = 1.5–3.0 mm, very thin. Incomplete, 20% preserved. Molded Judean floral. Barag and Hershkovitz Type D I–III. Second half of the first century C.E., about 50–66 C.E. Shoulder and underside fragment. On shoulder: single oak tree branch with 3 complete oak leaves, 2 unpreserved leaves, and 4 complete acorns in high relief. Wide ridge (1.5 mm) around filling-hole (which is 3.0 mm from ridge). Filling-hole preserved arc = 1.1 cm. Handle aperture, diameter 1.7 cm. Clay color: 7.5YR 4/0 dark gray. Slip color: 7.5YR 3/0 very dark gray. No visible inclusions. Finely levigated fabric. Hardness: 3.5. Hard fired. Good condition. No burning. Mold-made. Parallels, shape-type: Caesarea Maritima (Sussman 2008: 221–22, nos. 26–27).

73. Reg. no. 85.1520.1L.2. Not shown. GD = 4.5 cm. Incomplete, 5% preserved. Molded Judean floral. Barag and Hershkovitz Type D I–III. Second half of the first century C.E., about 50–66 C.E. Shoulder, underside, and handle fragment. Complete loop handle with two parallel grooves. Handle aperture, diameter 1.7 cm. Clay color: 10YR 7/1 light gray. Slip color: 7.5YR 3/0 very dark gray. No visible inclusions. Voids: round, poorly sorted (MF, VFS). Hardness: not measured. Fair condition. No burning. Mold-made. Parallels, shape-type: Caesarea Maritima (Sussman 2008: 221–22, nos. 26–27).

Type 8. Monolychnoi, Ring-Lamps with Elongated Nozzles with Circular Heads (cat. nos. 74–76)

Three nozzles characterized by a circular head with a broad, somewhat elongated neck were found at Sepphoris. They represent regionally manufactured Palestinian copies of Italian forms. The three nozzles (cat. nos. 74–76) included in this study are like those of the polylychnoi and "Augustan" (e.g., Broneer Type XXII) traditions in that they possess fairly elongated necks with stylized volutes. The heads of the Sepphoris nozzles, however, are circular and differ from the typically splayed nozzles of the Augustan group. The volutes of the Type 8 lamps are nearly completely integrated into the neck of the nozzle, unlike the flaring volutes of the Augustan nozzles that make up more of the shape of the body rather than serving as surface ornamentation. The fact that the volutes have become more integrated into the body of the lamp as a decorative device indicates a relatively later date for these lamps. A single band of two narrow, closely juxtaposed ridges outlines the circular nozzle-heads of the Sepphoris examples. Two opposing double concentric circles in high relief and positioned at the base of the nozzle-head represent the tips of the two stylized volutes composed of curving bands of two narrow, juxtaposed ridges that decorate each side of the nozzle and meet at the center of its neck (cat. nos. 74–75). A pointed and concentric *V*-shaped pattern decorates the bottom center of the nozzle neck (cat. nos. 74–75). The lamps to which cat. nos. 74 and 75 originally belonged may have been intended for hanging; the intricate triangle pattern decorating their nozzle-undersides indicates that they were meant to be seen from below. The elongated nozzles ensured maximum illumination, if indeed they were hanging lamps. Or the elongated nozzles may have belonged to nonhanging, bi-nozzled lamps of local or regional manufacture (suggested by their respective fabrics) that resemble Roman Imperial, Broneer Type XXI form (see, e.g., Sussman 2008: 225–26, especially no. 51).

The clay of these examples ranges from orange to a dull or dull-yellow orange. Catalog no. 76 is coated in a red slip, while others have none. Low to very low frequencies of medium or very fine sand grains of red hematite and/or white carbonaceous inclusions occur (cat. nos. 74–75). The average hardness is 2.4 on the Mohs scale. In contrast to the many varieties with spatulate heads, no parallels of nozzles with round nozzle-heads and degenerate volutes similar to ours have been located. The incompleteness of the examples makes further research difficult. Based on the existing evidence, a second-century C.E. or slightly later date for these examples seems appropriate, for they have features similar to the di- and trilychnoi of Jerash, Jordan dating to this period.

74. Reg. no. 84.1094.7L.8. L = 5.3 cm. W = 3.8 cm. Incomplete, 20% preserved. Monolychnos with round nozzle-heads. First century C.E. Circular, steeply sloping nozzle with broad neck. Single band of two narrow, closely juxtaposed ridges outline circular nozzle-head. Two opposing double concentric circles in positive relief decorate base of nozzle-head. Two curving bands, each consisting of two narrow aligned ridges stem from each circle and join together at the center of the nozzle-neck. Pointed, concentric *V*-shaped pattern decorates bottom center of nozzle-neck. Small circular wick-hole. Clay color: 5YR 6/6 orange. Hardness: 2.0. Inclusion(s): red, hematite (MS, low frequency). Good condition. Mold-made.
75. Reg. no. 84.1094.7L.9. L = 4.5 cm. W = 3.7 cm. Incomplete, 20% preserved. Monolychnos with round nozzle-head. First century C.E. Gently sloping, circular nozzle with broad neck. Single band of two narrow, juxtaposed ridges outline circular nozzle-head. Two opposing double concentric circles in positive relief decorate base of nozzle-head. Two curving bands, each consisting of two narrow juxtaposed ridges stem from each circle and join together at the center of the nozzle-neck. Pointed, concentric *V*-shaped pattern decorates bottom of nozzle. Small circular wick-hole. Well-executed ornamentation. Clay color: 10YR 7/3 dull yellow-orange. Hardness: 3.0. Inclusion(s): red, hematite (VFS, very low frequency); white (VFS, very low frequency). Good condition. Mold-made.
76. Reg. no. 85.1113.1L. GD = 8.1 cm. Incomplete, 10% preserved. Monolychnos. First century C.E. Complete nozzle-head and most of nozzle-bridge. Round nozzle-head (3.5 cm diameter) with elongated nozzle-bridge. Wide volutes run along sides of nozzle-bridge and end in very small coil, one on each side. Single band of two narrow, juxtaposed ridges outline circular nozzle-head and enclose flat circular space.

Wick-hole, 1.9 cm diameter. Expertly executed ornamentation and elegant design. Clay color: 5YR 6/6 reddish yellow. Slip color, evenly applied on exterior and interior surfaces of lower half and none applied on interior surface of upper half: 2.5YR 5/8 red. Hardness: 3.0. Medium–hard fired. Inclusion(s): gray (CSi, medium frequency). Very finely levigated clay. Good condition. Salt film covers exterior surface. Mold-made.

Type 9. Polylychnos, Rectangular Box-Like Lampstand (cat. no. 77)

A remnant of a rectangular lampstand was found at Sepphoris. Only one of presumably several nozzles is preserved. Portions of the surviving rectangular body indicate that the lampstand did not belong to the ring-lamp variety.

77. Reg. no. 84.1068.15L.10. L = 8.0 cm. W = 5.1 cm. Incomplete, 40% preserved. Polylychnoi. First century C.E. Complete elongated nozzle with broad neck and semi-round head. Plain neck. Tapered sides create spatulate impression. Medium-sized circular wick-hole. Closely juxtaposed raised ridges surround wick-hole. Two parallel ridges extend across base of neck perpendicular to nozzle-head. Clay color: 7.5YR 7/4 dull orange. Slip color: 10R 4/8 red (exterior only). Hardness: 2.5. Inclusion(s): gray (MS-CS, very low frequency), red, hematite (FS, very low frequency); muscovite mica (CSi, very low frequency). Good condition. Mold-made.

Type 10. Palestinian Discus (cat. nos. 78–142)

The Palestinian discus lamp type corresponds to Kennedy Type 5 and is considered a Palestinian version of Broneer XXV and Loeschcke VIII (Broneer 1930: 85). In fact, it corresponds to a very specific version of the Broneer XXV type and should not be confused with the numerous other round-bodied-with-discus types found in Roman Palestine and Arabia that are also versions of Broneer XXV (see, e.g., cat. nos. 143–44, which represent Syrian discus versions found at Sepphoris). The Palestinian discus type is easily distinguished from the other versions by its thin walls, fabric type, slipping, design execution, and motif types. Palestinian discus lamps enjoyed a widespread popularity in Roman Palestine and Arabia, probably because of the rich repertoire of motifs, which appealed to a mass-market clientele, depicted on their concave disks. They provided the medium par excellence by which religious themes and mythological subjects could easily be disseminated.

The chief feature of this lamp type, however, is a wide, round central discus, which provided the lamp-mold maker space for the depiction of a range of diverse images, including floral patterns, architectural features, religious symbols, mythological characters, erotic scenes, and deities (e.g., the Helios depiction on the Nabratein lamp of this type; Lapp 2009: lamp pl. B.13, photo 48). In order to facilitate the introduction of fuel into the discus lamps' shallow and relatively small oil chambers, the lamp users often intentionally broke out the discus center to create a larger filling-hole. This likely explains the partially missing discus of the Helios lamp found at Nabratein. Petrographic thin-section and trace-element analyses indicate that these lamps were manufactured with a clay rich in calcium carbonate (Lapp 1997a:

169–71). At least two clay sources were quarried for this type, one possibly located in Palestine in or near Beth Shean/Scythopolis and a second in southern Syria (Lapp 1997a: 174–79). Future analyses are necessary to determine the exact location of the clay sources. These lamps date from the first to third centuries C.E. (Lapp 1997a: 39; da Costa 2003: 1.50; Sussman 2008: 230, last third of the first to third centuries C.E.; Hadad 2002: 20, Type 7, Variant 1, late first [after 70]–second century C.E.).

Palestinian discus lamps are characterized by a round body with a small curved nozzle. Two incisions often mark each side of the nozzle on its underside. Lamps of this group typically have a round, mold-made body, with a small semicircular nozzle. The central, shallow concave discus is either plain or, more often than not, as already noted, decorated with an array of faunal, floral, mythological, and other motifs. One to three prominent ridges surround the central reservoir. The shoulder is rather wide and slightly sloping and is frequently decorated with a single band of evenly spaced, consecutive ornamental elements, such as a horseshoe (or ovolo), acanthus leaf, triangle, or dart pattern. Consecutively impressed acanthus leaves, ovals, or darts decorate the shoulders of many examples of this type. Although the shoulders of some of these lamps in the Sepphoris corpus are plain (cat. nos. 98, 110, 120–21, 127, 131, 141), many are decorated with one or more impressed ornamental elements: ovoli (maize kernel or horseshoe-shaped devices; cat. nos. 78, 85, 99, 103, 104, 112, 113, 115, 123, 124), acanthus leaves (cat. nos. 87, 129), and darts (cat. no. 94).

Many of these lamps have plain shoulders (cat. nos. 123–42). Two double-axe devices, one positioned in high relief on each side of the shoulder, decorate the shoulders of many Palestinian discus lamps and are unique to this type. This characteristic distinguishes them from other lamp versions of the Broneer XXV Type manufactured elsewhere in the Empire (Rosenthal and Sivan 1978: 85). Double axes occur on the shoulders of two discus lamps from Nabratein (Lapp 2009: lamp pls. B.13, C.12) and on the shoulders of several found at Sepphoris (e.g., cat. nos. 78, 82–83, 85, 99–100, 102, 112–15, 123, 127, 142). A considerable number of discus lamps have missing central disks. Lamp makers' marks are found on the bases of Palestinian discus lamps from Sepphoris (e.g., cat. nos. 124, 132; fig. 6).

Two opposing open volutes, remnants of the once flaring volutes that belonged to the overall shape of the lamp body of the Hellenistic to Early Roman periods, now merely decorate the shoulder of the lamp at the nozzle end. The base is flat and is frequently marked by several widely incised circles, thus imitating a lamp maker's mark (*planta pedis*); Greek initials commonly occur on the base. The initial may indicate the lamp maker or craftsperson who carved the lamp mold (especially a mold maker's mark). Lamps of this group no longer possess a handle. The clay is orange (or a hue thereof) and the slip, black or brown. The appearance of these lamps in Jerusalem and at Beth Sheʿarim confirms their date from as early as the second half of the first century C.E. and continuing through the third century C.E. (Rosenthal and Sivan 1978: 85; Broneer 1930: 36).

At Beth Shean, Hadad (2002: 16, 20, nos. 19–24, Variant 1) describes examples of the Palestinian discus group as "Type 7, Variant 1" and "imported" but does not provide any explanation about their origin. Archaeometric analyses indicate that at

103 104

Fig. 3. Mythological themes: images of Europa riding the bull on Palestinian discus lamps.

least two separate sources were exploited for clays used to manufacture this type, one source of which may be located in the area of Beth Shean and Pella (Lapp 1997a: 161–62, 174–79). Thus lamps belonging to Hadad Type 7, Variant 1 (Hadad 2002: 16, 20, nos. 19–24) do not represent imports but, rather, locally made lamps, as their petrographic characteristics further indicate. As Hadad (2002: 20, nos. 25–29) suggests, lamps of the Type 7, Variant 2 group are provincial forms that I would argue also originate from the Beth Shean and Pella area.

Palestinian discus lamps are typically coated in one of the following slips: brown (7.5YR 4/3), very dark grayish brown (10YR 3/2), red (2.5YR 5/6), reddish brown (5YR 5/4), or a bichrome combination with a red discus (10R 4/6) and a very dark gray shoulder (7.5YR 3/1; for examples, see Lapp 1997a: 236–40). Petrographic and dcp analysis of select Palestinian examples from Sepphoris indicate the use of a carbonaceous clay; the fabrics of 20 discus lamp samples collected from sites in Israel and Jordan were found to contain relatively high quantities of CaO, a combined total average of 28.50% (table 11 in chap. 4). By comparison, the calcium carbonate content of the bilanceolate (41.61%) and Beit Nattif versions (44.83%) was considerably higher (table 11).

Palestinian discus lamps and fragments represent the main lamp type found at Sepphoris, particularly in Area 84.2. Many of them were found in the fill, L84.1068, of a cistern located several meters east of a paved east–west street. Because of the quantity and variety of Palestinian discus lamps found at Sepphoris, 64 are included in this study (cat. nos. 78–142).

Discus Motif Groups

A wide range of pictorial themes occurs on the central disks of the Palestinian discus lamps from Sepphoris. The scenes can be categorized into the following groups: (1) floral, such as a rosette, a single maple-leaf motif, and a kantharos with floral arrangement (cat. nos. 78–87); (2) faunal including a bird of prey, bull, conch, dolphin, and lion (cat. nos. 88–93); (3) human body parts, especially human heads and legs (cat. nos. 94–98); (4) military themes, such as a charioteer and horseman (cat. nos. 99–101); (5) mythological characters, including an emperor-god, Europa riding a bull (see fig. 3), Helios, and Medusa (cat. nos. 102–11); (6) bi-associative motifs, specifically Dionysos or a satyr and a chalice (cat. no. 112); and (7) erotic scenes (cat. nos. 114–22).

Floral (cat. nos. 78–87). Several varieties of floral motifs occur on the Sepphoris examples. One variety includes a device consisting of four broad wispy or veined leaves alternating with four highly stylized pointed leaves. Motifs of rosettes with tightly spaced, radiating petals with pointed or rounded tips also occur on the disks. Other disks bear a floral motif of a single three-pronged maple-leaf composed of smaller leaves; the stem of the leaf is bisected by a horizontal bar (or stylized ribbon) where it curls upward into two symmetrical leaves with pointed petals. One discus bears the only representation of a kantharos with a floral arrangement in the corpus. Parallels of these various discus scenes appear at sites throughout Israel and Transjordan.

78. Reg. no. 84.1068.7L.4. L = 5.9 cm. W = 3.6 cm. Incomplete, 20% preserved. Palestinian discus. Late first to third centuries C.E. Discus and shoulder fragment. On discus: a raised rosette with juxtaposed, radiating petals with pointed tips. Shallow concave discus whose sides extend onto the shoulder forming a single prominent ridge. Broad, slightly sloping shoulder (Loeschcke VIIIb). On shoulder: a single band of deeply impressed broad, horseshoe devices with bulbous kernels; remnant of a single curl of an open double volute depicted in positive relief; remnant of a single double-axe device in positive relief. Clay color: 2.5YR 7/6 orange. Slip color: 10R 4–5/8 red, exterior and interior surfaces. Hardness: 2.0. Inclusion(s): white (FS, very low frequency). Good (+) condition. Parallels, shape-type: Beth Shean / Scythopolis (Hadad 2002: 16, 20, nos. 19–24, Variant 1).
79. Reg. no. 84.1068.6L.3. L = 6.5 cm. W = 3.1 cm. H = 2.1 cm. Incomplete, 40% preserved. Palestinian discus. Late first to third centuries C.E. Nozzle, shoulder, base, and discus fragment. On discus: a floral motif of rosette with tightly spaced, radiating petals with rounded tips. Single rounded prominent ridge around discus. Narrow, slightly sloping shoulder. Slightly sloping, rounded nozzle. On shoulder: a single band of tightly spaced impressed horseshoe devices with bulbous inner kernels. Two opposing open double volutes depicted in negative relief below the base of the nozzle. Several raised vertical strikes rendered between volutes. Flat base. Clay color: 5Y 8/4 pale yellow. Slip color: 5YR 4/3 dull reddish brown, exterior and interior surfaces. Hardness: 3.0. Inclusion(s): red, hematite (FS, medium frequency); white (FS, low frequency). Burning around wick-hole. Fair condition. Mold-made. Parallels, shape-type: Beth Shean/Scythopolis (Hadad 2002: 16, 20, nos. 19–24, Variant 1).
80. Reg. no. 84.1068.5L.2. L = 7.2 cm. W = 7.4 cm. H = 2.3 cm. Nearly complete, 90% preserved. Palestinian discus. Late first to third centuries C.E. On discus: a rosette composed of tightly spaced, radiating petals with broad and rounded tips. Missing discus. Deeply concave discus surrounded by three major rounded ridges of varying widths. Narrow, slightly sloping shoulder. On shoulder: a single band of closely spaced bulbous horseshoe (= ovule) devices. Broken nozzle. Flat base defined by single incised circle. Clay color: 10YR 8/3 light yellow-orange. Slip color: 5YR 3/6 dark reddish brown, exterior surface only. Hardness: 3.0. Inclusion(s): white (FS, very low frequency; VCS, very low frequency). Burning around wick-hole. Fair condition. Mold-made. Parallels, shape-type: Beth Shean/Scythopolis (Hadad 2002: 16, 20, nos. 19–24, Variant 1).

81. Reg. no. 84.1068.12L.3. L = 3.0 cm. W = 2.9 cm. Incomplete, 5% preserved. Palestinian discus. Late first to third centuries C.E. Discus fragment. On discus: a radiating, linear floral pattern. A thin single ridge encircles a small filling-hole within the discus. Clay color: 2.5Y 8/4 pale yellow. Slip color: 5YR 3/1 brownish black, exterior surface only. Hardness: 4.0. Inclusion(s): gray (MS, very low frequency); white (CS, very low frequency). Fair condition. Mold-made. Parallels, shape-type: Beth Shean/Scythopolis (Hadad 2002: 16, 20, nos. 19–24, Variant 1).
82. Reg. no. 84.2172.3L. L = 7.9 cm. W = 5.9 cm. H = 2.2 cm. Incomplete, 70% preserved. Palestinian discus. Late first to third centuries C.E. Discus, shoulder, nozzle, and base fragment. On discus: a floral motif of four broad veined leaves alternating with four highly stylized pointed leaves radiate from a common center composed of three raised concentric rings; a small central filling-hole. A single rounded ridge surrounds discus. Flat curved nozzle. Narrow, steeply sloping shoulder (Loeschcke VIIIb). On shoulder: remnant of a single double-axe device in positive relief on shoulder. Four raised strikes rendered beneath nozzle. Broad single ring base defined by two incised parallel circles with flat surface. Clay color: 7.5YR 7/4 dull orange. Slip color: 10R 4/4 reddish brown, nozzle and base only; 10YR 3/1 brownish black, discus and shoulder only; 10R 4/3 reddish brown, interior surface only. Hardness: 3.0. Good (–) condition. Mold-made. Parallels, shape-type: Beth Shean/Scythopolis (Hadad 2002: 16, 20, nos. 19–24, Variant 1); Salamis or Curium (Bailey 1988: pl. 58. Q2308), Tubas (Kennedy 1963: pl. 21.499) (exact), provenance unknown (Rosenthal and Sivan 1978: 89, no. 366) (exact); Adler Collection (Adler 2004: 65, no. 391).
83. Reg. no. 84.1068.3L.20. GD = ud. Incomplete, 20% preserved. Palestinian discus. Late first to third centuries C.E. Discus, shoulder, and nozzle fragment. Shallow concave discus whose sides extend onto shoulder, create a single, well-rounded prominent ridge. On discus: remnant of a floral motif composed of a wide leaf alternating with two stylized pointed leaves in positive relief. Broad, slightly sloping shoulder. On shoulder: a double axe in positive relief. One complete open double volute rendered at the base of the nozzle in high relief. Clay and slip colors: undetermined. Inclusion(s): undetermined. Fair condition. Mold-made. Parallels, shape-type: Beth Shean/Scythopolis (Hadad 2002: 16, 20, nos. 19–24, Variant 1).
84. Reg. no. 84.1068.3L.19. Not shown. L = 7.7 cm. W = 7.4 cm. H = 2.2 cm. Nearly complete, 95% preserved. Palestinian discus. Late first to third centuries C.E. Discus, shoulder, and nozzle fragment. Missing discus. Shallow concave discus whose sides extend onto shoulder, create a single, well-rounded prominent ridge. On discus: a floral motif composed of four broad and wispy leaves alternating with four highly stylized pointed leaves in positive relief. Broad, slightly sloping shoulder. On shoulder: two opposing double-axe devices in positive relief. Two opposing open double volutes are rendered at the base of the nozzle in relief. Five raised strikes between the two volutes are depicted directly beneath the wick-hole. A single raised and rounded ring base. On base: lamp maker's mark of Greek initial *eta* (H). Clay color: 2.5YR 7/6 orange. Slip color: 7.5YR 3/4 dark brown. Burnish: exterior surface traces. Hardness: 2.5. Inclusion(s): white (FS, low frequency; VCS, very low frequency). Burning around nozzle. Fair condition. Mold-made. Parallels, shape-type: Beth Shean/Scythopolis (Hadad 2002: 16, 20, nos. 19–24, Variant 1).
85. Reg. no. 84.2172.3L.3. L = 7.7 cm. W = 6.9 cm. Incomplete, 40% preserved. Palestinian discus. Late first to third centuries C.E. Discus, shoulder, and nozzle fragment. Shallow concave discus with a single filling-hole. Two opposing open double volutes in positive relief at base of nozzle. Four raised strikes rendered between

volutes. On discus: a floral motif composed of three-pronged leaf consisting of smaller leaves; a horizontal bar/stylized ribbon bisects a leaf stem where it curls upward into two symmetric leaves with pointed petals. A single, well-rounded ridge surrounds the discus. Narrow, slightly sloping shoulder. On shoulder: a single band of tightly spaced, deeply impressed bulbous horseshoe devices depicted on shoulder around the discus. Remnant of a single double axe in positive relief. Lamp maker's mark: Greek letter *eta* inside the base. Bichrome. Clay color: 2.5Y 8/4 pale yellow. Slip color: 1R 3/2 dark reddish brown, discus only; 7.5YR 3/1 brownish black, shoulder only; 2.5YR 4/2 grayish red, interior surface only. Hardness: 2.0. Inclusion(s): white (VCS, very low frequency); magnetite (MS-VCS, low frequency). Good condition. Mold-made. Parallels, shape-type: Beth Shean/Scythopolis (Hadad 2002: 16, 20, nos. 19–24, Variant 1); Jerusalem (B. Mazar 1971: 9, no. 3) (exact).

86. Reg. no. 84.1068.11L.1. L = 3.4 cm. W = 3.1 cm. Incomplete, 3% preserved. Palestinian discus. Late first to third centuries C.E. Discus medallion fragment. On discus: an incomplete floral motif of a three-pronged leaf composed of smaller leaves; a horizontal bar bisects leaf stem where it curls upward into two symmetrical pointed leaves; a remnant of a small filling-hole. Clay color: 7.5YR 8/4 light yellow-orange. Slip color: 10R 4/6 red, exterior and interior surfaces. Inclusion(s): red, hematite (FS, high frequency); magnetite (?)(FS, low frequency); muscovite mica (FS, medium frequency). Fair condition. Mold-made. Parallels, shape-type: Beth Shean/Scythopolis (Hadad 2002: 16, 20, nos. 19–24, Variant 1).

87. Reg. no. 84.1068.3L.17. L = 6.8 cm. W = 6.0 cm. Incomplete, 35% preserved. Palestinian discus. Late first to third centuries C.E. Discus and shoulder fragment. Shallow concave discus. On discus: a kanthoras containing a floral arrangement. A single filling-hole inside discus. A single well-rounded ridge surrounds discus. Narrow, steeply sloping shoulder. On shoulder: a single band of connected impressed darts encircle discus; remnant of volute curl (?) in positive relief. Clay color: 2.5Y 7/3 light yellow. Slip color: 10YR 3/1 brownish black, shoulder only; 5YR 4/2 dull reddish brown, discus and interior surfaces only. Hardness: 2.0. Fair condition. Mold-made. Parallels, shape-type: Beth Shean/Scythopolis (Hadad 2002: 16, 20, nos. 19–24, Variant 1); Abila (Fuller 1987: 476, fig. 102); Debaal (Hajjar 1965: pl. 20.F:374a) (exact); The Adler Collection (Adler 2004: 65, no. 390).

Faunal (cat. nos. 88–93). Palestinian discus lamps decorated with motifs of the following fauna were found at Sepphoris: conch (cat. no. 88), bull (cat. no. 89), camel (cat. no. 90), dolphin (cat. no. 91), lion (cat. no. 92), and bird of prey (cat. no. 93). Such faunal subjects frequently appear in the iconographic milieu of Roman art in the eastern provinces from the first to fourth century C.E. and represent local imitations of similar themes current in contemporary Roman art. The clay ranges from orange to dull or light yellow-orange.

88. Reg. no. 84.2193.1L.1. L = 5.7 cm. W = 7.3 cm. H = 1.9 cm. Incomplete, 30% preserved. Palestinian discus. Late first to third centuries C.E. Discus, shoulder, and base fragment. Shallow concave discus. On discus: an expansive conch with curled foot. Discus surrounded by broad, prominent, well-rounded ridge. Slightly sloping shoulder. On shoulder: a single band of consecutive, tightly spaced, degenerate impressed darts; remnant of a single circular curl of an impressed open volute directly below missing nozzle. Single ring base. Clay color: 10YR 8/3 light yellow-orange. Slip

color: 7.5YR 3/1 brownish black, exterior surface only; 2.5YR 4/6 reddish brown, exterior surface of base only. Hardness: 3.0. Burning on interior surface. Fair condition. Mold-made. Parallels, shape-type and discus motif, conch: Caesarea Maritima (Sussman 2008: 231, no. 83); Dor (Stern 1987: 78); shape-type: Beth Shean/Scythopolis (Hadad 2002: 16, 20, nos. 19–24, Variant 1).

89. Reg. no. 84.1068.7L.6. L = 3.3 cm. W = 2.3 cm. Incomplete, 3% preserved. Palestinian discus. Late first to third centuries C.E. Discus medallion fragment. On discus: a rearing bull in half profile and its head in frontal view. Clay color: 2.5YR 7/6 orange. Slip color: 10R 4/8 red. Hardness: 3.0. Inclusion(s): red, hematite (VFS, high frequency); white (VFS, high frequency); magnetite (VFS, high frequency). Fair condition. Mold-made. Parallels, shape-type: Beth Shean/Scythopolis (Hadad 2002: 16, 20, nos. 19–24, Variant 1).

90. Reg. no. 84.2051.3L.2. L = 3.3 cm. W = 2.0 cm. Incomplete, 3% preserved. Palestinian discus. Late first to third centuries C.E. Discus medallion fragment. On discus: a sitting camel with wide hind quarters and raised tail. Clay color: 5YR 7/4 dull orange. Slip color: 2.5YR 5/1 reddish gray. Hardness: 3.0. Good condition. Mold-made. Parallels, shape-type: Beth Shean/Scythopolis (Hadad 2002: 16, 20, nos. 19–24, Variant 1).

91. Reg. no. 84.1068.3L.32. L = 5.3 cm. W = 4.4 cm. Incomplete, 5% preserved. Palestinian discus. Late first to third centuries C.E. Discus medallion fragment. On discus: a wavy, double-forked tail of a dolphin. Clay color: 7.5YR 8/4 light yellow-orange. Slip color: 10R 4/8 red, exterior surface only. Hardness: 4.0. Inclusion(s): gray (granule, very low frequency); white (FS, low frequency). Poor condition. Mold-made. Parallels, shape-type: Beth Shean/Scythopolis (Hadad 2002: 16, 20, nos. 19–24, Variant 1); unknown provenance (Rosenthal and Sivan 1978: 88, no. 261) (close); Caesarea Maritima (Sussman 2008: 271, no. 80) (close).

92. Reg. no. 84.1068.14L.4. L = 3.4 cm. W = 2.5 cm. Incomplete, 5% preserved. Palestinian discus. Late first to third centuries C.E. Discus medallion fragment. On discus: a downward leaping lion with raised tail whose body is depicted in half profile and its head in frontal view. A remnant of a single filling-hole inside the discus. Clay color: 2.5YR 6/8 orange. Slip color: 2.5YR 4/8 reddish brown. Hardness: 4.0. Inclusion(s): white (VFS, medium frequency; VCS, very low frequency). Good condition. Mold-made. Parallels, shape-type: Beth Shean/Scythopolis (Hadad 2002: 16, 20, nos. 19–24, Variant 1).

93. Reg. no. 84.1068.3L.18. Not shown. L = 8.0. W = 6.9 cm. Incomplete, 70% preserved. Palestinian discus. Late first to third centuries C.E. Discus, nozzle, and shoulder fragment. Slightly sloping shoulder. Broken rounded nozzle. Single rounded ridge surrounds discus. A single ring base. On discus: a large bird of prey, probably an eagle, rendered in half profile, facing wick-hole. On shoulder: a single band of loosely spaced medium-sized horseshoe pattern with very small inner kernels impressed; degenerate symmetrically opposed double volutes directly below nozzle; five raised strikes depicted between volutes. Portion of discus, shoulder, and base are missing. Clay color: 7.5YR 3/1 light yellow-orange. Slip color: 7.5YR 3/1 brownish black (exterior surface only) and 10R 3/4 dark red. Hardness: 3.0. Inclusion(s): white (MS-CS, medium frequency). Burning on nozzle. Fair (–) condition. Mold-made. Parallels, shape-type: Beth Shean/Scythopolis (Hadad 2002: 16, 20, nos. 19–24, Variant 1).

Human Body Parts (cat. nos. 94–98). Representations of human heads (cat. nos. 94–96) and legs (cat. nos. 97–98) appear on several discus fragments from Sepphoris.

94. Reg. no. 84.2172.3L. L = 3.7 cm. W = 4.2 cm. Incomplete, 10% preserved. Palestinian discus. Late first to third centuries C.E. Discus and shoulder fragment. On discus: a human bust in half profile and in positive relief. A single ridge surrounds discus. Bichrome. Clay color: 7.5YR 8/4 light yellow-orange. Slip color: 2.5YR 3/4 dark reddish-brown, discus and interior surfaces; 2.5YR 3/1 dark reddish gray (shoulder only). Hardness: 3.0. Burning on interior surface. Fair (+) condition. Mold-made. Parallels, shape-type: Beth Shean/Scythopolis (Hadad 2002: 16, 20, nos. 19–24, Variant 1).
95. Reg. no. 84.1068.3L.26. Not shown. L = 5.2 cm. W = 2.0 cm. Incomplete, 5% preserved. Palestinian discus. Late first to third centuries C.E. Discus fragment. On discus: two juxtaposed human heads in positive relief; right head in 3/4 profile; left head in 1/2 profile; standing spiked hair style on both heads; raised leaf pattern between wings (?); heads primitively executed. On shoulder: a single band of impressed horseshoe devices. A single ridge surrounds discus. Clay color: 10YR 8.3 light yellow-orange. Slip color: 7.5YR 1.7/1 black, exterior and interior surfaces only. Hardness: 3.0. Fair condition. Mold-made. Parallels, shape-type: Beth Shean/Scythopolis (Hadad 2002: 16, 20, nos. 19–24, Variant 1).
96. Reg. no. 84.1068.10L.5. L = 3.2 cm. W = 2.5 cm. Incomplete, 5% preserved. Palestinian discus. Late first to third centuries C.E. Discus fragment. On discus: a large, elongated human head with full wavy hair and beard. Filling-hole remnant inside discus. Clay color: 5YR 8/4 pale orange. Slip color: 2.5YR 3/6 dark reddish brown. Hardness: 3.0. Inclusion(s): red, hematite (MS–CS, high frequency); white (MS, low frequency). Fair condition. Mold-made. Parallels, shape-type: Beth Shean/Scythopolis (Hadad 2002: 16, 20, nos. 19–24, Variant 1).
97. Reg. no. 84.1068.7L. Not shown. L = 3.3 cm. W = 1.5 cm. Incomplete, 3% preserved. Palestinian discus. Late first to third centuries C.E. Discus fragment. On discus: two extended legs with feet in half profile. Clay color: 7.5YR 7/4 dull orange. Slip color: 10R 4/6 red. Hardness: 3.0. Inclusion(s): white (VFS, low frequency). Good condition. Mold-made. Parallels, shape-type: Beth Shean/Scythopolis (Hadad 2002: 16, 20, nos. 19–24, Variant 1).
98. Reg. no. 84.2172.3L. Not shown. L = 5.5 cm. W = 5.4 cm. Incomplete, 5% preserved. Palestinian discus. Late first to third centuries C.E. Discus, shoulder, and nozzle fragment. On discus: two raised legs, the right one in half profile and the left in full frontal view; cloak tail between legs; leg stance and cloak suggest a centurion. On shoulder: remnant of complete, open double volute in positive relief is depicted on the shoulder, and thus the upper curl (wick-hole end) is greater in size than the lower. Nozzle-bridge: three (?) impressed horseshoe devices between symmetric volutes. Plain rounded and sloping shoulder. Round nozzle. A small single filling-hole inside discus. No ridge surrounds discus. Edge of discus extends onto shoulder. Clay color: 7.5YR 8/4 light yellow-orange. Slip color: 10R 3/4 dark red. Hardness: 4.0. Inclusion(s): red, hematite (MS, very low frequency). Poor condition. Mold-made. Parallels, shape-type: Beth Shean/Scythopolis (Hadad 2002: 16, 20, nos. 19–24, Variant 1).

Military Themes (cat. nos. 99–101). Three Palestinian discus lamp fragments portraying military subjects were found at Sepphoris. Two depict charioteers (cat. nos. 99–100) and the third, a military horseman draped in a toga (cat. no. 101). A parallel to the charioteer on cat. no. 99 was found at Antipatris (Neidinger 1982: 166, pl. 23.10). Lamps decorated with military schemes are quite rare in Roman Palestine and Arabia and tend to be imports from Italian workshops. An example is a fragment from Masada depicting the backside of a gladiator (*hoplomachus*; Bailey 1994: 81, no. 145, fig. 23). That the Sepphorean examples occur on regionally manufactured Palestinian discus lamps indicates that military schemes are part of the iconographic repertoire in the region. The charioteers and horseman of the Sepphoris sherds do not wear the characteristic helmets and armor or hold shields suggestive of gladiators as portrayed on the Masada fragment or on lamps from Ephesus (Büyükkolanci 2002: 94–95; Ladstätter 2002) and London (Campbell 2007: 635, fig. 2). The only other type of military theme occurring on locally manufactured lighting devices is the plan of a Roman fort that decorates the nozzle-bridge of an early fourth-century lamp from Samaria–Sebaste (Gichon 1972: 38–58; Dilke 1985: 148 n. 16, fig. 27; Gichon 1993; but cf. Lapp 1997a: 101–2, figs. 80–82).

99. Reg. no. 84.1068.9L.5. L = 7.1 cm. W = 3.6 cm. Incomplete, 25% preserved. Palestinian discus. Late first to third centuries C.E. Discus and shoulder fragment. On discus: Charioteer with the remnant of a galloping horse (including hind legs, buttocks, raised tail, and portion of body) and human figure standing upright and holding reins. Scene direction: left to right. On shoulder: a single band of medium-sized impressed horseshoe devices decorates shoulder; remnant of a single raised double axe. Nozzle base: no raised volutes. Remnant of round nozzle and wick-hole. A single rounded ridge surrounds discus. A small filling-hole within discus. Clay color: 2.5Y 8/4 pale yellow. Slip color: 10YR 3/1 brownish black, exterior and interior surfaces. Hardness: 3.0. Inclusion(s): white (MS, very low frequency). Poor condition. Mold-made. Parallels, shape-type: Beth Shean/Scythopolis (Hadad 2002: 16, 20, nos. 19–24, Variant 1). Parallels, shape-type and discus motif: Adler Collection (Adler 2004: 65, no. 392).

100. Reg. no. 84.1068.3L.21. L = 6.8 cm. W = 5.1 cm. Incomplete, 30% preserved. Palestinian discus. Late first to third centuries C.E. Discus and shoulder fragment. On discus: charioteer leading a galloping horse. Scene faces nozzle. Direction of movement: left to right. A single rounded ridge surrounds discus. A single small filling-hole inside discus. On shoulder: a single band of low and bulging impressed horseshoe devices; trace of a single raised double axe. Slightly sloping shoulder. Clay color: 7.5YR 7/3 dull orange. Slip color: 2.5YR 5/8 bright brown, exterior and interior surfaces. Hardness: 2.0. Poor condition. Mold-made. Parallels, shape-type: Beth Shean/Scythopolis (Hadad 2002: 16, 20, nos. 19–24, Variant 1). Parallels, shape-type and discus motif: Adler Collection (Adler 2004: 65, no. 392).

101. Reg. no. 84.1068.15L.5. L = 3.5 cm. W = 2.8 cm. Incomplete, 5% preserved. Palestinian discus. Late first to third centuries C.E. Discus medallion fragment. On discus: a draped horseman in 3/4 profile rides a harnessed horse in 1/2 profile. A single filling-hole within discus. Clay color: 7.5YR 7/4 dull orange. Slip color: 2.5YR 3/6 dark reddish brown, exterior only. Hardness: 2.0. Inclusion(s): red, hematite (FS, low frequency); white (FS-MS, medium frequency). Fair (-) condition. Mold-made.

Parallels, shape-type: Beth Shean/Scythopolis (Hadad 2002: 16, 20, nos. 19–24, Variant 1).

Mythological Themes (cat. nos. 102–11). A representation of an emperor god appears on a discus fragment (cat. no. 102). Depictions like this of a seated emperor god supported by a staff abound in the Mediterranean region during the Greco-Roman period. The theme of Europa riding a bull (Zeus) appears on two Palestinian discus lamp fragments found at Sepphoris (cat. nos. 103–4; fig. 3). An exact parallel of this motif appears on a discus lamp from Dor.[3] Five discus lamp medallions decorated with a Medusa were unearthed at Sepphoris (cat. nos. 105–9). Four have bulbous cheeks and flaring wavy hair (cat. nos. 105–8). Radiating bands in high relief surround the head of a Medusa (cat. no. 108) similar to a head on a Palestinian discus lamp from Caesarea Maritima (Sussman 2008: 270, no. 72). Catalog no. 109 bears a Medusa head with a highly Egyptianized headpiece. Medusa is a commonly depicted god throughout the ancient Mediterranean region. The sun-god Helios is depicted on two Palestinian discus fragments (cat. nos. 110–11). Helios is a Greek deity who was popular during the Hellenistic and Roman periods and is found in other Roman provinces of the Mediterranean. Despite the incompleteness of the motif in each example, the surviving curved and radiating points undoubtedly depict Helios's crown of radiant sunlight.

102a. Reg. no. 84.1068.6L.2. L = 8.1 cm. W = 6.9 cm. H = 1.9 cm. Incomplete, 75% preserved. Palestinian discus. Late first to third centuries C.E. Discus and shoulder fragment. On discus: an emperor god reclining on a flat surface and supported by a staff. On shoulder: a single band of small, round impressed horseshoe devices; two symmetrical, degenerate double open volutes below wick-hole; two raised double axes. Pattern faintly impressed on shoulder. A single ridge surrounds a discus that extends onto the shoulder. Round wick-hole. Small filling-hole within discus. Single ring base. Clay color: 2.5Y 8/4 pale yellow. Slip color: 2.5YR 4/2 grayish red. Hardness: 2.0. Inclusion(s): white (MS, medium frequency); black (FS, very low frequency). Poor condition. Mold-made. Parallels, shape-type: Beth Shean/Scythopolis (Hadad 2002: 16, 20, nos. 19–24, Variant 1).

102b. Reg. no. 84.2192.1L.1. L = 6.0 cm. W= 4.0 cm. Nearly complete, 90% preserved. Palestinian discus. Broken nozzle and discus partially missing. On surviving discus: rays of Helios's crown. On shoulder: two double axes in relief. On nozzle: double volutes. Flat concentric ring base. No lamp maker's mark. Clay color: 5YR 7/4 dull orange. Slip color: 10R 4/6 red (exterior). Hardness: 3.0. Inclusion(s): Limestone (MS, very low frequency). Condition: Good. Mold-made. Parallels: Nabratein (Lapp 2009: 271–72, lamp pl. B.13); Rosenthal and Sivan 1978: 86, nos. 347–48; Kennedy 1963: pl. 21.502).

103. Reg. no. 84.1068.10L.1. L = 6.3 cm. W = 4.3 cm. Incomplete, 35% preserved. Palestinian discus. Late first to third centuries C.E. Discus and shoulder fragment. On discus: a rearing bull with raised tail, in half profile; Europa, in frontal view, riding the bull and grasping a taut cloth that forms a bow over her head. Discus scene oriented toward nozzle. Discus scene, direction of movement: left to right. A small filling-

3. For an exhaustive examination of ancient texts on the myth of Europa and the artistic representation of Europa in antiquity, see Lecomte 2001.

hole within discus. A single rounded ridge surrounding the discus. On shoulder: a single band of tightly spaced, large, impressed horseshoe devices. Slightly sloping shoulder. Clay color: 2.5Y 8/4 pale yellow. Slip color: 10R 3/3 dark reddish brown, exterior surface only. Hardness: 2.0. Fair condition. Mold-made. Parallels, shape-type and discus motif: Aphek/Antipatris (Neidinger 1982: 161, no. 9 = pl. 22.1, very similar; also Kochavi 1989: 18); shape-type: Beth Shean/Scythopolis (Hadad 2002: 16, 20, nos. 19–24, Variant 1); Adler Collection (Adler 2004: 65, no. 394); IAA Collection (Sussman 2012: no. 370).

104. Reg. no. 84.1068.7L.1. L = 7.0 cm. W = 7.3 cm. H = 2.0 cm. Nearly complete, 90% preserved. Palestinian discus. Late first to third centuries C.E. Discus, shoulder, and base fragment. Missing nozzle. On discus: in positive relief, Europa riding a rearing bull without a tail. Scene faces nozzle. Direction of scene movement: left to right. A small single filling-hole within discus. A single rounded ridge surrounding discus. Slightly sloping shoulder. On shoulder: a single band of medium-sized bulging, inconsistently spaced, impressed horseshoes. Remnants of symmetrically opposing open double volutes at nozzle base. Single ring base. Clay color: 2.5Y 8/4 pale yellow. Slip color: 7.5YR 3/1 brownish black. Burnish: traces on exterior surface. Hardness: 2.0. Inclusion(s): white (VFS, very low frequency). Burning around nozzle. Poor condition: Poor. Mold-made. Parallels, shape-type: Beth Shean/Scythopolis (Hadad 2002: 16, 20, nos. 19–24, Variant 1); Adler Collection (Adler 2004: 65, no. 394); IAA Collection (Sussman 2012: no. 370).

105. Reg. no. 84.1296.22L.9. L = 2.7 cm. W = 3.1 cm. Incomplete, 5% preserved. Palestinian discus. Late first to third centuries C.E. Discus medallion. On medallion: a raging Medusa head with bulbous cheeks and flaring wavy hair in positive relief. Clay color: 7.5YR 8/2 light yellow-orange. Slip color: 10YR 3/1 brownish black, traces on exterior surface only. Hardness: 2.0. Good condition. Mold-made. Parallels, shape-type: Beth Shean/Scythopolis (Hadad 2002: 16, 20, nos. 19–24, Variant 1).

106. Reg. no. 84.1068.92L. GD = 3.0 cm. Incomplete, 5% preserved. Palestinian discus. Late first to third centuries C.E. Discus medallion. On medallion: a Medusa head with bulbous cheeks and flaring wavy hair in positive relief. Clay color: 7.5R 8/2 light yellow-orange. Slip color: 10YR 3/1 brownish black, traces on exterior surface only. Hardness: 2.0. Good condition. Mold-made. Parallels, discus motif, Medusa: Beth Shean/Scythopolis (Hadad 2002: 16, 20, no. 24, Variant 1); published: Lapp 1996: 220, no. 113.

107. Reg. no. 84.1068.7L.8. L = 2.1 cm. W = 2.1 cm. Incomplete, 5% preserved. Palestinian discus. Late first to third centuries C.E. Discus medallion. On medallion: Medusa head with bulbous cheeks and curly hair in positive relief. Clay color: 2.5YR 6/6 orange. Slip color: 2.5YR 3/6 dark reddish brown, exterior surface only. Hardness: 3.0. Inclusion(s): red, hematite (VFS, low frequency); white (VFS, low frequency). Good condition. Mold-made. Parallels, shape-type: Beth Shean/Scythopolis (Hadad 2002: 16, 20, nos. 19–24, Variant 1).

108. Reg. no. 84.1068.3L.1. L = 3.6 cm. W = 2.8 cm. Incomplete, 5% preserved. Palestinian discus. Late first to third centuries C.E. Discus medallion. On medallion: Medusa with bulbous cheeks and full curly hair in positive relief; a single rounded ridge encircles Medusa followed by radiating bands composed of two ridges in relief. Clay color: 7.5YR 7/4 dull orange. Slip color: 2.5YR 6/8 orange. Hardness: 3.0. Inclusion(s): red, hematite (FS-VFS, low frequency); white (VFS-CS, medium frequency). Good condition. Mold-made. Parallels, shape-type and discus motif, Medusa: Caesarea Maritima

(Sussman 2008: 230, no. 72); Galilean tomb (Lapp 1996: 220, no. 111); unknown provenance (Rosenthal and Sivan 1978: 40, no. 159) (similar); shape-type: Beth Shean/Scythopolis (Hadad 2002: 16, 20, nos. 19–24, Variant 1).

109. Reg. no. 84.1068.3L.33. L = 2.5 cm. W = 2.1 cm. Incomplete, 5% preserved. Palestinian discus. Late first to third centuries C.E. Discus medallion. On medallion: Medusa head with highly Egyptianized headdress in positive relief. Clay color: 10YR 8/3 light yellow-orange. Slip color: 2.5YR 4/2 grayish red. Hardness: 2.5. Inclusion(s): white (VFS, low frequency). Poor condition. Mold-made. Parallels, shape-type: Beth Shean/Scythopolis (Hadad 2002: 16, 20, nos. 19–24, Variant 1); Mampsis (Negev 1971: pl. 27.D) (similar); near Beth Shean (C. A. Kennedy 1963: pl. 22.508); unknown provenance (Rosenthal and Sivan 1978: 88, nos. 357–58; Bailey 1988: pl. 58, figs. 63, 131) (similar).

110. Reg. no. 84.1063.3L. Not shown. L = 4.0 cm. W = 2.4 cm. Incomplete, 3% preserved. Palestinian discus. Late first to third centuries C.E. Discus and shoulder fragment. On discus: remnants of four radiating points of the crown of Helios. A single flat ridge surrounds the discus. A slightly sloping plain shoulder. Clay color: 5YR 7/4 dull orange. Slip color: 10R 4/6 red, exterior surface. Hardness: 3.0. Inclusion(s): white (MS, very low frequency). Good condition. Mold-made. Parallels, shape-type and discus motif: Maresha (Oren and Rappaport 1984: pl. 14.A) (exact); Nabratein (Lapp 2009: 271–72, lamp pl. B.13 (exact); unknown provenance (Israeli and Avida 1988: 32, fig. 42; 33, fig. 45; Rosenthal and Sivan 1978: 86, nos. 347–48, C. A. Kennedy 1963: pl. 21.502) (exact); IAA Collection (Sussman 2012: no. 338); shape-type: Beth Shean/Scythopolis (Hadad 2002: 16, 20, nos. 19–24, Variant 1).

111. Reg. no. 84.1068.3L.28. Not shown. L = 4.4 cm. W = 2.7 cm. Incomplete, 3% preserved. Palestinian discus. Late first to third centuries C.E. Discus and shoulder fragment. On discus: remnant of radiating points of Helios crown. On shoulder: a single band of medium-sized, bulbous, impressed horseshoe devices with very small inner kernels; remnant of a double axe in positive relief. Clay color: 2.5YR 6/8 orange. Slip color: 10R 4/8 red, exterior and interior traces only; 10YR 2/1 black, exterior traces only. Hardness: 3.0. Inclusion(s): red, hematite (MS, very low frequency); white (FS-CS, very low frequency); black, magnetite (FS, very low frequency). Poor condition. Mold-made. Parallels, shape-type: Beth Shean/Scythopolis (Hadad 2002: 16, 20, nos. 19–24, Variant 1); parallels, shape-type and motif: IAA Collection (Sussman 2012: no. 338).

Bi-associative Motifs (cat. nos. 112–13). The decorated disks of most Palestinian discus lamps show a single motif, but several discus fragments have two separate motifs that are thematically bi-associated (cat. nos. 112–13; for additional examples, see also Lapp 1997a: 131). The bust of a satyr or of Dionysos facing a double-handled chalice to his left appears on one fragment (cat. no. 112). The chalice found on cat. no. 113 also belongs to the same type and is situated in the proper position to suggest a second similar fragment of this type. Another example of a bi-associative motif is a crane facing a crab that decorates a Palestinian discus lamp in the Schloessinger Collection (Rosenthal and Sivan 1978: 89, no. 365) and in the *Akademisches Kunstmuseum* at the University of Bonn (Hübinger 1993: 125, pl. 30.236). A further example of the crane and crab composition is portrayed on a Palestinian discus lamp from Caesarea Maritima (Sussman 2008: 231, 270, no. 74; for other examples, see Sussman 2012: nos. 453, 456, 464). I suggest that this unique juxtaposition of two stylistically separate subjects forms a thematic relationship between each character: the satyr or Dionysos drinks wine held in a chalice and, similarly, cranes eat crabs. A lamp from Tubas belonging to the bi-associative group has a representation of a man's head, probably a farmer, facing a sheath of wheat (C. A. Kennedy 1963: 99, no. 501, pl. 21; but Kennedy sees the depiction differently: the profile of a female head facing a pillar).

112. Reg. no. 84.1068.6L.7. L = 6.0 cm. W = 5.7 cm. Incomplete, 30% preserved. Palestinian discus. Late first to third centuries C.E. Discus and shoulder fragment. On discus: a representation of a large head of a satyr, or Dionysos, with full wavy hair and beard in positive relief is juxtaposed to a double-handled chalice. A single rounded ridge surrounds the discus. A single small filling-hole is located within the discus. On shoulder: a single band of consecutive and incised horseshoe devices surround discus. One complete double axe rendered on shoulder in positive relief. Clay color: 2.5Y 8/4 pale yellow. Slip color: 10YR 3/2 brownish black (exterior). Hardness: 3.5. Inclusion(s): white (FS-CS, low frequency); black (FS-CS, low frequency); muscovite mica (?)(CSi, very low frequency). Good condition. Mold-made. Parallels, shape-type and discus motif: unknown provenance (Rosenthal and Sivan 1978: 88, no. 359) (exact); Parallels, shape-type: Beth Shean/Scythopolis (Hadad 2002: 16, 20, nos. 19–24, Variant 1).

113. Reg. no. 84.1068.7L.11. L = 4.1 cm. W = 3.2 cm. Incomplete, 5% preserved. Palestinian discus. Late first to third centuries C.E. Discus and shoulder fragment. A single rounded ridge surrounds discus. On discus: a representation of a double-handled chalice in high relief. On shoulder: a single band of consecutively incised horseshoe devices decorate shoulder around discus; a partial double-axe device in positive relief. Clay color: 2.5Y 8/4 pale yellow. Slip color: 10YR 3/2 brownish black (exterior only). Hardness: 2.5. Inclusion(s): white (VFS, low frequency); magnetite (?) (VFS, low frequency); muscovite mica (?)(CSi, very low frequency). Carbon: Interior traces. Poor condition. Mold-made. Parallels, shape-type and discus motif: unknown provenance (Rosenthal and Sivan 1978: 88, no. 359) (exact); shape-type: Beth Shean/Scythopolis (Hadad 2002: 16, 20, nos. 19–24, Variant 1).

Erotic Imagery (cat. nos. 114–22). Erotic imagery appears on nine Palestinian discus lamps and fragments recovered on the western summit at Sepphoris (fig. 4; cat. nos. 114–22). Five types of lovemaking imagery suggest the use of different molds for their manufacture (Lapp 1997a: 96–98). A common symplegma scene includes a single, unclothed male kneeling behind a female. His back is slightly arched and his right arm is bent behind his posterior (cat. nos. 117–18). His left arm extends to the buttocks of the second figure, whose arms and lower legs support her body. A close parallel of this erotic scene appears on a discus lamp from the necropolis at Beth Sheʿarim (Avigad 1976: pl. 70.3). Two human figures engage in other sexual activities (cat. nos. 114–16). Erotic lamps have also been found at Abila, Ashkelon, Dor, Gerasa, and Jerusalem (Akeldama; Lapp 1997a: 94–96 nn. 34, 38; Lapp 2001b: 63, 65–66).

Clay oil lamps were the chief tangible medium upon which erotic scenes were portrayed and circulated in Roman Palestine and Arabia. Unlike the wide range of various erotic activities and significant quantity of sexual scenes found on material culture and brothel frescoes in Italy and elsewhere in the Roman Empire (Clarke 2003: 118–32, figs. 1–107), the archaeological evidence does not suggest widespread artistic expression of lovemaking or pornography in Roman Palestine and Arabia. Taking into account the popularity of erotic art throughout the Greco-Roman world, the occurrence of erotic lamps in Roman Palestine and Arabia is not extraordinary (Lapp 1996: 220–21, no. 114). The use of erotic lamps may have had some association with brothels and may represent evidence for the existence of brothels in the cities—and even villages (see discussion in D. Kennedy 2007: 177–80)—of Galilee, the Decapolis, and elsewhere in Roman Palestine. The presence of erotic lamps at Sepphoris may mean the existence of prostitution in the city.

Erotic scenes do not appear on several lamp types manufactured in Roman Palestine and Arabia, including the Darom, the Beit Nattif, and the Samaritan. This may be explained in part by their respective morphologies: even though motifs do occur on the nozzle and/or shoulder of lamps belonging to these forms, the lack of concave disks means that there is insufficient space to portray a complete erotic scene with two or more participants. Yet even though there is enough surface area to depict a single erotic motif, such as a phallus, no such representations occur on these lamps. On morphological grounds, the absence of erotic scenes on Caesarean discus lamps is a bit more difficult to explain, given the fact that their large central disks provide plenty of room to do so.

A more plausible explanation for the absence of pornographic scenes on these lamp forms may be a Jewish, Samaritan, or Christian rejection of erotic themes considered pagan. Rather, the Caesarean discus and Beit Nattif lamps typically portray Jewish or Christian symbols (e.g., a cross, a fisherman, or a menorah). Samaritan lamps depict images either of Samaritan religious objects (e.g., the Samaritan Torah Shrine and menorah) or of geometric patterns. And Darom lamps display a rich repertoire of motifs from daily life (e.g., baskets, boats, flora, bird cages, and even lamps). Thus the absence of erotic scenes from the repertoire of motifs used on these lamps may have been a deliberate choice; the lamp makers fashioned lamps with motifs or symbols to accommodate the cultural tastes and religious sensitivities of their Christian, Jewish, and Samaritan clientele.

The repertoire of erotic scenes on lamps made in Roman Palestine and Arabia is relatively limited, and the scenes are mild compared with the variety of extreme sexual practices depicted on lamps elsewhere in the Empire. For example, no fellation scene, such as is pictured on a first-century C.E. Italian lamp (Cantarella 1992: 98), has been found on lamps unearthed in the Holy Land. Absent, too, are heterosexual copulation images showing sex between two elderly persons, grotesques, or infants; such scenes occur on lamps from Greece and Cyprus (Bailey 1989: 64–65). The absence of homosexual scenes on locally manufactured lamps may reflect Jewish, Christian, and Samaritan prohibitions of homosexual and bisexual behavior; but one must keep in mind that, in comparison with heterosexual-copulation imagery on clay lamps, homosexual or bisexual scenes on lamps manufactured elsewhere in the Empire are also rare. Additionally, no lamps have been recovered in Israel and Jordan that portray a sexual organ, such as a phallus, as do such lamps found elsewhere: for example, an Early Roman lamp excavated at Herculaneum depicts a phallus with wings (Clarke 2003: fig. 63).

Erotic scenes depicted on lamps from Roman Palestine and Arabia can be grouped into three main categories: (1) heterosexual, (2) bestial, and (3) mythological. Heterosexual copulation between two adult humans is the most common type. Pornographic lamps showing a male kneeling behind a female have been found at northern sites, such as Beth Sheʿarim (Avigad 1976: pl. 70.3; Lapp 1997a: fig. 73) and Sepphoris (cat. no. 117), perhaps suggesting local manufacture in Galilee. A second group of heterosexual copulation shows a male grasping the upraised and bent left or right leg of a female lying on her back; he is positioned between her legs. Both participants rest on a sofa. Oil lamps portraying this type of scene have been found at the southern coastal city of Ashkelon (Lapp 2001b: 62–63) and at Sepphoris (cat. no. 114).

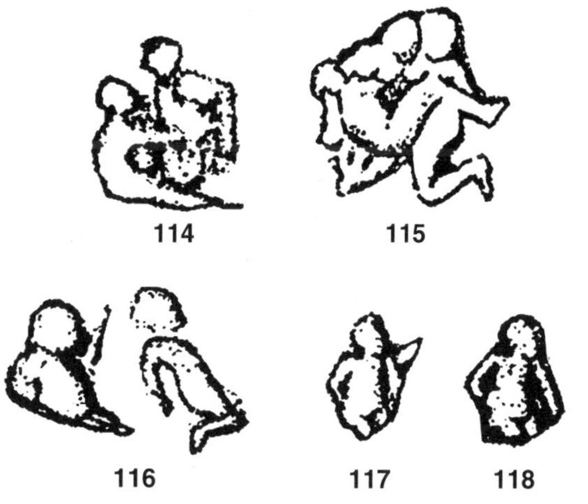

Fig. 4. Heterosexual erotic scenes on Palestinian discus lamps.

A less common representation portrays a male figure resting upright on a sofa with a female facing him. A complete discus lamp depicting this scene was found at Sepphoris (cat. no. 115). Another type of heterosexual symplegma scene on oil lamps from the Holy Land depicts a female figure lying on a sofa with both her legs fully extended and bent upright; with slightly bent knees, a prostrate male grasps her legs and seduces her. Several lamps found at Ashkelon portray this type of heterosexual erotic scene (Stager 1991: 47; Lapp 1997a: figs. 75–76; Lapp 2001b: 62). The most risqué of the heterosexual erotic scenes shows a female figure in a commanding, dominant sexual position: she is rendered in full frontal view squatting onto her male lover who lies horizontally below her in side view. The scene occurs on a lamp of probable local manufacture recovered from a tomb in the Decapolis city of Abila (Fuller 1987: 477, fig. 103, A; Lapp 1997a: fig. 78). Only a few clay lamps with erotic scenes of the bestial category have been found in Roman Palestine. The bestial type of erotic scene generally depicts a bull in congress with a woman lying on her back. Several such lamps have been discovered at the northern coastal city of Dor (Lapp 1997a: 95–96 n. 38).

The mythological rape of Leda by Zeus, who transforms himself into a swan is depicted on several lamps found in Transjordan and is the only type of mythological-erotic scene occurring in the region. That two lamp fragments portraying Leda and the swan were recovered among carved lamp molds from a pottery workshop at Gerasa /Jerash in Jordan (Illife 1945: 24, nos. 149–50) suggests that mythological-type erotic lamps were probably manufactured there.

114. Reg. no. 84.1068.6L.6. L = 7.0. W = 4.2 cm. Incomplete, 30% preserved. Palestinian discus. Late first to third centuries C.E. Discus and shoulder fragment. On discus: two

human figures, one sitting in an upright position and the other resting on top of the first with legs bent. On shoulder: a single band of consecutively incised horseshoe devices; a single double-axe device is depicted in positive relief on the shoulder; a trace of an incomplete raised open volute on the shoulder. Clay color: 2.5Y 8/4 pale yellow. Slip color: 10YR 3/1 brownish black. Hardness: 3.0. Inclusion(s): white (CS, very low frequency; exterior surface only). Poor condition. Mold-made. Parallels, shape-type: Beth Shean/Scythopolis (Hadad 2002: 16, 20, nos. 19–24, Variant 1); cf. collection of the IAA (Sussman 2012: no. 383).

115. Reg. no. 84.1068.3L.2. L = 8.4 cm. W = 7.1 cm. H = 2.3 cm. Complete. Palestinian discus. Late first to third centuries C.E. On discus: a representation of two human figures engaging in sexual intercourse, one lying on a flat surface with bent legs projected upward and upper body rendered in 3/4 profile supported vertically by extended arms, the other bent prostrate in 1/2 profile in front of the prostrate figure. On shoulder: a single band of consecutively incised horseshoe devices; two raised double axes are rendered in positive relief. Two raised open double volutes are depicted directly below the nozzle-bridge. Lamp maker's mark of the Greek ligature *eta rho* impressed in the base. Clay color: 2.5Y 8/4 pale yellow. Hardness: 3.5. Inclusion(s): white (FS, very low frequency; exterior surface only). Good condition. Mold-made. Parallels, shape-type: Beth Shean/Scythopolis (Hadad 2002: 16, 20, nos. 19–24, Variant 1). Published: Meyers, Netzer, and Meyers 1992: 23, lower right photograph; Lapp 1996: 220–21, no. 114.

116. Reg. no. 84.1068.14L.2. L = 6.5 cm. W = 4.6 cm. Incomplete, 30% preserved. Palestinian discus. Late first to third centuries C.E. Discus and shoulder fragment. On discus: an erotic scene between two human figures. Two kneeling, unclothed figures, the first one facing the posterior of the second, are depicted in 1/2 profile in positive relief. On shoulder: a single band of consecutive, tightly positioned incised horseshoe devices. A single ridge encircles the discus. Clay color: 5YR 8/4 pale orange. Slip color: 5YR 3/4 dark reddish brown and 7.5 YR 1.7/1 black (exterior). 10R 3/6 dark red (interior). Hardness: 3.0. Poor condition. Mold-made. Parallels, shape-type: Beth Shean/Scythopolis (Hadad 2002: 16, 20, nos. 19–24, Variant 1).

117. Reg. no. 84.1068.7L.9. L = 4.3 cm. W = 3.2 cm. Incomplete, 10% preserved. Palestinian discus. Late first to third centuries C.E. Discus and shoulder fragment. On discus: an unclothed male human figure with slightly arched back and two arms, one bent behind his posterior and the left extended, in positive relief. As rendered on other similar erotic poses involving two figures, the left arm extends to the buttocks of a second figure whose arms and lower legs support its body. On shoulder: a single band of consecutive and impressed horseshoe devices. Clay color: 10YR 8/3 light yellow-orange. Slip color: 10YR 4/1 brownish gray and 7.5YR 5/6 bright brown, exterior only. 10YR 5/2 grayish yellow-brown (interior). Hardness: 3.0. Inclusion(s): white (MS-CS, medium frequency). Fair condition. Mold-made. Parallels, shape-type: Beth Shean/Scythopolis (Hadad 2002: 16, 20, nos. 19–24, Variant 1).

118. Reg. no. 84.1068.9L.4. L = 4.3 cm. W = 3.1 cm. Incomplete, 10% preserved. Palestinian discus. Late first to third centuries C.E. Discus and shoulder fragment. On discus: a representation of a single unclothed human male figure with slightly arched back and two arms, the right one bent behind his posterior and the left extended, in positive relief. Left arm extends to the buttocks of a second figure, whose arms and lower legs support its body. On shoulder: a single band of consecutive impressed horseshoe devices. Clay color: 10YR 8/4 light yellow-orange. Slip color: 2.5YR 3/6

dark reddish brown. Hardness: 3.0. Inclusion(s): red, hematite (VFS, very low frequency); white (VFS, low frequency). Fair (+) condition. Mold-made. Parallels, shape-type and discus motif: Beth Sheʿarim (Avigad 1976: pl. 70.3) (exact).

119. Reg. no. 84.1068.6L.5. L = 6.9 cm. W = 3.3 cm. Incomplete, 30% preserved. Palestinian discus. Late first to third centuries C.E. Discus and shoulder fragment. On discus: two frontally facing human figures. On shoulder: a single band of consecutively incised horseshoe devices; a trace of a single double-axe device; a remnant of a single, incomplete curling volute. Clay color: 2.5Y 8/4 pale yellow. Slip color: 10R 3/2 dark reddish brown. Hardness: 2.0. Inclusion(s): white (MS, medium frequency). Poor condition. Mold-made. Parallels, shape-type: Beth Shean/Scythopolis (Hadad 2002: 16, 20, nos. 19–24, Variant 1).

120. Reg. no. 84.1068.14L.3. L = 5.9 cm. W = 4.4 cm. Incomplete, 25% preserved. Palestinian discus. Late first to third centuries C.E. Discus and shoulder fragment. On discus: a representation of two unclothed human figures, one sitting with his right arm resting on his lap and left arm extending to the buttocks of the second squatting figure, whose arms extend forward and upward in 1/2 profile. Plain shoulder. A single ridge surrounds the discus. Clay color: 2.5Y 8/4 pale yellow. Slip color: 2.5YR 4/3 dull reddish brown. Hardness: 3.0. Poor condition. Mold-made. Parallels, shape-type: Beth Shean/Scythopolis (Hadad 2002: 16, 20, nos. 19–24, Variant 1).

121. Reg. no. 84.1068.7L. Not shown. L = 3.3 cm. W = 2.2 cm. Incomplete, 3% preserved. Palestinian discus. Late first to third centuries C.E. Discus and shoulder fragment. On discus: a representation of two raised human heads rendered in 1/2 profile, facing and touching one another is depicted. Plain shoulder. A single ridge surrounds the discus. Clay color: 2.5Y 8/4 pale yellow. Slip color: 10YR 3/1 brownish black (exterior only). Hardness: 3.0. Inclusion(s): white (MS, very low frequency). Fair condition. Mold-made. Parallels, shape-type: Beth Shean/Scythopolis (Hadad 2002: 16, 20, nos. 19–24, Variant 1).

122. Reg. no. 84.1068.3L. Not shown. L = 2.2 cm. W = 1.2 cm. Incomplete, 3% preserved. Palestinian discus. Late first to third centuries C.E. Discus fragment. On discus: a probable erotic scene of a single figure in a lying upright position supported by a bent left arm. Clay color: 5YR 6/6 orange. Slip color: 10R 3/6 dark red, exterior and interior surfaces. Hardness: 3.0. Inclusion(s): white (MS, medium frequency). Good condition. Mold-made. Parallels, shape-type: Beth Shean/Scythopolis (Hadad 2002: 16, 20, nos. 19–24, Variant 1).

Discus Medallions

Discus medallions were a by-product of lamp breakage in antiquity; they were fragments of the concave discus of round-bodied lamps (e.g., Broneer Types XXI–XXV) that resulted from either the intentional breakage of the discus or from natural processes. Two types of discus medallions occur in the archaeological record: the round variety resembling a medallion, and the angular discus "chip" version (Lapp under review). Many scholars have debated the issue of the deliberate removal of the central discus of lamps (e.g., see overview in Tal and Bastos 2012: 104–14; also Sivan 1982: 115–16), but the function of the medallions themselves is not well understood and is rarely discussed.

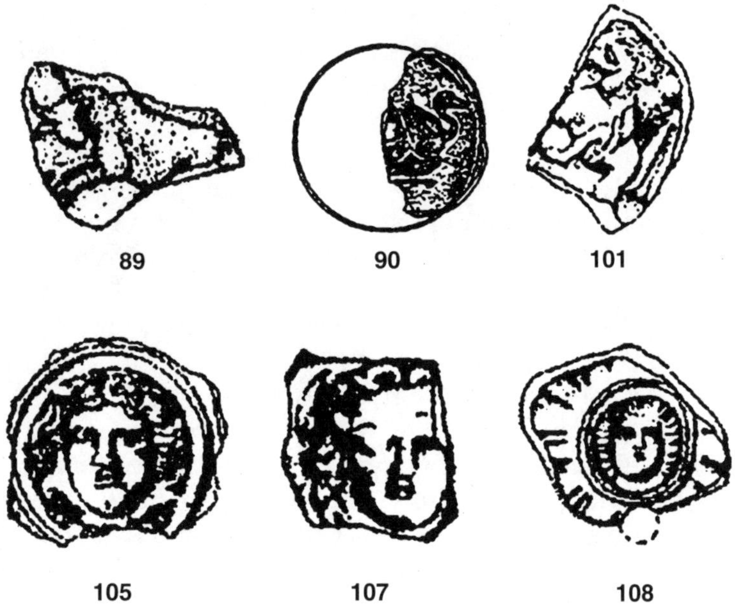

Fig. 5. Discus medallions used as photo amulets or gaming pieces.

Discus medallions and round-bodied lamps with missing disks have been found at numerous sites throughout the ancient Mediterranean. At Sepphoris, for example, several medallions were recovered on the western summit: they are decorated with a maple leaf (cat. no. 86), a bull (cat. no. 89), a camel (cat. no. 90), a soldier riding a horse (cat. no. 101), and most commonly, the face of Medusa (cat. nos. 105–9). The medallions come from lamps belonging to the Palestinian Discus type. At Pergamon in Asia Minor, numerous discus medallions were unearthed, many of which depict Medusa, erotic scenes, animals, and fowl (Heimerl 2001: pls. 18.773–76, 778–831; 19.832–95).

Given the wealth of symbolism, iconographic elements, and mythological motifs occurring on the ubiquitous discus lamps of the late antique Mediterranean region, it is possible that, once the disks were broken out, the resulting medallion was kept and served as an amulet (Lapp in press b; see fig. 5). Archaeological evidence for such a practice is known from Byzantine Nubia, where medallions were cut from earthenware magic bowls and served either as amulets or gaming pieces (Żurawski 1992: 96–97, figs. 8–9).[4] These medallions bear either magical names or images of a profiled human face circumscribed with Greek characters forming the names of either Romaios or Persos (Żurawski 1992: 96). Although this sixth-century Christian Nubian practice postdates the missing discus phenomenon, it nonetheless may point to an earlier, possibly Roman-period origin.

Distribution

The Palestinian discus lamp was widely popular throughout Roman Palestine and among the Decapolis cities of Transjordan. Several have even been found as

4. For a discussion of amulets and magical bowls in late antiquity, see Naveh and Shaked 1985.

far south as the Nabatean port city of Roman Aqaba (e.g., reg. nos. 12618, 27279, 49244 in Lapp forthcoming a). At a minimum, discus lamps have been found at some 53 Palestinian and Arabian sites (see Lapp 1997a: 39–44, figs. 21–22). Significant deposits have been uncovered, for example, at Dor (Rosenthal-Heginbottom 1995: 244–46, fig. 5.22:1, photo 5.56), in the Mercantile Quarter at Antipatris (Neidinger 1982: 160–66, nos. 9–39, pls. 22–23), in the western summit of Sepphoris (Lapp 1996: 220–21, nos. 112–14), in a villa room at Ashkelon (Stager 1991: 45–47), and in the Akeldama Tomb in Jerusalem (Avni and Greenhut 1996: 84–85, fig. 4.9:1–2, 4). In addition to being found at Sepphoris, Palestinian discus lamps and fragments have been found at the four synagogue sites of the Meiron Excavation Project: Gush Ḥalav (Meyers and Meyers 1990: 158–59, pl. A.9–10, 12, 16), Khirbet Shemaʿ (Meyers, Kraabel, and Strange 1976: pl. 8.9:1–6, 8, 13, 14), Meiron (Meyers, Strange, and Meyers 1981a: pls. 9.16:1–3, 4–8, 10; 9.17), and Nabratein (Lapp 2009: lamp pls. B.5–6, 11–13; C.8, 11–13). Lamps of this type were also found at Nazareth, Capernaum, Pella, Abila, Tel Halif, Jericho, Jerusalem, Ramat Rahel, Beit Nattif, Gezer, Tell en-Naṣbeh, Rosh Haʿayin, Samaria, Silet edh-Dahar, Huqoq, Beth Sheʿarim, Beth Shean, Gerasa, Ḥorvat Uza, and Debaal, and in the Judean Desert (Rosenthal and Sivan 1978: 89–90).

Of all the lamp types manufactured in Roman Palestine, Palestinian discus lamps are the ones most often found outside Israel and Jordan. For example, several have been unearthed in association with the Sanctuary of Apollo at Tyre in Lebanon (Marchand 1996: 64–65, nos. 64–67). Numerous examples have been found on Cyprus (Oziol 1977: 184–89, pl. 31.548–55). Only three, however, come from as far west as Athens (published under the heading "Various Imported Lamps" with no further identification of provenance; see Perlzweig 1961: 84, cat. nos. 132–33, pl. 5). The variety of findspots indicates that Palestinian discus lamps were used in villages, towns, and cities with Roman, Jewish, and mixed populations (Lapp 2001c: 297). Palestinian discus lamps were chiefly manufactured in the late first to third centuries C.E. (Rosenthal and Sivan 1978: 85; Lapp 1997a: 39–44; da Costa 2003: 1.50).

The Palestinian discus lamps and fragments found at Sepphoris suggest that the Jewish community there was not isolated from the cosmopolitan cities of the Jordan Valley. Discus lamps commonly depict images regarded as pagan. Furthermore, recent preliminary archaeometric analyses suggest that the lamp type likely originated in or near the city of Beth Shean/Scythopolis in the northern Jordan Valley, thus indicating contact with that region (Lapp 1997a: 114–79). Hayes describes the Palestinian discus lamp type as a "South Syrian/North Palestinian" form and cautions that a northern Palestinian source for this group should not be dismissed (Hayes 1981: 86). The widespread occurrence of this type throughout Roman Palestine, far exceeding its occurrence at sites in Phoenicia, Syria, and elsewhere in the Roman Empire, corroborates Hayes's suggestion of a production center in Roman Palestine. Archaeometric analysis further supports the case for a Palestinian origin. Comparative dcp-oes and petrographic thin-section analyses suggest that at least two clay sources were exploited in the manufacture of this type: selected discus lamp fragments recovered from several cities of the Decapolis, along the coast, and in Galilee were found to exhibit similar chemical compositions to lamps unique to Palestine—the bilanceolate and Beit Nattif forms—indicating a Palestinian origin for at least one group of discus lamps (Lapp 1997a: 154–57, 159–62, 388–89, figs. 115–16; Lapp 2001c: 297). In order to confirm the whereabouts of the one or more workshops that manufactured Palestinian discus lamps, other evidence—such as molds, kilns, and wasters—must be found.

123. Reg. no. 84.1068.3L.16. L = 8.6 cm. W = 7.2 cm. H = 2.1 cm. Nearly complete, 95% preserved. Palestinian discus. Late first to third centuries C.E. Shallow concave discus whose sides extend onto the shoulder, thus forming a single prominent rounded ridge. Missing discus. Slightly sloping curved nozzle. Narrow slightly sloping shoulder. On shoulder: a single band of closely juxtaposed bulbous deeply impressed horseshoe devices with large inner kernels; two double-axe devices in positive relief; two opposing open double volutes, between which are three (?) raised vertical strokes. Single ring base. Clay color: 7.5YR 8/3 light yellow-orange. Slip color: 10R 5/6 red and 5YR 3/1 brownish black. Hardness: 3.0. Inclusion(s): white (FS-CS, medium frequency). Fair condition. Mold-made. Parallels, shape-type: Beth Shean/Scythopolis (Hadad 2002: 16, 20, nos. 19–24, Variant 1).
124. Reg. no. 84.1068.15L.3. L = 8.8 cm. W = 7.6 cm. H = 1.8 cm. Nearly complete, 95% preserved. Palestinian discus. Late first to third centuries C.E. Missing discus. Narrow, slightly sloping shoulder. Two ridges with wide, flat surfaces encircle deep discus. On shoulder: a single band of deeply impressed, consecutive horseshoe devices. Flat, round nozzle. On each side of the nozzle base: two nodules (stylized nail-heads). Slightly raised circular platform base. On base: impressed *planta pedis*. Clay color: 5YR 7/4 dull orange. Slip color: 2.5YR 3/6 dark reddish brown. Hardness: 3.5. Inclusion(s): white (CS-VCS, low frequency). Burning on nozzle. Fair condition. Mold-made. Parallels, shape-type: Beth Shean/Scythopolis (Hadad 2002: 16, 20, nos. 19–24, Variant 1).
125. Reg. no. 84.1068.12L.2. L = 7.8 cm. W = 7.4 cm. Incomplete, 40% preserved. Palestinian discus. Late first to third centuries C.E. Discus and shoulder fragment. Plain discus with steep sides that extend onto the shoulder, forming a single rounded ridge. On discus: a small filling-hole. Narrow, slightly sloping shoulder. On shoulder: a single band of consecutive, well-executed bulbous impressed horseshoe devices. Clay color: 7.5 YR 8/4 light yellow-orange. Slip color: 2.5YR 3/4 dark reddish brown, exterior and interior surfaces. Burnish: traces on exterior surface. Hardness: 3.0. Inclusion(s): white (FS, very low frequency). Fair condition. Mold-made. Parallels, shape-type: Beth Shean/Scythopolis (Hadad 2002: 16, 20, nos. 19–24, Variant 1); unknown provenance (Israeli and Avida 1988: 33, no. 46).
126. Reg. no. 84.1068.7L.7. L = 5.8 cm. W = 5.4 cm. Incomplete, 20% preserved. Palestinian discus. Late first to third centuries C.E. Discus and shoulder fragment. Five ridges surround broad discus. Small central filling-hole. Narrow, slightly sloping shoulder. On shoulder: trace of single band of connected, consecutively rendered bulbous horseshoe devices decorate shoulder. Clay color: 5YR 8/4 pale yellow. Slip color: 10R 3/4 dark red. Hardness: 3.0. Inclusion(s): white (FS, very low frequency); quartz (VFS, medium frequency). Good condition. Mold-made. Parallels, shape-type: Beth Shean/Scythopolis (Hadad 2002: 16, 20, nos. 19–24, Variant 1).
127. Reg. no. 84.1209.2L. L = 8.6 cm. W = 7.1 cm. H = 2.0 cm. Nearly complete. Palestinian discus. Late first to third centuries C.E. Plain discus, most of which is missing. Single broad ridge surrounds discus. On shoulder: two double-axe devices; raised

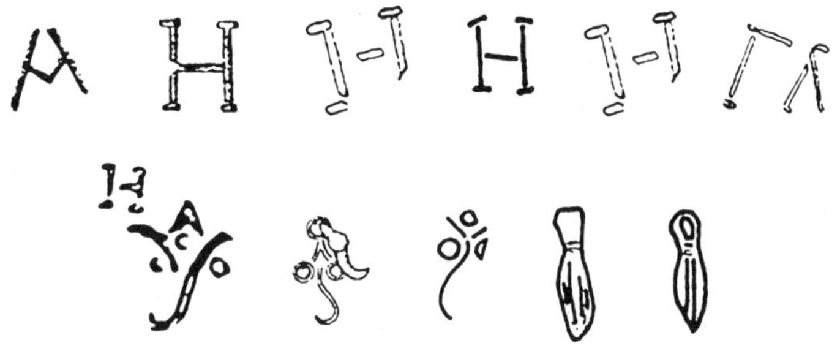

Fig. 6. Lamp makers' marks on bases of Palestinian discus lamps.

open double volutes at base of nozzle; four raised strikes between volutes. Round nozzle. Slightly sloping, plain shoulder. Double concentric ring base in which outer ring is wider than the inner. Underside: two parallel incised lines where shoulder meets nozzle base. Clay color: 5YR 7/6 orange. Slip color: 2.5YR 3/4 dark reddish brown. Hardness: 3.0. Inclusion(s): white (VC, very low frequency). Burning on nozzle. Fair condition. Mold-made. Parallels, shape-type: Beth Shean/Scythopolis (Hadad 2002: 16, 20, nos. 19–24, Variant 1).

128. Reg. no. 84.1094.8L.1. L = 8.5 cm. W = 7.3 cm. H = 1.9 cm. Nearly complete, 95% preserved. Palestinian discus. Late first to third centuries C.E. Description: whole discus lamp with missing disk. Round, level nozzle. Broad, level then slightly sloping shoulder. Shallow discus whose sides extend onto the shoulder forming a rounded prominent ridge. Two opposing open double volutes depicted in positive relief on the shoulder at the base of the nozzle. Remnants of raised vertical strokes between volutes rendered directly below the nozzle. Single band of closely juxtaposed, consecutive horseshoe devices are impressed on shoulder around the discus. Two double-axe devices in positive relief decorate shoulder. Remnants of indiscernible motif on discus. Circular, slightly raised platform base. Clay color: 5Y 8/3 pale yellow. Slip color: 2.5YR 3/2 dark reddish brown. Hardness: 3.0. Inclusion(s): white (FS-MS, low frequency). Burning on nozzle. Poor condition. Mold-made. Parallels, shape-type: Beth Shean/Scythopolis (Hadad 2002: 16, 20, nos. 19–24, Variant 1).

129. Reg. no. 84.1068.12L.1. L = 10.1. W = 8.8 cm. H = 2.5 cm. Incomplete, 80% preserved. Palestinian discus. Late first to third centuries C.E. Discus, shoulder, nozzle, and base fragment. Missing discus. Rounded nozzle. Shallow discus encircled by three rounded ridges, the outer two of equal width, and the innermost, most narrow. Narrow, steeply sloping shoulder. On shoulder: a single band of unevenly spaced impressed darts. Two small impressed circles on each side of the nozzle base. Slightly raised platform base. Clay color: 7.5YR 8/4 light yellow-orange. Slip color: 5YR 3/2 dark reddish brown, exterior only; 10R 4/6 red, interior surface only. Hardness: 2.0. Inclusion(s): white (VFS, very low frequency). Good condition. Mold-made. Parallels, shape-type: Beth Shean/Scythopolis (Hadad 2002: 16, 20, nos. 19–24, Variant 1).

130. Reg. no. 84.1068.9L.1. Not shown. L = 8.7 cm. W= 7.5 cm. H = 2.0 cm. Complete. Palestinian discus. Late first to third centuries C.E. Plain discus. Round nozzle. Missing discus. Single ridge surrounding discus extends onto shoulder. On shoulder: two double-axe devices in high relief. Two open double volutes are rendered at base of nozzle on the shoulder. Single ring base. Bichrome but poorly executed. Clay color: 10YR 7/3 dull yellow-orange. Slip color: 10R 4/8 red, exterior surface of discus and base only; 10YR 3/1 brownish black, remnant of exterior surface of body. Hardness: 3.5. Inclusion(s): white (MS, low frequency; discernible on exterior surface

only). Burning around wick-hole. Fair condition. Mold-made. Parallels, shape-type: Beth Shean/Scythopolis (Hadad 2002: 16, 20, nos. 19–24, Variant 1); Huqoq (Ravani 1961: figs. 3–4.23; 5, respectively) (exact).

131. Reg. no. 84.1068.6L.1. L = 8.2 cm. W = 7.2 cm. H = 1.8 cm. Nearly complete, 95% preserved. Palestinian discus. Late first to third centuries C.E. Broken round nozzle. Discus sides extend onto shoulder, forming a single rounded ridge. Plain, slightly sloping, broad shoulder. On shoulder: two double-axe devices in positive relief. Two open raised double volutes appear at the base of the nozzle on the shoulder. Four raised "strokes" are depicted between the volutes. Single rounded (?) ring base. On discus: remnant of indiscernible motif. Clay color: 2.5Y 8/4 pale yellow. Slip color: 10YR 2/1 black, exterior only. Hardness: 3.5. Inclusion(s): white (FS, very low frequency; exterior surface only). Fair (–) condition. Mold-made. Parallels, shape-type: Beth Shean/Scythopolis (Hadad 2002: 16, 20, nos. 19–24, Variant 1).

132. Reg. no. 84.1068.3L.14. L = 7.7 cm. W = 5.7 cm. H = 2.0 cm. nearly complete, 55% preserved. Palestinian discus. Late first to third centuries C.E. Shallow concave discus whose outer edge extends onto the shoulder. Missing discus. On shoulder: trace of a single double-axe device in positive relief; a single band of loosely spaced alternating and impressed acanthus leaf-and-horseshoe pattern. Trace of a single, raised, open double volute on shoulder. Remnants of three raised strikes rendered at base of nozzle. Slightly sloping shoulder. On base: lamp maker's mark, Greek ligature *upsilon chi*. Single ring base. Clay color: 2.5Y 7/3 light yellow. Slip color: 5YR 3/2 dark reddish brown (exterior and interior surfaces). Hardness: 2.0. Inclusion(s). white (FS, very low frequency). Burning on interior surface. Poor condition. Mold-made. Parallels, shape-type: Beth Shean/Scythopolis (Hadad 2002: 16, 20, nos. 19–24, Variant 1).

133. Reg. no. 84.1057.3L.12. L = 6.2 cm. W = 2.2 cm. Incomplete, 15% preserved. Palestinian discus. Late first to third centuries C.E. Discus and shoulder fragment. On discus: indiscernible motif. Slightly sloping shoulder. On shoulder: a single band of closely spaced, well-executed impressed acanthus leaf-darts; a single raised open double volute. Clay color: 2.5Y 8/4 pale yellow. Slip color: 10YR 3/1 brownish black. Hardness: 3.0. Inclusion(s): white (VFS, very low frequency); muscovite mica (CSi, low frequency). Fair condition. Mold-made. Parallels, shape-type: Beth Shean/Scythopolis (Hadad 2002: 16, 20, nos. 19–24, Variant 1).

134. Reg. no. 84.1094.6L.1. Not shown. L = 7.9 cm. W = 7.1 cm. H = 2.2 cm. Nearly complete, 95% preserved. Palestinian discus. Late first to third centuries C.E. Missing discus. Three ridges surround shallow discus. Narrow, slightly sloping shoulder. On shoulder: a single band of deeply impressed, bulbous, unconnected horseshoe devices. Level, curved nozzle. Single ring base. Clay color: 5YR 7/4 pale orange. Slip color: 10R 4/6 red, exterior surface only. Burnish: discus only. Hardness: 2.0. Inclusion(s): white (CS, low frequency). Burning on nozzle. Poor condition. Mold-made. Parallels, shape-type: Beth Shean/Scythopolis (Hadad 2002: 16, 20, nos. 19–24, Variant 1).

135. Reg. no. 84.1068.3L.7. L = 8.5 cm. W = 7.1 cm. H = 2.1. Nearly complete, 85% preserved. Palestinian discus. Late first to third centuries C.E. Discus and shoulder fragment. On discus: a remnant of indiscernible motif. Missing discus. Sides of discus extend onto shoulder, creating a single, rounded prominent ridge. Wide, slightly sloping shoulder. On shoulder: a single band of repeated, poorly executed and loosely spaced impressed horseshoe devices; two opposing, open double volutes in

positive relief at the base of the nozzle. Single ring base. Slightly sloping wick-hole in side view. Clay color: 5YR 7/4 dull orange. Slip color: 7.5YR 2/1 black to 2.5 YR 4/6 red-brown, exterior and minor interior surface traces. Hardness: 3.5. Inclusion(s): white (FS, low frequency). Burning around wick-hole. Poor condition. Mold-made. Parallels, shape-type: Beth Shean/Scythopolis (Hadad 2002: 16, 20, nos. 19–24, Variant 1).

136. Reg. no. 84.1068.3L.4. L = 8.6 cm. W = 7.3 cm. H = 2.0 cm. Nearly complete, 95% preserved. Palestinian discus. Late first to third centuries C.E. Discus and shoulder fragment. Missing discus. Prominent well-rounded ridge surrounds discus. Round nozzle. Narrow, steeply sloping shoulder. Single band of consecutively impressed bulbous horseshoe devices. Very shallow discus. On discus: remnant of floral (?) motif. Double ring base. Traces of single curls of opposing volutes depicted in positive relief directly beneath base of nozzle. Four raised strokes depicted between curls. Clay color: 10YR 8/4 light yellowish orange. Slip color: 10R 4/6 red, exterior and interior surfaces. Hardness: 3.0. Burning around wick-hole. Fair condition. Mold-made. Parallels, shape-type: Beth Shean/Scythopolis (Hadad 2002: 16, 20, nos. 19–24, Variant 1).

137. Reg. no. 84.1068.3L.9. L = 8.3 cm. W = 4.4 cm. H = 2.0 cm. Incomplete, 35% preserved. Palestinian discus. Late first to third centuries C.E. Discus and shoulder fragment. Missing discus. Part of shoulder and nozzle missing. Flat, curved nozzle. Narrow, slightly sloping shoulder. On shoulder: a single band of consecutive juxtaposed bulbous impressed horseshoe devices; a single double axe; an open double volute left of the nozzle base. Shallow, plain concave discus whose sides extend onto the shoulder, forming a single rounded prominent ridge. Small, slightly raised single ring base. Clay color: 2.5R 6/6 orange. Slip color: 2.5YR 4/8 reddish brown. Hardness: 3.0. Inclusion(s): red, hematite (FS, medium frequency); white (FS-CS, high frequency). Burning on nozzle. Fair condition. Mold-made. Parallels, shape-type: Beth Shean/Scythopolis (Hadad 2002: 16, 20, nos. 19–24, Variant 1).

138. Reg. no. 85.3406.2L. Not shown. GD = 3.7 cm. Incomplete, 3% preserved. Palestinian discus. Late first to third centuries C.E. Discus fragment. On discus: an erotic scene, to the left, a male on his knees against (to the right) a female on her knees. Filling-hole remnant, 0.5 cm diameter. Clay color: 7.5YR 7/4 pink. Slip color: 2.5YR 4/8 red. Slip splashes on part of shoulder and on interior surface. Inclusions: gray (LF, VFS); white (VLF, FS); red (VLF, FS). Hardness: 3.0. Fair condition. Mold-made. Parallels, shape-type: Beth Shean/Scythopolis (Hadad 2002: 16, 20, nos. 19–24, Variant 1).

139. Reg. no. 84.1179.6L. Not shown. GD = 5.4 cm. Incomplete, 10% preserved. Palestinian discus. Late first to third centuries C.E. Base fragment. Single ring around flat base. On base: lamp maker's initials in Greek: *beta alpha*: BA (reversed). Bichrome. Hard fired. Inclusions: white (FS, MF). Fair condition. No burning. Mold-made. Parallels, shape-type: Beth Shean/Scythopolis (Hadad 2002: 16, 20, nos. 19–24, Variant 1).

140. Reg. no. 95.1051.13L.7. Not shown. GD: ud. Incomplete, 30% preserved. Palestinian discus. Late first to third centuries C.E. Base fragment. Single ring (5.0 mm wide) delineates circular flat base. On base: lamp maker's initial and trademark, in Greek: *alpha;* rosette with four preserved circles and two diagonal lines. Two parallel grooves delineate where nozzle meets underside. Clay color: 5YR 6/6 reddish yellow. No slip. Hard fired. Inclusions: white (CS, very low frequency); orange, quartz sand,

translucent, sub-rounded, well sorted (VFS, very high frequency). Good condition. No burning. Instrument marks, swirls on interior surface used to press clay into mold. Mold-made. Parallels, shape-type: Beth Shean/Scythopolis (Hadad 2002: 16, 20, nos. 19–24, Variant 1).

141. Reg. no. 84.1069.1L. L = 7.3 cm. W = 6.0 cm. H = 2.0 cm. Nearly complete, 95% preserved. Palestinian discus. Late first to third centuries C.E. Curved nozzle. Broken discus with indiscernible motif. Slightly sloping shoulder. Plain shoulder. On shoulder: trace of single double-axe device in positive relief; two open double volutes at nozzle base; faint raised strokes appear between the volutes. Steeply sloping discus demarcated by a deeply incised line. Raised circular platform base. Clay color: 7.5YR 8/4 light yellow-orange. Slip color: 2.5YR 5/6 bright brown, exterior traces. Hardness: 2.0. Inclusion(s): severe surface encrustation and weathering inhibits proper identification. Burning around wick-hole. Poor condition. Mold-made. Parallels, shape-type: Beth Shean/Scythopolis (Hadad 2002: 16, 20, nos. 19–24, Variant 1).

142. Reg. no. 84.1056.3L.4. L = 8.9 cm. W = 8.2 cm. H = 2.9 cm. Nearly complete, 90% preserved. Palestinian discus. Late first to third centuries C.E. Discus, shoulder, and base fragment. Plain discus. Plain shoulder. Broad rounded single ridge surrounds the discus. Slightly sloping shoulder. On shoulder: two double-axe devices in positive relief; two nodules imitating stylized nails on nozzle base. Single ring base. Clay color: 5YR 7/4 dull orange. Slip color: 5YR 6/6 orange, exterior traces only. Hardness: 3.0. Inclusion(s): gray (FS-CS, high frequency); red, hematite (FS-CS, high frequency). Burning around wick-hole. Poor condition. Mold-made. Parallels, shape-type: Beth Shean/Scythopolis (Hadad 2002: 16, 20, nos. 19–24, Variant 1

Type 11. Syrian Discus (cat. nos. 143–44)

Two Syrian discus lamp fragments were found at Sepphoris: a rosette decorates the central discus of one example (cat. no. 143), while a menorah with seven branches occurs in molded relief on the concave discus of the other (cat. no. 144; see also Lapp 1996: 222, no. 117). The menorah is depicted with curved, rope-like branches supported by a curved tripod base. Two similar menorahs decorate Syrian discus lamps found at Caesarea Maritima (Sussman 2008: 234–35, nos. 107, 113). No Jewish symbols (e.g., *lulav*, shofar, and incense shovel) that flank the menorahs of other examples of lamps of this type (see Sussman 2008: 235, nos. 104, 105) appear on the Sepphoris fragment. An incised rope band surrounds the central discus of the lamp and serves as a decorative border between the discus and the plain shoulder. No slip covers the lamp. Syrian discus lamps with similar menorahs with rope-like branches have been unearthed at Qaṣrawet in northern Sinai (Oren 1982: pl. 29.B) and at the glass-making center of Jalame in Lower Galilee (MacDonnell 1988: 133, fig. 6-6.83–84, pl. 6-3). A Syrian discus lamp fragment decorated with a menorah flanked by two *lulavim* was also recovered from the Lerna Tunnel Fountain in Corinth, Greece (reg. no. L-69-36, unpublished, personal on-site examination at the Corinth Archaeological Museum by the author on January 31, 1997). The nozzle-bridge of the Corinth example is decorated with three impressed circles, a characteristic feature of this lamp type.

In Roman Palestine, examples have also been found at Shiqmona (Elgavish 1994: 115, fig. 89) and at Sumaqa on Mt. Carmel (Kingsley 1999: fig. 13.5–7). At

Caesarea Maritima, 75 fragments of this lamp type were found, most of which were recovered with coins dated to the third/fourth–sixth centuries C.E. in areas CC and KK (Sussman 2008: 234–35). Sussman rightly observes that the menorahs decorating many of these lamps suggest marketing of this lamp type to Jewish customers in Syria and the Diaspora (Sussman 2008: 235). A cross rendered on another example suggests that this Syrian discus type was made with a Christian clientele in mind (Waliszewski 1995: 305–6, pls. 35–36). The rosette on the first Sepphoris example (cat. no. 143) would have appealed to a mass market of customers with diverse religious affiliations (Lapp 2007: 372–77). The discovery of the Sepphoris lamp fragment on the western summit indicates contact with Jewish communities in Syria. That Syrian discus lamps were also unearthed at Antioch (Waagé 1941: Type 49b+d 135, 141; Dobbins 1977: 405, 521; Modrzewska-Marciniak 1988: pl. 14.7–8) may indicate interaction between the Jewish residents of Sepphoris and Antioch, where a synagogue and inscriptions point to the existence of a community (M. H. Williams 1998: 1.19, 24, 29–31, 64–66, 86, 113; 2.15, 62, 79–80; 3.19; 4.4; 5.9, 28–29, 52, 97–98; 7.7, 24). Sussman dates this lamp type to the third–fourth centuries C.E. (Sussman 2008: 235).

143. Reg. no. 84.4001.11L. L = 5.8 cm. W = 3.8 cm. Incomplete, 30% preserved. Syrian discus. Loeschke Type VIII. Discus and shoulder fragment. Third–fourth centuries C.E. Wide, slightly sloping plain shoulder. Single, flat surface ridge with incised diagonal cross-hatches surrounds discus. On discus: a rosette with wide, radiating "pie slice"-shaped petals with rounded tips. Incised, double concentric circles decorate outer space between the intersection of the two petals. Clay color: 5YR 6/6 orange. Hardness: 3.0. Inclusion(s): red, hematite (FS, low frequency). Excellent condition. Mold-made. Parallels, shape-type: Caesarea Maritima (Sussman 2008: 234–36, nos. 104–17).

144. Reg. no. 85.1093.1L.2 (IAA 95-3896). GD = 6.0 cm. Incomplete, 40% preserved. Syrian discus. Loeschke Type VIII. Third–fourth centuries C.E. Discus and shoulder fragment. On discus: remnant of a menorah with seven semicircular branches in molded relief: three complete branches (left), two incomplete (right), and one branch and upper central shaft (missing). Branches decorated with incised rope pattern. Complete semicircular tripodal base. No cross-bar. Incised rope pattern around discus. Two filling-holes, one on each side of menorah shaft. Clay color: 5YR 6/4 light reddish brown. Inclusion(s): white (FS, low frequency). Hardness: 3.0. Hard fired. No burning. Good condition. Mold-made. Parallels, shape-type: Shiqmona (Elgavish 1994: 115, fig. 89, dated to the fourth century C.E.); Sumaqa (Kingsley 1999: fig. 13.5–7); shape-type and discus motif, menorah: Sussman 2008: 235, nos. 107, 113; Corinth, Greece, Lerna Tunnel Fountain (reg. no. L-69-36, unpublished, personal on-site examination at the Corinth Archaeological Museum by the author on January 31, 1997). Published: Lapp 1996: 222, no. 117.

Type 12. Jebel Jofeh (cat. nos. 145–54)

Ten examples of lamps belonging to the Jebel Jofeh type were found at Sepphoris (cat. nos. 145–54). The term *Jebel Jofeh* refers to a tomb located in the ancient Decapolis city of Philadelphia, today's Amman, Jordan, where a significant quantity of this type of lamp was recovered (Harding 1950: 81–94). However, they also occur in significant numbers at Gerasa, their confirmed place of origin (Kehrberg 2001: 232; Hadad 2002: 22–24) and where some were manufactured at the hippodrome workshop (Kehrberg 2001: 232–33, JH18, fig. 3.d). Jebel Jofeh lamps are characterized by a round body and a shoulder decorated with an impressed herringbone pattern. They also feature a large, central filling-hole surrounded by one to three prominent, well-rounded ridges and a narrow, slightly sloping shoulder. A band of herringbone (cat. no. 146) or a mixture of herringbone and circles (cat. no. 149) typically surrounds the filling-hole. Some examples may have a plain ledge around the filling-hole (cat. no. 145).

The Sepphoris examples have a standing, somewhat acute knob handle that is often decorated with two incised parallel grooves on its top surface. The clay is commonly orange or a light yellowish orange. Lamps of this group bear a dark reddish slip. The base is flat. In many respects, the round body and curved nozzle (sometimes heart-shaped and decorated with two opposing incised circles) are reminiscent of the Roman discus lamp of the first–third centuries C.E. But the Jebel Jofeh herringbone patterning, enlarged central filling-hole, standing handle, and tall body represent common features characterizing lamps of the north in the Late Roman and Early Byzantine periods. Comparative examples of this lamp type have been dated from the third to fourth centuries C.E. (Hadad 2002: 22). Parallels have been found at Abila, Amman, Arraba, Beth Sheʿarim, Jordan Valley (east of Tubas), and Pella (Hadad 2002: 22–23). Hadad dates the Jebel Jofeh group—her Type 13—to the mid-third to mid-fourth centuries C.E. (Hadad 2002: 24).

The shoulders of some lamps belonging to this group are decorated with a combined pattern of herringbone and round globules (cat. nos. 143–44). In respect to some lamps, globules decorate the otherwise plain shoulder (cat. no. 145). Several of these lamps with globules have been found at Jebel Jofeh in Amman, and one example at Beth Sheʿarim. They were discovered in a context dating to the second half of the third century C.E. (Harding 1950: 81–94).

145. Reg. no. 84.2175.1L.1. L = 8.1 cm. W = 3.1 cm. H = 2.9 cm. Incomplete, 35% preserved. Jebel Joffeh. Round-bodied with deeply impressed herringbone and pronounced nodules. Jebel Jofeh, Hadad Type 13. Mid-third to mid-fourth centuries C.E. Shoulder fragment with large central filling-hole. A single prominent ridge surrounds filling-hole. Narrow discus with sloping sides and decorated with broad band of deeply impressed herringbone pattern. Narrow, steeply sloping shoulder. On shoulder: consecutive, evenly spaced large round nodules in positive relief. Clay color: 2.5YR 6/6 orange. Hardness: 2.0. Burning around nozzle, discus, and interior surface. Fair (–) condition. Mold-made. Parallels, shape-type: Amman (Harding 1950: pl. 25.51, 54, 65, 83) (similar); Caesarea Maritima (Sussman 2008: 238, no. 124); Pella (McNicoll and Smith 1980: fig. 5.1; similar shape-type and relief circular nodules on shoulder with herringbone pattern around filling-hole: da Costa 2003: 2.508, nos. 29, round-bodied no discus type, fugitive discus group, relief decoration subgroup 2); Beth Sheʿarim (Avigad 1976: pl. 70.8) (similar).

146. Reg. no. 84.2135.3L.2. L = 4.7 cm. W = 2.4 cm. Incomplete, 15% preserved. Jebel Jofeh. Round-bodied with deeply impressed herringbone and pronounced nodules.

Jebel Jofeh, Hadad Type 13. Mid-third to mid-fourth centuries C.E. Shoulder fragment with large central filling-hole. No ridge surrounds filling-hole. Narrow discus consists of two juxtaposed broad bands bearing impressed herringbone pattern. Narrow, gently sloping shoulder decorated with consecutive, evenly spaced, large round nodules (2.0 mm diameter) in positive relief. Clay color: 2.5YR 6/6 orange. Slip color: 10R 4/6 red, exterior and also interior traces. Hardness: 2.0. Inclusion(s): gray (MS, very low frequency); white (MS, very low frequency). Burning on interior surface. Good condition. Mold-made. Parallels, similar shape-type and relief circular nodules on shoulder with herringbone pattern around filling-hole: Pella (da Costa 2003: 2.508, no. 29, round-bodied no-discus type, fugitive discus group, relief decoration subgroup 2).

147. Reg. no. 84.2147.3L.1. L = 9.3 cm. W = 4.7. Incomplete, 25% preserved. Jebel Jofeh. Round-bodied with pronounced nodules. Jebel Jofeh, Hadad Type 13. Mid-third to mid-fourth centuries C.E. Shoulder fragment with poorly defined shallow discus. Narrow, very rounded shoulder decorated with consecutive, irregularly spaced round nodules in positive relief. Plain unpierced elongated knob handle. Clay color: 5YR 6/6 orange. Hardness: 2.5. Inclusion(s): gray (VFS-FS, low frequency); white (FS-MS, very low frequency). Burning on interior surface. Poor condition. Mold-made. Parallels: Amman (Harding 1950: pl. 25:31) (exact).

148. Reg. no. 84.2027.3L.2. L = 8.6 cm. W = 7.0 cm. H = 2.7 cm (with handle = 3.2 cm). Nearly complete, 90% preserved. Jebel Jofeh. Bilanceolate transitional with round body and deeply impressed herringbone. Jebel Jofeh, Hadad Type 13. Mid-third to mid-fourth centuries C.E. Missing nozzle. Narrow, slightly sloping shoulder. Single rounded ridge encircles large, central filling-hole. Plain, slightly sloping ledge extends away from the filling-hole where it meets shoulder. On shoulder: a single band of deeply impressed diagonal hatches. Unpierced standing handle. Handle base: two deeply impressed parallel notches. Flat base. Clay color: 2.5YR 7/6 orange. Slip color: 10R 4/6 red, exterior only. Hardness: 2.0. Inclusion(s): gray (MS-CS, low frequency); white (CS-granule, very low frequency). Considerable burning around wick-hole. Poor condition. Mold-made. Parallels: Rehovot (Sussman 1969: pl. 28, fig. 5, first lamp, right; Jordan Valley, east of Tubas (C. A. Kennedy 1963: pl. 23.519), Pella (Smith 1973: pl. 65.507) (close); provenance unknown (Israeli and Avida 1988: 94, fig. 271) (close).

149. Reg. no. 84.5002a.2L.1. L = 8.0 cm. W = 3.8 cm. Incomplete, 35% preserved. Jebel Jofeh. Bilanceolate transitional with round body and deeply impressed herringbone. Jebel Jofeh, Hadad Type 13. Mid-third to mid-fourth centuries C.E. Shoulder and handle fragment. A broad rounded ridge surrounds large central filling-hole. Narrow, gently sloping ledge extends from around the filling-hole upward to the shoulder. A single band of impressed, consecutive, closely spaced double concentric circles decorates the inner rim. On shoulder: a single band of deeply impressed herringbone. Unpierced standing handle with two deeply incised parallel grooves on its top surface. Clay color: 5YR 7/6 orange. Slip color: 2.5YR 4/6 reddish brown (exterior). Hardness: 2.5. Inclusion(s): gray (MS, very low frequency). Fair condition. Mold-made.

150. Reg. no. 84.2147.4L.1. L = 9.3 cm. W = 7.2 cm. H = 2.9 cm (with handle = ca. 3.4 cm). Complete. Jebel Jofeh. Bilanceolate transitional with round body and deeply impressed herringbone. Jebel Jofeh, Hadad Type 13. Mid-third to mid-fourth centuries C.E. Part of nozzle missing. Rounded nozzle. Narrow, slightly sloping shoulder. Two

rounded ridges of equal width surround narrow sloping sides that lead to a third rounded ridge of similar width encircling a large, central filling-hole. On shoulder: a single band of deeply impressed herringbone and also on inner sloping sides. Unpierced standing knob handle with two incised parallel grooves decorating its upper surface. Clay color: 7.5YR 8/4 light yellow-orange. Slip color: 10R 3/4 dark red, exterior and interior surface traces. Hardness: 2.0. Inclusion(s): gray (FS, low frequency); white (FS, low frequency). Burning on nozzle, handle, and flange of filling-hole. Fair (+) condition. Mold-made parallels: Rehovot (Sussman 1969: pl. 28, fig. 5, first lamp, right); Jordan Valley, east of Tubas (C. A. Kennedy 1963: pl. 23.519; Pella (Smith 1973: pl. 65.507) (close); provenance unknown (Israeli and Avida 1988: 94, fig. 271) (close).

151. Reg. no. 84.2158.1L. L = 4.7 cm. W = 2.3 cm. Incomplete, 10% preserved. Jebel Jofeh. Bilanceolate transitional with round body and deeply impressed herringbone pattern. Jebel Jofeh, Hadad Type 13. Mid-third to mid-fourth centuries C.E. Shoulder fragment. Missing nozzle. Narrow, slightly sloping shoulder. Single rounded ridge encircles large, central filling-hole. Plain, slightly sloping ledge extends away from the filling-hole where it meets shoulder. Single prominent ridge surrounds outer edge of rim. Narrow, slightly sloping shoulder decorated with a single band of a deeply impressed herringbone pattern. Flat base. Clay color: 2.5YR 6/6 orange. Slip color: 10R 4/4 reddish brown (exterior surface only). Hardness: 3.0. Inclusion(s): gray (VFS, low frequency); white (VFS, low frequency). Burning around nozzle. Fair condition. Mold-made.

152. Reg. no 84.2183.3L.1. Not shown. L = 5.1 cm. W = 2.4 cm. Incomplete, 5% preserved. Jebel Jofeh. Bilanceolate transitional with round body and deeply impressed herringbone pattern. Jebel Jofeh, Hadad Type 13. Mid-third to mid-fourth centuries C.E. Shoulder fragment. Part of nozzle. Narrow, slightly sloping shoulder. Three rounded ridges encircle large, central filling-hole. Plain, slightly sloping ledge extends away from the filling-hole where it meets shoulder. On shoulder: single band of deeply impressed herringbone. Flat base. Mold-made. Clay color: 2.5YE 6/6 orange. Slip color: 10R 4/4 reddish brown, exterior and interior surface traces. Hardness: 3.0. Inclusion(s): white (CS, very low frequency). Burning around nozzle. Fair (–) condition.

153. Reg. no. 84.1000.00L.4. L = 7.3 cm. W = 6.3 cm. H = 3.0 cm. Incomplete, 55% preserved. Jebel Jofeh. Mid-third to mid-fourth centuries C.E. Body fragment. Narrow, steeply sloping shoulder. On shoulder: a single band of crudely executed impressed herringbone pattern. Very large central filling-hole. Unpierced knob handle with steeply sloping top. Two parallel incised lines decorate front of handle. Ring base consisting of numerous well-rounded concentric ridges. Clay color: 5YR 6/6 orange. Hardness: 3.0. Inclusion(s): gray (FS-CS, low frequency); red, hematite (FS, low frequency); white (FS-VCS, low frequency); quartz (CSi-VFS, very high frequency). Extensive burning on interior surface. Poor condition. Mold-made. Parallels: Capernaum (Loffreda 1974: 92, photo 25.8–9).

154. Reg. no. 84.2174.1L.2. L = 5.3 cm. W = 6.5 cm. H = 2.7 cm. H, with handle = 4.2 cm. Incomplete, 55% preserved. Jebel Jofeh. Mid-third to mid-fourth centuries C.E. Body fragment of lamp with large central filling-hole. Narrow, gently sloping shoulder devoid of ornamentation. Tall, broad rectangular-shaped, unpierced handle. Ring base consisting of numerous, closely juxtaposed, well-rounded ridges. Mold-made. Clay color: Munsell 5YR 7/4 dull orange. Hardness: Mohs 2.5. Inclusion(s): chert

(MS, low frequency). Limestone (FS-MS, low frequency). Quartz (VFS, very high frequency). Poor condition. Parallels: none.

Type 13. Northern Stamped (cat. nos. 155–61)

Northern stamped lamps are characterized by a shoulder with deeply stamped, simple geometrical patterns, such as circles and wavy lines. The shoulder is generally wide and slightly sloping. The ovoid body contains an almond-shaped, shallow discus, inside of which a filling-hole is often medium in size. The discus is most often plain but in some examples may bear a simple decorative device in low relief. The surface of this lamp is not coated in a slip. A groove (or two) generally delineates the discus. Short knob handles, plain or decorated with an impressed leaf or line are commonly associated with this lamp type. The base is flat and undecorated. The nozzle with a small wick-hole is integrated into the body of the lamp. Sepphoris cat. no. 155 encompasses a generally massive, heavy appearance with crudely fashioned ornamentation. The pattern of cat. no. 159 resembles a stylized pinecone, similar to that found on a northern stamped lamp from Nahariya (Sussman 1989: 39, 56, no. 42a–b). The clay color of the Sepphoris examples ranges from orange and dull orange to light yellow-orange. White and hematite are the most common inclusions. Quartz and a black mineral, possibly magnetite, also occur in several of the lamps. The average hardness is 3.0 on the Mohs scale. The Sepphoris examples possess features characteristic of this lamp type: pale calcium carbonate–rich fabrics, oval shapes, deeply impressed decorations, and unslipped surfaces. Four types of the northern stamped form (Sussman Types 1, 4–6) were identified. Sepphoris examples (cat. nos. 155, 159) belong to Sussman Type 6 lamps dating to the end of the third–fifth centuries C.E. (Sussman 1989: 42).

Like the lamps of the northern stamped Type 6 group, the Type 1–4 lamps have an ovoid body with an almond-shaped discus. Similarly, the broad and slightly sloping shoulder is decorated with simple floral or geometrical devices. The base is flat and usually undecorated, and a knob handle is common. The nozzle with a small wick-hole is integrated into the body of the lamp. Types 1–4 distinguish themselves from Type 6 by a more precise and carefully executed ornamentation. Their overall appearance is more orderly and fine, certainly not massive like the Type 6 variety. Some northern stamped lamps—such as the two wall lamps in the Maʿayan Barukh Collection (Sussman 1989: 33–34, 56, nos. 23, 28)—were intended to be hung on a wall; the angle of the pierced handles of these lamps enabled them to be hung by a hook.

The northern stamped lamp type is a derivative of the *Firmalampe* ("Factory Lamp"; Sussman 1989: 24). *Firmalampen* were widely manufactured in Italy, Gaul, and Germania, but examples found in Aqaba suggest local Palestinian or Arabian production (Lapp 2012: fig. 3.50, table 3). Or they may have originated from a Phoenician workshop (Hayes 1981: 88; Sussman 1989: 24, 26). Several lamp bases found at Lahav (Tell Halif), for example, are stamped with the lamp maker's signature, *Damwn*, which indicates such an origin (Gofna and Sussman 1974: fig. 4, lamp nos. 7, 9). Should the signature reflect the name of the lamp-mold carver, however, it is possible that the molds were carved in Phoenicia and purchased and used by lamp makers at sites such as Sepphoris in Galilee. Given the widespread occurrence of this lamp type in Galilee, a second production center, if not the primary one, was probably there. Taking into consideration Adan-Bayewitz's observation that many of the known lamp workshops in Roman Palestine and the lamp molds found in this region were excavated in cities (Adan-Bayewitz 1995: 180), likely candidates for the manufacture of northern stamped lamp molds include Ptolemais, Tyre, and Sepphoris, which are located in Phoenicia and Galilee, where the greatest concentrations of this group occur.[5]

Although I admit that the deeply impressed patterns of the northern stamped type tend to be crudely executed and that the geometrical and floral decorations are carelessly applied, this style should not be viewed as "poor workmanship" indicative of northern manufacture (Sussman 1989: 49). Rather, the shoulder patterns of the northern stamped lamp type are reflective of a local Galilean folk art that should be accepted as a fully legitimate style of artistic expression. These lamps provide evidence that the lamp maker(s) of the northern stamped type intended to distinguish their lamps from those manufactured in other workshops. Further, the quality of lamp workmanship is a measure of the skill of the individual lamp-mold carver and of the lamp maker, not of a region. As a northern region, Galilee is home to urban centers such as Sepphoris, where superb workmanship is in evidence in a number of archaeological finds, including the Dionysiac and synagogue floor mosaics and the expertly fashioned "Sepphorean" discus lamp. Although the geometrical and floral patterns decorating northern stamped lamps would have appealed to a religiously mixed population and mass market, a menorah found impressed on the base of a lamp of unknown provenance (Sussman 1989: 43–44, no. 52.a, b) indicates that the lamp maker had target-marketed the distribution of lamps, or at least this example, to a Jewish clientele.

The most significant concentration of northern stamped lamps occurs in western Palaestina Secunda and southern Phoenicia. An impressed net pattern decorating the shoulders of northern stamped lamps found at Sepphoris (cat. nos. 155, 160) somewhat resembles those unearthed at Nabratein and Khirbet Shemaᶜ (Lapp 2009: lamp pl. C.15; Meyers, Kraabel, and Strange 1976: pl. 8.10:11). In addition to Sepphoris, northern stamped lamps of the type common in western and central Galilee have been found at Hanita, Lahav, Nahariya, Beth Sheᶜarim, Hagoshrim, Metulla, and Shiqmona (Sussman 1989: 22, 56). Northern stamped lamps have been recovered from Galilean synagogue sites, including Gush Ḥalav (Meyers and Meyers 1990: pl. B.4–7, 10, 12, 13–16). At Nabratein, one nearly complete lamp and three fragments belonging to the northern stamped type were recovered from the synagogue there (Lapp 2009: lamp pls. C.15; photo 49; C.10; B.8, 14). Outside Israel, northern stamped lamps have been discovered in Lebanon at Tyre (Marchand 1996: 65–67, nos. 69–73, 79–83). A complete northern stamped lamp, a version of Sussman's Type 6a, was recovered in the Agora at Athens, Greece, and published under

5. As of this writing, however, 20 complete lamp molds—more than from any other site in Israel—have been found at the small village or hamlet of Shikhin, 1.5 km from the Sepphoris citadel. See below, ch. 5 n. 3 (p. 184).

the heading "Various Imported Lamps" (Perlzweig 1961: 103, pl. 11.376). Several examples have even been uncovered as far south as the Fayoum in Egypt and as far north as Regensburg, Germany (Hayes 1980: 88, pl. 42.357). Sussman identifies seven types of northern stamped lamps, ranging in date from the third century C.E. to the beginning of the seventh century C.E.: Types 1 and 2, second–third centuries C.E.; Types 3 and 4, third century C.E.; Type 5A–B, no date provided; Type 5C, third–fourth centuries C.E.; Type 6E, end of the third into the fifth century C.E.; Type 7A, B1, C, D, E1, E2, unclear dating; Type 7B2, fifth–sixth centuries C.E.; and Type 8, sixth or beginning of the seventh century C.E. (Sussman 1989: 28, 36, 42, 47, 49, 54–55).

Archaeological evidence indicates that northern stamped lamps were manufactured for ritual use as symbol-offerings, meaning the actual lamps in body and form symbolized light and were deposited unused by members of the funerary party or sanctuary participants as symbols of light. Numerous unused northern stamped lamps were recovered from the Sanctuary of Pan at Banias (Berlin 1999: 40), indicating that they were used to provide symbolic illumination, not actual light. Because the Sepphoris examples of this lamp type were not recovered from a burial context, it is not possible to determine whether they functioned as light-symbols. That they were found in the western summit indicates that they were used to provide actual illumination in a daily life setting.

155. Reg. no. 85.1156.1L.1. GD = ud. Incomplete, 10% preserved. Northern stamped. Sussman Type 6A. Third–fifth centuries C.E. Ovoid body. Almond-shaped shallow concave discus. On discus: two complete stamped flower blossoms comprising six triangular petals. Single groove surrounds discus. Medium-sized filling-hole (0.9 cm diameter) located within discus. Broad, slightly sloping shoulder. On shoulder: deeply impressed "net" pattern alternating with crudely executed circle devices. Complete, pointed, cone-shaped unpierced handle decorated with a single impressed groove. Clay color: 10YR 8/4 very pale brown. No slip. Hardness: undetermined. Inclusion(s): gray, sub-angular (FS, medium frequency; CSi, high frequency). Good condition. Finger impressions on interior surface. Mold-made.
156. Reg. no. 84.2077.1L.2. L = 6.1 cm. W = 3.6 cm. Incomplete, 25% preserved. Northern stamped. Version of Sussman Type 4, third century C.E. Shoulder and discus fragment. Two wide, rounded, prominent, closely juxtaposed ridges surround remnant of shallow discus. Remnant of medium-sized filling-hole. Broad, slightly sloping shoulder. On shoulder: carelessly applied, impressed flower (daisies) devices. A single pinched handle with impressed groove. Clay color: 5YR 8/4 pale orange. Hardness: 2.0. Inclusion(s): gray (FS-MS, medium frequency); red, hematite (FS, low frequency); white (VFS-FS, medium frequency). Fair condition. Mold-made. Parallels, shape-type: Caesarea Maritima (Sussman 2008: 237–38, nos. 121–23).
157. Reg. no. 84.1152.2L.1. L = 9.3 cm. W = 6.6 cm. H = 2.7 cm. Complete. Northern stamped. Version of Sussman Type 5B, third century C.E. Pear-shaped body with rounded nozzle. Medium-sized filling-hole positioned within an almond-shaped shallow discus. Broad, slightly sloping shoulder bearing impressed, evenly spaced, consecutive flower-blossom devices. Carelessly rendered incised lines containing repeated hatch marks delineate discus and channel. Unpierced knob handle decorated on top surface with a single, centered, incised line. Flat base. Clay color: 5YR

7/6 orange. Hardness: 3.0. Inclusion(s): red, hematite (VFS-MS, medium frequency); white (VFS, very low frequency). Burning on body and around filling-hole. Good condition. Mold-made. Parallels, shape-type: Caesarea Maritima (Sussman 2008: 237–38, nos. 121–23); Beth Sheʿarim (Avigad 1976: pl. 70.17) (close); unknown provenance (Rosenthal and Sivan 1978: 111, no. 448) (close).

158. Reg. no. 84.1056.3L.2. L = 8.9 cm. W = 6.6 cm. H = 3.9 cm. Complete. Northern stamped. Version of Sussman Type 5B, third century C.E. Ovoid body. Medium-sized filling-hole within almond-shaped, shallow discus. Broad, slightly sloping shoulder. On shoulder: crudely executed, carelessly stamped geometric devices. A single "pinched" handle with impressed central groove. Flat base. Clay color: 5YR 8/4 pale orange. Slip color: 5 YR 6/4 dull orange. Hardness: 3.0. Inclusion(s): gray (MS, low frequency; VCS, low frequency); red, hematite (FS-MS, medium frequency); black, magnetite (?)(FS, low frequency). Burning around nozzle. Poor condition. Mold-made. Parallels, shape-type: Caesarea Maritima (Sussman 2008: 237–38, nos. 121–23).

159. Reg. no. 84.2092.5L.2. L = 8.8 cm. W = 5.7 cm. H = 2.8 cm. Complete. Northern stamped. Sussman Type 6B. Third–fifth centuries C.E. Ovoid body. Deeply impressed pinecone pattern, crudely executed. Medium-sized filling-hole located within almond-shaped shallow discus. Single groove surrounds discus. Broad, slightly sloping shoulder. On shoulder: a random assortment of wavy lines and round depressions. Sharply curved, unpierced handle decorated with a single incised line. Nozzle with small wick-hole integrated into lamp body. Flat, undecorated base. Clay color: 7.5YR 8/4 light yellow-orange. Hardness: 3.0. Inclusion(s): red, hematite (FS, very low frequency); white (FS-CS, medium frequency); black, magnetite (?) (VFS-FS, medium frequency). Considerable burning around wick-hole. Good condition. Mold-made. Parallels: Beth Sheʿarim (Avigad 1976: pl. 70.25; Avigad 1956: pl. 18.9) (similar); provenance unknown (Rosenthal and Sivan 1978: 111, no. 448; Bailey 1988: pl. 59.Q:2314) (similar).

160. Reg. no. 84.1001.15L.1. L = 5.1 cm. W = 3.0 cm. Incomplete, 15% preserved. Northern stamped. Sussman Type 5A, no date provided, though probably third–fourth centuries. Shoulder fragment. Medium-sized filling-hole encircled by a single prominent ridge. Broad, very slightly sloping shoulder. On shoulder: an elaborate "net" pattern and circles in negative relief. On nozzle-bridge: an incised, stylized palm branch (a diagnostic feature of the Type 5A lamps) extending vertically from central filling-hole. Clay color: 10YR 8/3 light yellow-orange. Hardness: 3.0. Inclusion: white (VFS, low frequency); quartz (?)(CSi, very high frequency). Fair condition: Fair. Mold-made. Parallels: Khirbet Shemaʿ (Meyers, Kraabel, and Strange 1976: pl. 8.10:11) (exact).

Sussman Type 1: Nozzle-Bridge with T-Device (cat. no. 161)

The chief characteristics of the Sussman T-device northern stamped type lamp include three prominent ridges surrounding a large central filling-hole, a narrow, rounded, steeply sloping shoulder, and a pear-shaped body whose nozzle is more prominent and far less integrated into the body of the lamps. The shoulder is either plain or decorated with simple geometrical devices such as circles or horseshoe motifs. The face of the nozzle is regularly decorated with two incised perpendicular lines in the shape of the letter *T*. Two incised circles often decorate each side of the wick-hole. Like the lamps belonging to the two Galilean types, the lamps of this group possess a knob handle and a flat base. Sometimes the base is decorated with concentric incised rings. The T-device lamps are prevalent in the north of Israel, particularly at Beth Sheᶜarim, Capernaum, Pella, and Meiron; they date to the third–fourth centuries C.E. A nearly complete example of the T-device lamp type from Sepphoris (cat. no. 161) corresponds to Sussman Type 1 (Sussman 1989: 26, nos. 3–6; see also Sequence A in da Costa 2003: 1.92). It is made from an orange clay covered with a light brown slip and containing white inclusions ranging from a fine to medium sand size.

161. Reg. no. 84.1173.4L.1. Not shown. L = 8.1 cm (incomplete). W = 6.2 cm. H = 2.5 cm. Incomplete, 45% preserved. Northern stamped. Sussman Type 1. Second to third century C.E. Body fragment. Pear-shaped body with large filling-hole. Prominent triple ridges around central filling-hole. Plain, narrow, rounded, steeply sloping shoulder. Elongated nozzle decorated with T-device, incised lines, and two opposing circles. Flat base with concentric incised rings. Clay color: 5YR 7/6 orange. Slip color: 2.5YR 5/8 bright brown (exterior only). Hardness: 3.0. Inclusion: white (FS-MS, very high frequency). Good (-) condition. Mold-made. Parallels: Magdala (Loffreda 1976: fig. 5.21); Meiron (Meyers, Strange, and Meyers. 1981: pl. 9.17:6) (close); Beth Sheᶜarim (Avigad 1976: pl. 70.9–11) (close); see also Sussman 1989: 26, nos. 3–6.

Type 14. Beit Nattif (cat. nos. 162–66)

Beit Nattif lamps were one of the most sophisticated lamp forms manufactured in Roman Palestine. Their precise motifs and patterning indicate the use of expertly and exquisitely carved molds, making these lamps among the great luxury fine wares of the ancient Mediterranean. While menorahs flanked by other Jewish ritual objects such as the *lulav* (palm frond) and shofar decorate the nozzle-bridges of many Beit Nattif lamps, Christian crosses are portrayed on other examples. This indicates that the lamps were intended for marketing to a Jewish and Christian clientele. Such molded symbols also suggest the use of this lamp type as an apotropaic symbol and as symbols of light, especially when deposited unused in a funerary context. The primary Beit Nattif types are easily identifiable by their pale-white chalky fabrics with uniformly applied red slip on the upper half with patches on their bases. The northern versions, however, are fashioned in an orange fabric and are typically unslipped. The most significant archaeological deposits of Beit Nattif lamps were recovered from Beth Guvrin, particularly from the lower chamber of the *sacellum* of the amphitheater and from tombs of the south cemetery. Beit Nattif lamps were also found in the western summit at Sepphoris, the synagogue at Hammath Tiberias, Beth Shean, Pella, and the Red Sea port of Roman Aila.

In 1930, carved white lamp molds were recovered from two cisterns at Beit Nattif, located in the Shephelah, thus suggesting this village as the site for their manufacture (Baramaki 1936: pl. 12). An assemblage of clay lamps characterized by their precise form and carefully executed ornamentation was also recovered (Baramki 1936: 5–6, 7, pls. 6.8–15; 7; 10.1–25; 11.1–16).

In recent years, a large quantity of Beit Nattif lamps has been recovered from Jewish and Christian tombs of the south cemetery at Beth Guvrin (Magness 2008:

figs. 5.7:1–3; 5.9:124–26).[6] Excavations at several sites in the north and Transjordan have also yielded substantial numbers of high quality Beit Nattif lamps (Sepphoris: Lapp 1991: 125–27, figs. 111–16; Beth Shean/Scythopolis: Hadad 2002: 26, 30–35, nos. 74–113; Pella: da Costa 2003: 1.68; 2.515–25, nos. 86–135; Gadara: Lapp 1997a: 240, 259, table 2, nos. BNG8, BNG9, BNG10). A workshop at Beth Shean/Scythopolis may also have produced this lamp type.[7]

Magness (2008: 129) observes that at nearby Beth Guvrin most of the lamps found in the south cemetery belong to one of three types associated with the workshop at Beit Nattif, which she dates to the second half of the third to fifth centuries C.E.: round lamps with decorated discus, lamps with a bow-shaped nozzle, and ovoid lamps with large filling-hole (Magness 2008: 129–30). Although scholars identify a northern version of this type (Rosenthal and Sivan 1978: 105; Hadad 2002: 26, 30–35, Type 17, nos. 74–113; Magness 2008: 130), petrographic thin-section and trace-element analyses of selected Beit Nattif fabrics collected from northern sites are inconclusive as to this version's origin of manufacture (Lapp 1997a: 145–66, pl. 7 [2], figs. 115–16, 119–20, tables 1–3, 5–7; da Costa 2003: 1.16–17, 19, 68–76, 235–36, 259–62).

Primary Beit Nattif, Southern Versions

The Beit Nattif lamp type received its name from the numerous lamps found in Cistern II at Beit Nattif (Bethorous) located southeast of Jerusalem in the Judean Desert. Rosenthal and Sivan categorize this lamp type as "Lamps with Bow-Shaped Nozzle" (Rosenthal and Sivan 1978: 104). The group also corresponds to Elgavish's Types 34 and 37 and Kennedy's Types 11, 12, and 14. Beit Nattif lamps are often identifiable by their ornate geometric and/or floral patterning on shoulder and nozzle-bridge, a degenerate spatulate-shaped nozzle whose contours are incorporated into the body of the lamp, a pyramid-shaped handle, and a large filling-hole. The nozzle appears pinched on each side and, as Rosenthal and Sivan note, produces the effect of volutes (Rosenthal and Sivan 1978: 104). The spatulate-shaped nozzle, rounded oil chamber, and large filling-hole are reminiscent of the same features exhibited by the Herodian type.

The small, nonfunctional pyramidal handle is often decorated with concentric triangles, but some handles are plain. Loop handles are rare. Typical designs on the shoulder of the lamps' body include herringbone, single guilloche, and circle devices. Patterns commonly found on the nozzle-bridge include: a single floral motif consisting of one to four stylized pinecones, alternating ropes and circles, grape clusters, simple geometrical patterns, and alternating acanthus leaves with lines and circles. Kanthoroi, architectural features such as two columns supporting an arch, and palm trees with two birds flanking their trunks also belong to the repertoire of decorative elements on the nozzle-bridges. Religious symbols on the nozzle-bridge

6. I thank Amos Kloner and Jodi Magness, who permitted me to examine a number of these finds at Beth Guvrin in the summer of 1993.

7. I appreciate the kindness of Shulamit Hadad for permitting me to examine the bow-shaped lamps from Scythopolis.

of this lamp type include menorahs and crosses and have been found in the two cisterns at Beit Nattif, the south cemetery at Beth Guvrin, the synagogue at Hammath Tiberias, the Temple Mount in Jerusalem, and Pella (Beit Nattif: Baramki 1936: pl. 10.24; pl. 11.5; Beth Guvrin: Oren and Rappaport 1984: 114–53, pls. 10–18; Magness 2008: 143, fig. 5.9:1; Hammath Tiberias: Dothan 1983: 65, fig. 5.B:2, 4 [second row, right], pl. 24; Pella: da Costa 2003: 2.522, nos. 117–18; 700, no. 92). A prominent ridge surrounds the filling-hole. Beit Nattif lamps have a ring base. The lamp motif on the Beth Shean and Beth Guvrin fragments can either be interpreted simply as an image of a lamp, or, as I believe particularly with respect to the Beth Guvrin example from the south cemetery (Magness 2008: 144, fig. 5.10:3), as a symbol of light.

Rosenthal and Sivan (1978: 105) distinguish two typological groups, one earlier and the other later. They observe that, whereas the earlier forms have "an elegant shape with a nozzle well-proportioned in relation to the body . . . and the relief is clear," the later specimens "tend to have a broad and often badly shaped nozzle and a crude handle . . . often made from poor molds." Rosenthal and Sivan (1978: 105) point to several lamps from Gezer as examples of the later type. Da Costa (2003: 1.68–69), however, challenges their distinction between "southern" and "northern" forms.

Israeli and Avida also identify two groups of Beit Nattif lamps. Examples belonging to their first group have a developed angular nozzle with prominent concavities and a large central filling-hole encircled by a single convex ridge (Israeli and Avida 1988: 116). The nozzle and shoulder of this Beit Nattif subtype are elaborately decorated with recurrent patterns of geometrical designs. A particularly distinct device found on the nozzle of this lamp is a complex pattern of combined braids and knots. A single band of a herringbone pattern, concentric circles, or guilloche surrounds the filling-hole on the shoulder. The collective appearance of the rich ornamentation is reminiscent of repousée or granulated jewelry (Israeli and Avida 1988: 116). The second group proposed by Israeli and Avida (1988: 116) contains smaller lamps that are decorated with simple geometrical patterns. Unlike the lamps of the first group, those of the second are rounder, their filling-holes are very large, and their nozzles are relatively small. The dense ornamentation on these lamps is less clearly defined than that of the first group. Most characteristic of this second group is a rope or herringbone pattern, a single band of which extends horizontally across the neck of the nozzle between the nozzle and the filling-hole (Israeli and Avida 1988: 116).

Beit Nattif Round-Bodied with Decorated Discus and Flat Shoulder

The Beit Nattif round lamps with decorated discus and flat shoulders are characterized by a round body with a decorated sunken discus. Geometrical devices (e.g., swastika), floral patterns (e.g., grapes attached to spiraling vines), and animals (e.g., peacock, human face) ornament the disk. Unlike the motifs rendered on the Palestinian discus lamp group, the decorative figures are oriented so that they can be viewed from the handle end of the lamp rather than from the nozzle end. Like the bow-shaped lamps from Cistern II, a curved band of herringbone pattern decorates the nozzle of this round lamp type. The nozzle is slightly rectilinear. The standard type of handle is pyramid shaped and is decorated with concentric triangles in molded relief. A single filling-hole is positioned at the center of the discus. The base of this lamp type is flat. Rosenthal and Sivan (1978: 104) note that the clay fabric ranges from light green to buff.

Beit Nattif Ovoid Type

Ovoid lamps were found in Cistern I at Beit Nattif. The chief diagnostic features of this type are an ovoid body with a large filling-hole, a short nozzle, and a pyramidal or knob-shaped handle. The walls of this type are thin and, because they are hard fired, clink when tapped. They are fashioned in a light brown, yellowish-brown, or reddish-brown clay. The upper half is coated with a red slip that is fired brown or black (Magness 2008: 130). As Magness (2008: 121) observes, the motifs that typically decorate the nozzle-bridge and shoulder are similar to those decorating the primary version of the Beit Nattif lamp variety. As an example, she cites a wreath across the nozzle-bridge and a scroll on the shoulder. Several basic geometrical motifs decorating the shoulders of some ovoid lamps have been overinterpreted. For example, circles with dots have been interpreted as shields (Sussman 1982: 84, no. 136; 109, no. 204) and hooks as either *gamma*s or a military structure (Gihon 1993: 190–93). A motif on the nozzle-bridge of a Samaritan lamp has been interpreted as the plan of a Roman camp (Gihon 1972: 43–44, figs. 1, 4) but is actually a geometrical design (a square with quadrants) commonly found on lamps of this type (see Lapp 1997a: figs. 80–82). Magness (1992: 161–65; 2008: 130) dates this group to the third–fourth centuries C.E. (cf. Rosenthal and Sivan 1978: 99).

Lamp-Pattern Samples for Customers

Several complete upper parts belonging to round-bodied lamps were recovered from Cistern I at Beit Nattif. However, they apparently were never intended for use as lamps. Because they show no evidence of burning and are not found at any other site, Rosenthal and Sivan (1978: 103) have posited a plausible explanation for their purpose: they served as samples for customers of the lamp maker's workshop. The hodgepodge of different motifs and patterns rendered on the disks and shoulders of these lamp-pattern samples offered customers an array of designs from which to choose.

The diagnostic characteristics of the lamp samples are the flat shoulder and the unique treatment of its decoration. Two devices, usually raised squares marked with intersecting lines, divide the shoulder into two decorative zones: the handle side and the nozzle side. Each zone bears an ornamental pattern that is distinct from the other. Patterns include hatching, herringbone, or elaborate geometrical forms.

Examples of the lamp-pattern samples have been found only at Beit Nattif (Baramki 1936: pls. 6.1–3, 7; 7.2–8; 11.19, 23), so their absence at Sepphoris is not surprising. The numismatic finds associated with the lamp samples of Cistern II date them to the last quarter of the third century C.E. (Baramki 1936: 8–10). Rosenthal and Sivan (1978: 104) suggest a date for them in the second half of the third and first half of the fourth centuries.

Beit Nattif Secondary, Northern Versions

Over the years, lychnologists have postulated the existence of a northern version of the Beit Nattif lamp type (see Rosenthal and Sivan 1978; Israeli and Avida 1988; Hadad 2002; Magness 2008). Petrographic thin-section and trace-element analyses of select Beit Nattif lamp fabrics from Sepphoris, Gadara, Beth Shean, and Pella, however, revealed inconclusive results regarding the origin (Lapp 1997a: 158; da Costa 2003: 1.72). A preliminary count analysis revealed 31 Beit Nattif fragments unearthed at the Nabatean port of Roman Aila on the Red Sea (Lapp forthcoming). Until such molds are discovered, the exact location of the workshops that manufactured the northern version will continue to elude us. The chalky white fabrics of the Beit Nattif lamps found at Beth Guvrin are consistent with the chalky sedimentary geology of the area. They are distinct from the brown fabrics of the northern versions, indicating the probable exploitation of different clay sources. But did the workshop located at Beit Nattif also manufacture lamps using brown clays? If this is the case, the Beit Nattif lamps found in the north then originated from the Shephelah and not from a northern workshop.

Beit Nattif Lamps at Sepphoris (cat. nos. 162–66)

Five lamp fragments belonging to a northern version of the Beit Nattif type were found by the Duke excavations at Sepphoris (cat. nos. 162–66). Additional Beit Nattif lamps were also excavated by the Hebrew University team at the site (see Gärtner 1999: 52–54, pls. 10.40–45, 11.46–48), which, as pointed out by Hadad, are similar to the lamps from Cistern I at Beit Nattif; see Hadad 2002: 26 n. 76). This northern version corresponds to Hadad Type 17 (Hadad 2002: 26–35). Hadad reports that 25 complete, 13 nearly complete, and 122 fragments were found at Beth Shean (Hadad 2002: 26–35, nos. 74–113). She dates this lamp type to the fourth–fifth centuries C.E. (Hadad 2002: 35). The occurrence of both Darom and Beit Nattif lamps at Sepphoris indicates connections between this upper Galilean site and Judea, albeit in the Early and Late Roman periods, respectively.

Similar to the parallels of this lamp type found at Beit Nattif, Pella, Gezer, and Silet edh-Dhahr (Rosenthal and Sivan 1978: 105), those discovered at Sepphoris have the precise ornamentation common to this group—specifically, that of a tightly spaced band of herringbone on the shoulder (e.g., cat. no. 162) and sharply arched nozzles with stressed cavities. A close similarity in clay and slip colors is readily visible among the Beit Nattif lamps of the Sepphoris corpus. These lamps are commonly fashioned in an orange to dull yellowish-orange clay and are coated in a red (10R 5/8 red) or dark reddish-brown slip (2.5YR 3/6), which is a slip color commonly associated with the upper part of the Beit Nattif lamp. White inclusions are most common in these examples. Chert, hematite, and quartz are also present but in less abundance. The average hardness of these lamp fragments is approximately 2.3 on the Mohs scale.

Catalog no. 164 has a decorated square, pyramid-shaped handle, a deep and undecorated concave discus, a precisely executed band of nodules around the discus, and a radial-ribbed pattern on the shoulder—all elements of which are commonly associated with Beit Nattif lamps. Catalog no. 162 represents the site's most complete Beit Nattif lamp fragment. It belongs to the main group of Beit Nattif lamps as outlined by Avida and Israeli, because, like the members of this group, this fragment has a developed arched nozzle with stressed concavities, a precisely ornamented herringbone shoulder, a single floral motif (stylized pinecone) positioned on the nozzle-bridge, and a characteristic dark reddish-brown slip (see main group designated by Israeli and Avida 1988: 116). The ornate shoulder of cat. no. 163 bears the simple, precise herringbone pattern and concentric-circle device imitating silver jewelry characteristic of the Beit Nattif lamp.

All the Sepphoris examples of the Beit Nattif lamp type seem to belong to the northern version, though similar to those found in Cistern I at Beit Nattif. None exhibits the diagnostic pale white, chalky fabrics of those manufactured at Beit Nattif and found in significant quantities in the south cemetery at Beth Guvrin and the Red Sea port of Roman Aila. Further archaeometric analyses of the Sepphoris examples will be necessary to determine the location of their clay sources, and a southern source should not be ruled out. That they are similar in design may suggest that the lamp-mold carvers operating at Beit Nattif traded/sold their molds to lamp makers manufacturing the northern versions in workshops perhaps located in cities such as Beth Shean.

162. Reg. no. 84.2077.4L.2. L = 6.9 cm. W = 3.5 cm. Incomplete, 45% preserved. Beit Nattif. Second half of the third to fifth centuries C.E. Upper half body fragment. Stylized spatulated nozzle with small circular wick-hole. Raised pinecone motif directly beneath the wick-hole. Four opposing concentric circles depicted in relief on each side of the motif, followed by a fifth directly below the pinecone. On shoulder: a single band of tightly spaced, carefully executed impressed herringbone pattern. Narrow, slightly sloping shoulder. Small, circular, central filling-hole set within medium-sized basin defined by a broad, prominent ridge. Clay color: 7.5YR 8/4 light yellow-orange. Slip color: 2.5YR 3/6 dark reddish-brown, exterior and interior surfaces. Hardness: 3.0. Inclusion(s): gray (MS, low frequency); white (FS, very low frequency). Considerable burning around wick-hole. Poor condition. Mold-made. Parallels, shape-type: Caesarea Maritima (Sussman 2008: 238–39, no. 125); Jerusalem (Bagatti and Milik 1958: pl. 38, photo 120.8); Pella (da Costa 2003: 2.522, nos. 117–18, Beit Nattif type, group: 3 lozenges); provenance unknown (Rosenthal and Sivan 1978: 107, no. 437) (close).

163. Reg. no. 84.2077.4L.6. L = 3.4 cm. W = 2.0 cm. Incomplete, 15% preserved. Beit Nattif. Second half of the third to fifth centuries C.E. Discus and shoulder fragment. On discus remnant: an indiscernible motif. Remnant of a single rounded ridge that had encircled missing central filling-hole. On shoulder: a single band of tightly spaced, well-executed herringbone pattern in high relief. Single raised concentric circle design positioned at the end of the ornamental band. Clay color: 10 YR 7/3 dull yellowish orange. Slip color: 5 YR 4/4 dull reddish brown. Hardness:

2.0. Inclusion(s): white (FS, very low frequency). Good condition. Mold-made. Parallels, shape-type: Caesarea Maritima (Sussman 2008: 238–39, no. 125).

164. Reg. no. 84.2012.1L. L = 5.8 cm. W = 4.6 cm. Incomplete, 30% preserved. Beit Nattif. Second half of the third to fifth centuries C.E. Discus and shoulder fragment. Plain discus. Deeply concave discus with steeply sloping walls. Small, circular, central filling-hole. Discus encircled by a single band of consecutive, evenly spaced circles in positive relief. Two narrow raised bans enclose circles. On shoulder: raised, parallel curving "bars"; remnant of raised concentric-circle design. Pyramid-shaped handle, decorated with circular nodule and raised ridge. Clay color: 5YR 6/4 dull orange. Core color: 10YR 7/3 dull yellow-orange. Slip color: 10R 5/8 red, exterior surface; 10R 4/6 red, interior surface. Hardness: 2.0. Inclusion(s): gray (MS, very low frequency); quartz (MS, very low frequency). Fair condition. Mold-made. Parallels, shape-type: Caesarea Maritima (Sussman 2008: 238–39, no. 125).

165. Reg. no. 84.2069.1L.4. Not shown. L = 6.0 cm. W = 2.3 cm. Incomplete, 35% preserved. Beit Nattif. Second half of the third to fifth centuries C.E. Shoulder, nozzle, and discus fragment. Plain, deeply concave discus with steeply sloping sides. Discus encircled by a band of consecutive, evenly spaced circles in positive relief. Two narrow raised bands enclose the circles. On shoulder: raised, parallel curving "bars." Single narrow rounded ridge defines outer edge of shoulder. On nozzle-bridge: curved concentric "bars" in high relief. Clay color: 5YR 6/6 orange. Slip color: 10R 5/6 red, exterior surface only. Hardness: 2.0. Inclusion(s): gray (FS-MS, medium frequency); white (MS, very low frequency). Good condition. Mold-made. Parallels, shape-type: Caesarea Maritima (Sussman 2008: 238–39, no. 125).

166. Reg. no. 84.1353.1L. GD = 7.2 cm. Incomplete, 45% preserved. Beit Nattif with bow-shaped nozzle. Second half of the third to fifth centuries C.E. Upper body and shoulder fragment. Discus with ridge, 3.1 cm diameter. On nozzle-bridge: two bands of laurel between which is a wide, plain rounded line. On shoulder: consecutive evenly spaced circles imitating metal jewelry inlay in high relief. Below narrow sharply curved shoulder, a tapered ledge. Small pyramidal handle with four faces: face oriented toward filling-hole is decorated with an inverted *V* containing a single round nodule in high relief. Filling-hole, 1.0 cm diameter. Clay color: 10YR 7/3 very pale brown. Slip color: 2.5YR 5/6 red. Hardness: 3.5. Hard fired. Inclusion(s): gray, sub-rounded (FS, medium frequency); red, hematite, well sorted (CSi, very high frequency; and FS, low frequency). Fair condition. Two mended fragments. Burning around wick-hole. Mold-made. Parallels, shape-type: Caesarea Maritima (Sussman 2008: 238–39, no. 125).

Type 15. Gilead (cat. no. 167)

Gilead lamps such as the Sepphoris example (cat. no. 167) are characterized by a round or oval body. The nozzle is pronounced with indented sides. Consecutive chevrons typically decorate the shoulder. Shoulder sherds can be easily misinterpreted as northern versions of the Beit Nattif group. The shoulder of the Sepphoris example is similar to lamp JH27 found in Chamber E8 of the Jerash hippodrome (Kehrberg 2011: fig. 5d). Two opposing birds commonly decorate the nozzle-bridge. The lug handles are longer than broad, and tend to be rectangular. Kehrberg (2001: fig. 2) identifies 23 different handle types. Double ring bases without lamp makers' marks are common. The fabric of the Sepphoris example is orange (2.5YR 6/6), and its exterior and interior surfaces are coated with a reddish-brown (10R 4/4) slip.

Petrographic thin-section analysis conducted on a Gilead shoulder fragment from Jerash (and resembling JH30 in Kehrberg 2001: fig. 5g) indicates a distinct fabric of well-rounded and high sphericity quartz grains in a well-sorted matrix (Lapp 1997a: 134, 274, table 1; sample BNJ2, misidentified here as a Beit Nattif, northern version type, and rightly considered a Gilead lamp sherd by da Costa 2003: 1.71–72, 260, pl. 7.2). Initial findings of dcp-oes analysis of the same shows a high abundance of calcium (61.22%), but low levels of aluminum (7.05), silica (23.09%), and titanium (0.45%) (Lapp 1997a: table 7, sample BNJ2). The results are confirmed in a proton-induced X-ray emission (PIXE) analysis of a larger sampling that shows "a discrete though not particularly homogenous group" that is different from the lamps manufactured at Jerash (da Costa 2003: 1.99, 2.233–34; for an explanation of PIXE, see Rice 1987: 398).

Gilead lamps are rare and have a limited geographical distribution in the region of Gilead, thus the name (da Costa 2003: 1.97). The most significant deposit of this type is Chamber E8 of the Jerash hippodrome (Kehrberg 2001). Given the fact that examples have been found only at Amman, Jerash, Pella, and Umm Qeis (Bisheh 1972; da Costa 2003: 1.97), the discovery of the Gilead lamp fragment at Sepphoris is certainly noteworthy. At Jerash, Gilead lamps were recovered with coins dating to the second half of the fourth century to the early fifth century C.E., but the lamps were likely produced for the entirety of the fifth century C.E. (da Costa 2003: 1.97, 100). The Sepphoris Gilead lamp fragment was recovered from L84.1092, the fill inside cistern 84.1106; the fill contained a sixth-century C.E. coin of Anastasius I minted at Constantinople.

167. Reg. no. 84.1092.5L.1. L = 5.7 cm. W = 2.9 cm. Incomplete, 35% preserved. Gilead type. Fourth–fifth centuries C.E. (da Costa 2003: 1.97, 100). Shoulder and handle fragment. Broad, rounded shoulder. Square, pyramid-shaped handle. Central large filling-hole encircled by a prominent rounded ridge. On shoulder: a single band of tightly spaced, carefully executed impressed herringbone pattern. Two parallel vertical grooves at base of handle. Thick walls. Clay color: 2.5YR 6/6 orange. Slip color: 10R 4/4 reddish brown, exterior and interior surfaces. Hardness: 3.0. Inclusion(s): red, hematite (FS-MS, low frequency). Good condition. Mold-made. Parallels: Pella (Smith 1973: pl. 60.1040, exact; da Costa 2003: 1.97, no. 158); Jebel Joffeh (Bisheh 1972).

Type 16. Bilanceolate (cat. nos. 168–72)

Deeply impressed chevrons typically decorate the shoulder of bilanceolate lamps, so designated because of the two spear-shaped ends characteristic of this type. Chevrons and concentric circles also occur. Da Costa (2003: 2. Group 1, cat. nos. 37–73; Group 2, nos. 79–83; Group 3, nos. 74–78; Group 4, cat nos. 84–85) identifies four groups of bilanceolate lamps based on shoulder decoration: Group 1, chevrons; Group 2, circles; Group 3, circles and lines; and Group 4, relief decoration. Da Costa (2003: 1.58, fig. 4) also points out four chevron decoration schemes based on their respective orientation on certain examples of this lamp type. The variation in shoulder decoration may suggest the work of different mold carvers and lamp makers, a way to distinguish their products from those of another craftsman (see also da Costa 2003: 1.58). Circles may also decorate shoulders. Sometimes impressed concentric circles are intermingled with this chevron ornamentation (Sepphoris cat. nos. 227–28). Bilanceolate lamps have a flat ovoid body and a medium to large filling-hole, thus greatly facilitating the introduction of oil into the fuel chamber. A series of three to seven deeply incised grooves decorates the space between the wick-hole and filling-hole. The filling-hole itself is surrounded by a wide, well-rounded ridge. A deep groove typically surrounds the ridge. Some examples exhibit an impressed rope pattern enclosed by an outer incised line or an impressed rope pattern around the ring of the filling-hole and not enclosed by an outer incised line. Some are plain with no rope pattern.

A characteristic feature is the relatively flat shape of the lamp body—especially the handle, central filling-hole, and nozzle, which are all on a single horizontal plane. The handle is usually knob shaped or a triangular lug. Two and occasionally three parallel grooves decorate the handle; some examples have only one groove. These lamps commonly have a knob-shaped handle, projecting outward on a horizontal axis; it usually does not exceed the height of the lamp. The base is slightly convex. Lamp makers' marks, especially initials, are absent. A deeply impressed palm frond decorates the base of some examples. Another diagnostic feature is the integration of the nozzle into the body of the lamp.

The average length of the five Sepphoris examples is 7.8 cm, the width 5.6 cm, and the height 2.6 cm; their range in length is 7.4–8.1 cm, width 5.5–5.8 cm, and height 2.4–2.9 cm (Lapp 1997a: 52). The execution of patterns is generally poor but in some examples is sharp. Slip covers the upper half with dribbles on the underside. Makers of bilanceolate lamps used several different slip colors: dark reddish brown (10R 3/3), reddish brown (2.5YR 4/6 and 10R 4/4), dark red (10R 3/6), and red (10R 4/6). The fabrics of some examples are highly friable and exhibit flaking. They were low to moderately fired. Inclusions include black grits (VFS, LF)/white grits (FS, LF)/red grits (VFS, VLF) and white grits (MS, LF)/some examples with no grits; fabric: dull orange (5YR 6/4, 7/4, 7/6), dull yellow-orange (10YR 7/3, 8/3), and grayish brown (7.5YR 5/2). Some examples are characterized by impressed concentric circles around the filling-hole ridge, followed by a half-band of chevrons below. A rare pattern of only concentric circles or concentric circles intermixed with chevrons also occurs.

Although the bilanceolate lamp was most commonly used in the Byzantine period, it was first manufactured in the Late Roman period. These lamps are commonly dated to the third–fourth centuries C.E. (Rosenthal and Sivan 1978: 110–11). Da Costa (2003: 1.57, 66–67) dates them from the third to early fifth centuries C.E. and Hadad (Hadad 2002: 26) from the fourth to fifth centuries C.E.

Bilanceolate lamps are commonly found in northern Palestine (primarily in Galilee), the northern Jordan Valley, and Transjordan, with the most significant concentrations in northern Palestine, suggesting their manufacture in this area and their chief market. Bilanceolate lamps have been unearthed at Araba I/Gabara, Beth Sheʿarim, Beth Yerah/Philoteria, Capernaum, Hammath-Tiberias, Jalame, Nazareth, Sepphoris, Samaria, Abila, Gadara, Pella, and Beth Shean/Scythopolis (see parallels listed for cat. nos. 168–72). Eleven bilanceolate lamps and fragments were uncovered at Sepphoris (cat. nos. 149–59), a comparatively low number that may change when the other Sepphoris excavations publish their lamps. At Gadara, 690 examples were found, representing 38.9% of the lamps recovered at the Decapolis site (da Costa 2003: 1.57). At Beth Shean, Hadad (2002: 26–29, nos. 53–73) reports the discovery of 37 complete bilanceolate lamps, 16 almost complete, and 538 fragments. At Pella, 138 examples were found, representing 9% of the lamps found at this Decapolis site (da Costa 2003: 1.57). However, a few have been uncovered as far south as the Red Sea port of Roman Aila (Lapp forthcoming a: reg. nos. 46566, Area J; 11911, Area B). Bilanceolate lamps are surprisingly absent from deposits at Gerasa.

Petrographic analysis indicates that lamp makers exploited marl clay sources for the manufacture of bilanceolate lamps, marls characterized by a high abundance of foraminifera, primarily globeriginids and ostracod valves exhibiting secondary calcitic replacement (Lapp 1997a: 148–49, Petrographic Fabric Group 2; 261–63, for petrographic descriptions; pl. 6.2, photomicrograph of sample BS9 found at Sepphoris; Lapp 1997b: 525). That the morphological structure of the microfossils are preserved in the fabrics suggests low firing; the foraminifera represent original constituents of the clay fabrics. One would expect the microfossils' internal structures to disintegrate under high firing temperatures, and if used as temper, to be disarticulated and fractured from grinding during clay preparation. Occasional quartz sand inclusions exhibit high angularity and low sphericity, suggesting that this component was added by the lamp maker as temper (Lapp 1997a: 261–63). Foraminiferal clays are found in central and northern Israel as well as in Jordan, suggesting these areas as possible clay sources quarried for the manufacture of bilanceolate lamps (Goren 1995: 206–7, appendix A, nos. 5, 6, and 12; Lapp 1997a: 163–64; cf. da Costa 2003: 1.63). Bilanceolate lamp fabrics have a very high abundance of calcium carbonate and match chemically, suggesting the exploitation of one or more clay sources somewhere in central Israel (Lapp 1997a: 162–65).

Although no actual workshop for the manufacture of bilanceolate lamps has been discovered, clues about their possible origin have emerged. First, the type of clay source exploited for its manufacture was a foraminiferous-rich marl. Second, because the greatest concentrations of this lamp type have been found at Beth Shean, Pella, and Gadara, one or more marly clay sources should be located somewhere near these sites; and such clays do exist in the area. Preliminary petrographic and trace-element analyses indicate a provenance in the North Jordan Valley, probably in or near Scythopolis (Lapp 1997a: 162–65). The clays have similar chemical compositions. Analysis of 15 samples indicates relatively high abundances of calcium carbonate and iron: CaO 41.97%; SiO_2 33.42%; Al_2O_3 14.52%; and Fe_2O_3 5.67% (Lapp 1997a: 292, table 10). Building on my petrographic and chemical analysis of bilanceolate lamps from Galilee and Transjordan (Lapp 1997a: 148–49, Petrographic Fabric Group 2; 157-59; 261–63, for petrographic descriptions; pl. 6.2; tables 5, 7; figs. 116, 120) and applying multivariate statistical analysis to the resulting data, da Costa (2003: 1.63–65, figs. 5–6) distinguishes two main groups and six outliers within the bilanceolate type. Notably, all the Beth Shean samples are outliers. She posits that they are suggestive of a common heterogeneous source (da Costa 2003: 1.64). Grouping her samples by shoulder decoration, da Costa observes that, "allowing for the very limited number of samples, the result may even show that circles and chevrons might be produced by the same workshops, or at least by different workshops using the same clays" (da Costa 2003: 1.64).

Archaeological evidence indicates that bilanceolate lamps were made for ritual use as symbol-offerings, meaning the actual lamps in body and form symbolized light and were deposited unused by members of the funerary party or sanctuary participants as symbols of light (for a discussion on lamps as symbols, see Sussman 1981: 4). I agree with da Costa's observation (2003: vol. 1.ii, 356) that "there are indications that in funerary contexts lamps were increasingly used symbolically, and not actually lit." Da Costa (2003: vol. 1.ii, 356) argues that bilanceolate lamps in particular are too small to provide light for very long. Several bilanceolate lamps recovered from Tomb 39A at Pella show no traces of use, suggesting they were used as funerary offerings and not for their lighting (McNicoll et al. 1992: 143; Lapp 1997a: 211). I agree and am convinced that, in this context, they were deposited as photo amulets (Lapp in press b). Otherwise, why bring a lamp whose function is to provide light in a dark space such as a tomb and not light it? Unused bilanceolate lamps were also recovered from the Sanctuary of Pan at Banias (Berlin 1999: 40), again indicating that they were used as light-symbols and not light sources.

168. Reg. no. 84.1000.00L.3. L = 8.1 cm. W = 5.5 cm. H = 2.9 cm. Complete. Bilanceolate. Third to early fifth centuries C.E. Ovoid body. Medium-sized filling-hole surrounded by a plain, well-rounded prominent ridge. Broad slightly sloping shoulder. Deeply impressed rope pattern consisting of diagonal hatches surrounds the filling-hole ridge. On shoulder: two opposing unconnected bands of herringbone are impressed in a crude manner. Nozzle with small wick-hole incorporated into the lamp body. Four parallel, deeply impressed consecutive lines are rendered directly below the wick-hole. Cone-shaped handle projects horizontally and not higher than the lamp. Two faintly impressed parallel lines decorate handle. Flat base. Clay color: 7.5YR 7/6 orange. Slip color: 10R 3/6 dark red, exterior traces only. Hardness: 2.0. Burning around wick-hole. Good condition. Mold-made. Parallels: Abila (Fuller 1987: 472, fig. 98a–b) (exact); Capernaum (Loffreda 1976: photo 25.3, 11, 13); Jordan Valley, east of Tubas (C. A. Kennedy 1963: pl. 23.533) (exact); Maʾoz (Tzaferis 1982: fig. 11.3) (exact); Hammath Tiberias (Dothan 1983: 62–64, fig. 5: M, N, O, pl. 24.4 [upper left], 9); Hammat Gader (Coen-Uzzielli 1997: 320–22, 340–41, nos. 5–14, figs. 1–3, pls. 1.5–6, 2.1–8); Umm Qeis/Gadara (Nielsen, Anderson, and Holm-Nielsen 1993: 169–70, 194, nos. 341–343; da Costa 2003: 2. nos. 36, 46, 59, 74, 76–78, 84–85); Meiron (Meyers, Strange, and Meyers 1981: pl. 9.18:11–13) (exact); Pella (Smith 1973: pls. 60.6, 52–53, 56, 96; 65.35, 258; 78.288, 290, 372) (exact); da Costa 2003: 2. nos. 35, 37–45, 47–58, 60–73); Beth Sheʿarim (Avigad 1976: pl. 71.29); provenance unknown (Rosenthal and Sivan 1978: 111, no. 451) (exact); Museum of Art and History, Geneva (Chrzanovski 2011: 296–98, nos. 267–69).

169. Reg. no. 84.2184.5L.1. L = 5.1 cm. W = 3.4 cm. Incomplete, 20% preserved. Bilanceolate. Third to early fifth centuries C.E. Flat ovoid body fragment. Medium-sized

filling-hole surrounded by plain, well-rounded prominent ridge. Narrow, slightly sloping shoulder. Deeply impressed rope pattern consisting of diagonal hatches surrounds ridge of filling-hole. On shoulder: a single band of herringbone impressed in a crude fashion. Nozzle with small wick-hole is integrated into the body of the lamp. Four parallel, deeply impressed consecutive lines rendered directly beneath wick-hole. Clay color: 5YR 7/4 dull orange. Core color: 10YR 7/2 dull yellow-orange. Slip color: 2.5YR 5/6 bright brown, exterior and interior surfaces. Hardness: 3.0. Burning around wick-hole. Fair (–) condition. Mold-made.

170. Reg. no. 84.2003.5L. L = 6.1 cm. W = 5.2 cm. Incomplete, 20% preserved. Bilanceolate. Third to early fifth centuries C.E. Flat ovoid body fragment. Medium-sized filling-hole encircled by broad, well-rounded prominent ridge. Narrow, steeply sloping shoulder. Crudely executed herringbone pattern in low relief impressed on shoulder. Cone-shaped handle, projecting horizontally and not higher than the body. Handle decorated with three parallel incised lines. Clay color: 5YR 7/4 dull orange. Slip color: 10R 4/6 red, exterior surface only. Hardness: 3.0. Inclusion(s): surface too weathered for proper identification. Poor condition. Mold-made.

171. Reg. no. 84.1056.2L.2. L = 8.0 cm. W = 5.6 cm. H = 2.4 cm. Complete. Bilanceolate. Third–early fifth centuries C.E. Flat ovoid body. Medium-sized filling-hole surrounded by plain, well-rounded prominent ridge. Narrow, slightly sloping shoulder. Deeply impressed rope pattern consisting of diagonal hatches surrounds ridge of filling-hole. On shoulder: a single band of herringbone impressed in a crude fashion. Nozzle with small wick-hole is integrated into the lamp body. Knob-shaped handle, projecting horizontally and not higher than the lamp. Two impressed, elongated parallel grooves decorate handle. Flat base decorated with a stylized palm branch in negative relief. Clay color: 5YR 7/4 dull orange. Slip color: 10R 3/3 dark reddish brown. Hardness: 3.5. Inclusion(s): gray (VFS, very low frequency); red, hematite (VFS, very low frequency); white (VFS, very low frequency). Burning around nozzle. Good (–) condition. Mold-made.

172. Reg. no. 84.2077.4L.4. L = 9.0 cm. W = 3.7 cm. Incomplete, 40% preserved. Bilanceolate. Third to early fifth centuries C.E. Shoulder fragment. Flat ovoid body. Medium-sized filling-hole encircled by broad well-rounded ridge decorated with irregularly spaced impressed circles. Broad, slightly sloping shoulder. Crudely executed concentric circles of varying diameters impressed in a single band around the shoulder. Outer edge of shoulder delineated by a single band of deeply impressed herringbone pattern. Sharply curved handle decorated with two vertical lines in low relief. Clay color: 10YR 8/3 light yellow-orange. Slip color: 2.5YR 4/6 reddish brown, exterior and interior traces. Hardness: 3.0. Inclusion(s): gray (FS-MS, high frequency). Considerable burning on interior surface. Fair (–) condition. Mold-made. Parallels: Abila (Fuller 1987: 473, fig. 99a); Capernaum (Loffreda 1974: fig. 28.7) (close), Meiron (Meyers, Strange, and Meyers 1981: pl. 9.19:1) (similar); Pella (Smith 1973: pl. 78.386), Beth Sheʿarim (Avigad 1976: pl. 71.28) (exact); Khirbet Shemaʿ (Meyers, Kraabel, and Strange 1976: pl. 8.10:2–3) (close); provenance unknown (Rosenthal and Sivan 1978: 111, no. 450) (close).

Type 17. Galilean (cat. nos. 173–76)

The lamps of the Galilean type are characterized by a wide round body with a large filling-hole, which would have greatly facilitated the introduction of oil into the fuel chamber. The wide splayed nozzle tends to be rectangular in shape and is straight edged. The volutes have been absorbed into the nozzle-bridge and are delineated by two parallel lines in high relief that follow the nozzle's contours. The lamp fabrics include a range of colors, including varying hues of gray or brown. Some examples are covered with a red, reddish brown, or brown slip (for Munsell measurements of fabric and slip colors, see catalog entries). Handles are generally conical shaped, though some examples have standing "tongue" handles. In section, the shoulder slopes steeply. This contrasts with the curved shoulder of the "Sepphorean" discus type and therefore should not be identified as the latter form on this basis and for other reasons already presented. The wide shoulders allow the addition of patterns and motifs that are reflective of local Galilean folk art. Patterns include geometrical designs (e.g., ovoli, dots, radials), vegetal images (stylized grape clusters and spiraling vines), and zoomorphic forms (birds and human faces) (see examples in Hadad 2002: nos. 257–58, ovoli and dots; nos. 263–64, grape clusters and spiraling vines; and nos. 260, 265, birds and a human face, respectively).

Schematic human faces adorn the nozzle-bridges of several examples from Beth Shean (e.g., Hadad 2002: 56, nos. 260, 265). Hadad (2002: 56, nos. 261–62) astutely identifies two examples in which the entire lamp is fashioned in the shape of a schematic human face: the conical handle is the nose, the large filling-hole is the mouth, on the shoulder edge is the hair, on part of the nozzle-bridge is the beard, and the eyes are portrayed on the shoulder. It still must be determined whether these face lamps served some sort of apotropaic purpose in ritual or daily life for, to fill the lamp with fuel, the oil would be poured into the human's mouth and "guzzled" into the stomach. The depiction of human faces is rare on lamps in Roman Palestine and Arabia. A number of human-faced lamps are found in burial contexts, as at Beth Sheʿarim and Pella, suggesting that this decorative scheme may be connected with themes of afterlife or eternal life.

A common motif decorating the nozzle-bridge of this type is an inverted *V* in high relief. Although its identification remains uncertain, perhaps it is a stylized representation of a mountain, not unlike mountains found on Samaritan lamps (for examples of Mount Gerizim depicted on Samaritan lamps, see Sussman 1978). Other motifs decorating the nozzle-bridges include a palm tree and a face (Hadad 2002: nos. 264, 260). A chalice with five flutes decorates the nozzle-bridge of another lamp (cf. the interpretation of the motif as a five-branched menorah in Hadad 2002: 56, 59, no. 263). Some nozzle-bridges are plain. That no explicitly identifiable Jewish or Christian symbols decorate examples of this type suggests that the mold makers and lamp makers had no specific clientele in mind for these lamps. Rather, they were made with a mass market in mind because this lamp would appeal to both Jews and non-Jews. That these lamps enjoy a widespread distribution in northern Palestine and Transjordan suggests a production center located in this area, leading da Costa (2003: vols. 1.i and 2.120–28, nos. 195–214; probably Galilean, nos. 215–16; possibly Galilean, nos. 217–19) to refer to them as "Galilean" lamps. Perhaps the most significant deposit of such lamps was found by Mazar in Catacomb 24 at Beth Sheʿarim (B. Mazar 1973: 213, fig. 23.3); in one of two pits there, 280 of these Galilean lamps were recovered (Hadad 2002: 61). Lamps of this type date to the fifth–early sixth centuries C.E. (Hadad 2002: 61).

173. Reg. no. 84.4125b.1L.2. L = 11.3 cm. W = 7.3 cm. H = 2.95 cm. H, without handle = 3.5 cm. Complete. Galilean round-bodied with spatulated nozzle and large filling-hole. Fifth–early sixth centuries C.E. Round body. On nozzle-bridge: an inverted-*V* device containing three round nodules forming a triangle, perhaps portraying a stylized mountain of the type found on Samaritan lamps depicting Mt. Gerizim. Elongated square nozzle, 3.7 cm wide. Wick-hole, diameter 1.0 cm × 0.8 cm. Standing tongue handle. Large filling-hole, 3.5 cm diameter. Wide (0.6 cm) pronounced ridge around filling-hole. Beaded rope ridge around filling-hole. On shoulder: round nodules and inverted, random and pronounced *V* patterns. Prominent, slightly convex ring base (4.3 cm diameter). Clay color: 10YR 7/3 very pale brown. Slip color: 7.5YR 6/4 light brown. Hardness: undetermined. Inclusion(s): gray, sub-angular (VHF, VFS). Slight burning traces on nozzle. Poor condition. Heavy grayish salts or lime powder on surface. Three significant cracks. Mold-made.
174. Reg. no. 84.4121b.1L.2. L = 9.7 cm. W = 6.9 cm. H = 6.95 cm. Complete. Galilean round-bodied with spatulated nozzle and large filling-hole. Fifth–early sixth centuries C.E. Round body. On nozzle-bridge: an inverted-*V* device containing three round nodules forming a triangle topped with an additional round nodule: device perhaps portrays a stylized mountain of the type found on Samaritan lamps depicting Mt. Gerizim. Square, splayed nozzle. Wick-hole, 0.8 cm diameter. Filling-hole, 2.3 cm diameter. Wide bulbous shoulder. On shoulder: band of round nodules closest to filling-hole, a single ridge, and a band of stylized almonds and inverted *V*s filled with three round nodules. All decoration in molded high relief. Thumb-pressed, round concave base (2.5 cm diameter). Clay color: 7.5YR 7/4 pink. Slip color: 5YR 6/6 reddish yellow. Hardness: undetermined. Inclusion(s): gray, sub-rounded (MF, VFS); red hematite (LF, FS). Slight traces of burning on nozzle. Fair condition. Mold-made. Parallels, shape-type: Pella (da Costa 2003: 2.540, nos. 196–97).
175. Reg. no. 84.4125b.1L. Complete. Galilean round-bodied with spatulated nozzle and large filling-hole. Fifth–early sixth centuries C.E. Round body. On nozzle-bridge: an inverted-*V* device forming a triangle, perhaps portraying a stylized mountain of the type found on Samaritan lamps depicting Mt. Gerizim. Elongated square nozzle. Standing tongue handle. Large filling-hole. Rope-like ridge around filling-hole. On shoulder: round nodules and inverted, random, and pronounced *V* patterns. Prominent, slightly convex ring base. Clay color: 10YR 7/3 very pale brown. Slip color: 7.5YR 6/4 light brown. Hardness: undetermined. Inclusion(s): gray, sub-angular (VHF, VFS). Slight burning traces on nozzle. Poor condition. Mold-made.
176. Reg. no. 84.5002a.2L.1. L = 5.5 cm. W = 3.3 cm. Incomplete, 10% preserved. Galilean round-bodied with spatulated nozzle and large filling-hole. Fifth–early sixth centuries C.E. Spatulate nozzle fragment. Two raised ridges outline concavities. Two parallel raised lines decorate neck of nozzle. Small round wick-hole. Clay color: 7.5YR 8/4 light yellow-orange. Slip color: 10R 4/4 reddish brown. Hardness: 2.0. Inclusion(s): white (VCS, very low frequency). Burning around nozzle. Fair condition. Mold-made.

Type 18. Caesarean Discus (cat. nos. 177–81)

Limestone molds for lamp and figurine manufacture were found in the excavations at the port city of Caesarea Maritima. The molds were found in association with an apsidal building located near the Byzantine *praetorium* in Area CC57 (Sussman 1980: 79–79, pls. 15–16; 2008: 296–300, nos. 1–7). Sussman (1980: 76–79) identifies two types of lamps that were produced using these molds. Type 1 lamps are characterized by a large round body with a wide concave discus. The main diagnostic feature of this type is the linear style of the motifs portrayed on the disks, noticeably incised by freehand into the limestone mold (Sussman 1980: 77). The visual effect created by the graffiti-like "sketch" of the discus images of Caesarean round lamps differs from the more generic, standardized and factory-like motifs of the Palestinian discus type. A variety of motifs are depicted on the disks of this type: geometric patterns, human and animal figures, floral designs, aediculae, crosses, and menorahs. The short nozzle is wide and splayed. Two parallel lines in molded relief follow the outline of the nozzle. A triangle composed of double lines is rendered on the nozzle below the wick-hole and touches the ridge encircling the central discus. Curved radial lines commonly decorate the shoulder and create a ribbed appearance. A small pointed triangular handle is common. The base is generally flat. Type 1a lamps differ from the latter mainly in that the nozzle is longer and the discus less wide.

The distribution of this lamp type is most prominent in the central part of Israel. Caesarean discus lamps are found at the northern sites of Sepphoris and Khirbet Shemaʿ. They occur as far south as ʿEin Gedi (see, e.g., Sussman 1980: pl. 15.1a) and the Red Sea port of Roman Aila, where a fragment uncovered in Area O bears the trace of a human face (see Lapp forthcoming a: reg. no. 50711). Caesarean discus lamps are rare outside Byzantine Palestine, but two fragments have been found at Palaepaphos in Cyprus (Bezzola 2004: 142, nos. 601–2). Scholars propose several chronological ranges for lamps of this type: fourth century C.E. (Lapp 1997a: 47, Type 10); fourth–fifth centuries C.E. (Sussman 1980: 76); fourth–sixth centuries C.E. (Israeli and Avida 1988: 112); sixth–seventh centuries C.E. (Vine and Hartelius 2000: T26A); and fifth–sixth centuries C.E. (da Costa 2003: 1.177, 180–81). I propose a revised date of late fourth to sixth centuries C.E.

Lamp makers of the Caesarean discus type target marketed their products to both a Jewish and Christian clientele (Lapp 2007: 375). They exploited religious difference by manufacturing lamps bearing Jewish symbols for Jewish buyers and lamps with Christian symbols for a Christian market (Lapp 2007: 372–77). Until recently, Caesarea round lamps—known for their rich repertoire of Christian images (e.g., cross, Gospel aedicula, fisherman, and peacock)—were considered to have been made exclusively for a Christian population (Sussman 1980: 77 n. 4, 78; MacDonnell 1988: 133, fig. 6-6.86, pl. 6-4). Examples with Jewish symbols indicate that this lamp type was also popular in the Jewish community. A Caesarean discus lamp fragment with a menorah flanked by a *lulav* (palm frond) was discovered on the western summit at Sepphoris (Lapp 1996: 222, no. 118). By comparison, a simple floral pattern decorates the wide discus of a complete Caesarea discus lamp found in a niche of the ʿEin Gedi synagogue (Barag, Porat, and Netzer 1981: 112). A Torah Shrine with a drawn *parochet* (curtain) is portrayed on a Caesarean discus lamp from a tomb near Caesarea Maritima, further suggesting Jewish use of this lamp type (Siegelmann 1992: 65, fig. 4.1; Fine 1996: 171, cat. no. 69).

At Sepphoris, two lamp fragments and a spatulated mold-made nozzle were discovered (cat. nos. 132–34). The shoulder of this lamp type is curved and steeply sloping and is decorated with a single band of consecutive, evenly spaced, slightly curved radials, thus forming a ribbed pattern (cat. nos. 132–33). Like other lamps of this type, the three examples of the Sepphoris corpus are made of pale clay, particularly a dull or light yellowish orange (cat. nos. 132–33). Generally, Caesarean discus lamps, like the Sepphoris examples (cat. nos. 132–33) are not coated in a slip. Catalog no. 134, however, has a reddish-brown slip; this makes it a rare example for the Caesarean group, so it may belong to an altogether different contemporary northern lamp type. White represents the only inclusion type found in the Sepphoris examples. The grain size ranges from medium to very coarse sand and occurs at a very low frequency. The average hardness is 2.0 on the Mohs scale.

177. Reg. no. 84.1263.1L.2. L = 5.2 cm. W = 2.1 cm. Incomplete, 10% preserved. Caesarean discus. Late fourth to sixth centuries C.E. Shoulder fragment with discus remnant. Plain discus whose outer edge laps onto shoulder, forming a single prominent ridge. An additional ridge surrounds discus. Narrow, rounded and steeply sloping shoulder. On shoulder: a ribbed pattern of radials in positive relief. Clay color: 10YR 7/2 dull yellow-orange. No slip. Hardness: 2.0. Inclusion(s): white (MS, very low frequency). Good condition. Mold-made. Parallels, shape-type: Caesarea Maritima (Sussman 2008: 245–47, nos. 170–78; Sussman 1980: pl. 15.1, lamp mold) (similar); provenance unknown (Israeli and Avida 1988: 112, fig. 321[C]; Rosenthal and Sivan 1978: 125, nos. 513–14).

178. Reg. no. 84.1252.2L.1. L = 4.0 cm. W = 2.3 cm. Incomplete, 10% preserved. Caesarean discus. Late fourth to sixth centuries C.E. Shoulder fragment. Outer edge of discus extends onto the shoulder, thus forming a single prominent ridge. Two additional closely juxtaposed minor ridges encircle discus. On discus: a floral pattern consisting of branches in high relief. Narrow, rounded, and steeply sloping shoulder. On shoulder: a ribbed pattern in positive relief. Clay color: 7.5YR 7/4 dull orange. Hardness: 2.0. Fair condition. Mold-made. Parallels, shape-type: Caesarea Maritima (Sussman 2008: 245–47, nos. 170–78).

179. Reg. no. 84.1463.2L.2. GD = 5.7 cm. Incomplete, 20% preserved. Caesarean discus. Late fourth to sixth centuries C.E. Shoulder and discus fragment. Wide round discus. Narrow, steeply curving shoulder. On shoulder: some radials in high relief. On discus, oriented left: hind quarters with raised tail of lion in molded relief. Fabric color: 5YR 7/6. No slip. Inclusions: gray, well sorted (VHF, VFS). Hardness: undetermined. Fair condition. Heavy lime on exterior surface. No burning. Sepphorean workshop (?). Mold-made. Parallels, shape-type: Caesarea Maritima (Sussman 2008: 245–47, nos. 170–78).

180. Reg. no. 84.4142b.2L.8 (IAA 95-3897). GD = 7.0 cm. Incomplete, 25% preserved. Caesarean discus. Late fourth to sixth centuries C.E. Discus and shoulder fragment. Wide concave discus. On discus: an incomplete menorah with a rectangular tripod base; three branches (left) and four others (right) stem from the central shaft; remnant of a *lulav* flanks menorah on the left; on the upper right, a curved nodule may

be a trace of a shofar. Menorah and associated Jewish symbols fashioned in graffito-like, haphazard linear style in molded high relief. Two rounded ridges around discus. Narrow, curved and steeply sloping shoulder. On shoulder: a ribbed pattern of radials in positive relief. Conical unpierced and chipped handle. Clay color: 5YR 6/4 light reddish brown. No slip. Hardness: 3.0. Hard fired. Inclusion(s): white (FS, very low frequency). Good condition. Mold-made. Parallels, shape-type: Caesarea Maritima (Sussman 2008: 245–47, nos. 170–78). Published: Lapp 1996: 222, no. 118.

181. Reg. no. 84.1505.1L. Not shown. GD = 5.3 cm. Incomplete, 5% preserved. Caesarean discus. Late fourth to sixth centuries C.E. Discus and shoulder fragment. Plain concave discus. Prominent ridge around discus. On shoulder: evenly spaced radials or ribbing. Clay color: 10YR 6/1 light gray-gray. Slip color: 10YR 5/3 brown. Slip splashes on part of shoulder and on interior surface. Inclusions: gray, sub-angular (VHF, VFS-CSi). Hardness: 3.0. Good condition. Medium firing. No burning. Mold-made. Parallels, shape-type: Caesarea Maritima (Sussman 2008: 245–47, nos. 170–78).

Type 19. Galilean Discus (cat. nos. 182–88)

Galilean discus lamps are characterized by a wide, deeply concave discus typically decorated with Jewish symbols in high relief. These include the seven-branched menorah (cat. no. 184) and the Torah Shrine (cat. nos. 183–85). The discus images are fashioned in a linear, realist style, a graffito etched into a lamp mold by freehand. In this respect, the Sepphoris examples represent the Galilean equivalent of the Caesarea discus lamp group, also with haphazardly rendered linear motifs decorating the central disk. By manufacturing their own version of the popular Caesarean discus type, Galilean lamp makers provided their local clientele an alternative but similar lamp type. A distinguishing feature is the extension of the edge of the large broad discus above the shoulder, creating a wide, rounded, and prominent ridge. This differs from the Caesarean and earlier Palestinian discus types whose respective disks do not extend higher than the shoulder in section.

The Galilean discus lamps excavated at Sepphoris were intended primarily for a Jewish clientele. The Jewish symbols portrayed on them suggest that both the lamp-mold maker and lamp maker manufactured these objects with members of the local Sepphorean Jewish community in mind. For example, the vestige of a menorah with two, though incomplete branches out of the probable original seven is portrayed on cat. no. 184: an outer right, semicircular branch decorated with consecutive *V*-shaped hatches (not visible in the drawing) and a second branch adorned with consecutive small circles. A remnant of the elongated rectangular leg of the menorah base is preserved. A highly stylized Torah Shrine with closed cabinet doors flanks the menorah to the right. The doors are indicated as concentric squares and the aedicula as a triangle. Similar aediculae suggestive of Torah Shrines are found on cat. nos. 182–83.

At Jalame, a discus and shoulder fragment of a Galilean discus lamp were found below a hard clay floor (d:24) near coins of Arcadius and Honorius (MacDonnell 1988: 120, 136, fig. 6-7, pl. 6-4.101). MacDonnell (1988: 136) describes the motif on the discus as a "geometric design." Upon closer examination of the fragment's photograph, however, and rotating it counterclockwise by 45° (MacDonnell 1988: pl. 6-4.101), we find that the motif—a square (cabinet) topped by a triangle (aedicula)—appears to be a stylized Torah Shrine like that on the Sepphoris example (cat. no. 184). Like the Sepphoris example, the Torah Shrine on the Jalame sherd occurs on the edge of the discus, most of which is missing. Thus, it likely flanked a menorah in the original discus of the complete lamp. Similar to the Sepphoris fragment, scallops also decorate the shoulder of the Jalame sherd, differing only with respect to the inclusion of concentric circles in each scallop loop. MacDonnell (1988: 136) dates the Jalame example to the late fourth century C.E.

The round demi-dome gable and columns with spiraling flutes of the shrine portrayed on cat. no. 185 resemble images of the Torah Shrine found in ancient Jewish art (e.g., the synagogue mosaic at Hammath Tiberias), but in the absence of certain features, no definitive religious affiliation can be attributed to this shrine (see discussion in Lapp 1996: 221–22, no. 116). The motif decorating the discus of fragment cat. no. 186 is simply a shrine. Considerably thicker in section and exhibiting a different fabric, cat. no. 186 should not be confused with the Palestinian discus type.

Three branches of a highly stylized menorah are preserved on a Galilean discus fragment (cat. no. 187). The branches are fashioned as bent lines (similar to branch type no. 8, lower in Hachlili 1998: 329, fig. 7.14). Narrower, parallel horizontal lines decorated the branches, creating an intended wood-like texture effect. A single circle at each tip represents a stylized flame.

Of the examples found at Sepphoris, the complete conical handle of cat. no. 182 indicates this was the handle of choice for this lamp group. Galilean discus lamps are fashioned in a limited variety of clay fabric colors and associated slips, including: light gray (10YR 7/2) fabric with light reddish-brown (2.5YR 6/4) slip; or pink fabric (7.5YR 7/4) with light red (2.5YR 6/6) slip. Some examples exhibit a spongy fabric indicative of a high percentage of pores. Their fabrics tend to be gritty, which reflects a high abundance of inclusions, particularly orange, translucent, surrounded quartz sand. In this respect, Galilean discus lamps are similar to the Sussman Types 1 and 2 Samaritan group. The distribution of the Galilean discus type is limited to Galilee and is found only at a few sites, notably Sepphoris and Jalame. For this reason, they were probably manufactured in one or more workshops located in Galilee, perhaps even in Sepphoris, where the greatest numbers, though low, have been found. Until combined petrographic and trace-element analyses are conducted on samples of this group, their origin of manufacture will remain unknown. No bases have been found to determine whether lamp makers' marks were used.

Several Galilean discus lamp fragments (cat. nos. 183, 184, 187, 188) were found in L84.4142b, described by the excavators as unsealed fill in a ritual bath (L84.4290). The fill contained a bronze coin of an uncertain ruler, dated to 383–395 C.E. (Ariel and Bijovsky forthcoming, IAA 48596), and pottery dating mainly to the fourth–fifth centuries C.E. Based on morphological similarities with the Caesarean discus type (see table 1, pp. 5–8 above), and considering the late fourth-century date of the Jalame example (MacDonnell 1988: 136, no. 101), the Galilean discus lamps from Sepphoris's western summit date from the late fourth through the fifth centuries C.E.

182. Reg. no. 84.1351.1L. GD = 5.2 cm. Incomplete, 20% preserved. Galilean discus. Late fourth to fifth centuries C.E. Discus, shoulder, nozzle, and body underside fragment. Narrow, steeply curving shoulder. On shoulder: tendrils in high relief. On discus: the point of a gable of an aedicula, possibly a Torah Shrine fashioned in linear high relief. A prominent, rounded double ridge (0.3 cm wide) incorporated into the outer edge and direct continuation of a wide concave discus. Clay color: 10YR 7/3 very pale brown. Slip color: 5YR 6/6 reddish yellow. Inclusions: black (VHF, CSi); gray, sub-rounded (CS, LF). Hardness: not measured. Fair condition. Medium-hard firing. No burning. Mold-made

183. Reg. no. 84.4124b.10L. GD = 5.3 cm. Incomplete, 15% preserved. Galilean discus. Late fourth to fifth centuries C.E. Discus, shoulder, and underside fragment. Wide concave discus. On discus: remnant of an aedicula gable, possible Torah Shrine, two round nodules inside aedicula, probably stylized scrolls; to the right, a curl; thin parallel lines running between curl and aedicula may represent rays of light. Double ridge around discus comprises two rounded minor, thin (1.0 mm) ridges with groove (1.0 mm) between them. Plain, curved, and steeply sloping shoulder. Clay color: 2.5YR 4/0 dark gray. Slip color, traces: 2.5YR 5/0 gray. Inclusions: gray, sub-angular (VFS, high frequency). Hardness: 3.0. Medium to hard fired. Good (–) condition. Medium-hard firing. Burning, interior surface. Film of white lime powder covers exterior surface. Mold-made.

184. Reg. no. 84.4142b.2L.10. GD = 5.5 cm. Incomplete, 20% preserved. Galilean discus. Late fourth to fifth centuries C.E. Discus and shoulder fragment. Quite deeply concave discus. On discus: the remnant of a menorah flanked by a nearly complete Torah Shrine aedicula with closed cabinet doors indicated as concentric squares (to the right); below shrine, perhaps a remnant of the elongated rectangular leg of the menorah base; only part of outer semicircular right branch decorated with consecutive *V*-shaped hatches, and a trace of a second branch decorated with consecutive small circles. On shoulder: connecting semicircles topped with round nodules. All motif elements rendered in molded high relief. A single, wide (0.8 cm) rounded prominent ridge incorporated into the outer edge and direct continuation of a wide concave discus (preserved discus radial 4.5 cm). Wide, steeply sloping and curved shoulder. Clay color: 7.5YR 7/4 pink. Slip color: 2.5YR 6/4 light reddish brown. Slight traces of slip on shoulder. Inclusions: orange, translucent, sub-rounded quartz sand (VFS, HF); gray, sub-rounded (VFS, VHF); gray sub-angular (FS, LF). Very gritty fabric due to high abundance of inclusions. Possibly the same clay source or same type of clay used to manufacture lamps belonging to Samaritan, Sussman Types 1 and 2, and probably made in same workshop as no. 69. Hardness: Not measured. Fair condition. Film of white salts or lime covers exterior surface. Fragile. Minor flaking (of discus). Medium firing. No burning. Mold-made. Parallels, shape-type and Torah Shrine motif on discus: Jalame (MacDonell 1988: 120, 136, fig. 6-7.101).

185. Reg. no. 85.1060.1L.1. L = 4.5 cm. W = 7.5 cm. Incomplete, 30% preserved. Galilean discus. Late fourth to fifth centuries C.E. Discus, shoulder, and nozzle-bridge fragment. Plain, sharply sloping, narrow shoulder. On discus: a shrine with round architrave supported by two columns (left column missing) decorated with diagonal incised lines in high relief; a single band of consecutive raised nodules ornament architrave; beneath architrave, a demi-dome. Although no menorah or rendition of the Holy Ark exists between the columns of this architectural feature, the style is reminiscent of other examples associated with Jewish symbolism. Mold-made. Clay

color: 10YR 7/3 dull yellow-orange. Slip color: 5YR 6/4 dull orange, exterior surface only. Hardness: 3.0. Inclusion(s): gray (FS, high frequency); white (VCS, very low frequency); muscovite mica flakes (CSi, high frequency). Gritty. Burning traces, interior surface. Good condition. Parallel (exact): Sepphoris (Lapp 1991: fig. 205).

186. Reg. no. 84.1163.2L.1. L = 4.3 cm. W = 3.8 cm. Incomplete, 15% preserved. Galilean discus. Late fourth to fifth centuries C.E. Discus fragment. On discus: two plain columns supporting a curved line, suggesting a highly stylized shrine with round architrave. Clay color: 7.5YR 7/4 dull orange. Slip color: 2.5YR 5/6 bright brown, exterior surface only. Hardness: 2.0. Fair condition. Mold-made.

187. Reg. no. 84.4142b.2L.2. GD = 4.1 cm. Thickness = 0.3–0.8 cm. Incomplete, 10% preserved. Galilean discus. Late fourth to fifth centuries C.E. Discus and nozzle-bridge fragment. On discus: two branches of stylized menorah; the branches comprise bent lines, between which are thinner, parallel horizontal lines, creating wood texture; a single circle at each tip represents a highly stylized flame. Plain shoulder. A single, rounded prominent ridge incorporated into the outer edge and direct continuation of a wide concave discus. Relatively wide, steeply sloping and curved shoulder. Clay color: 10YR 7/2 light gray. Slip color: 2.5YR 6/4 light reddish brown. No visible inclusions. Spongy fabric indicative of pores. Hardness: not measured. Fair condition. Medium firing. No burning. Mold-made.

188. Reg. no. 84.4142b.3L.2. GD = 5.5 cm. Incomplete, 10% preserved. Galilean discus. Late fourth to fifth centuries C.E. Discus and shoulder fragment. Plain, shallow, concave discus. On shoulder: tightly spaced band of herringbone; between herringbone and ridge, a line of evenly spaced round nodules. A single, rounded prominent ridge incorporated into the outer edge and direct continuation of a wide concave discus. Relatively wide, steeply sloping curved shoulder. Clay color: 7.5YR 7/4 pink. Slip color: 2.5YR 6/6 light red. Inclusions: orange, translucent, sub-rounded quartz sand (FS, VHF); gray, sub-rounded (CS, LF). Very gritty fabric due to high abundance of inclusions. Possibly the same clay source or same type of clay used to manufacture lamps belonging to Samaritan, Sussman Types 1 and 2. Hardness: not measured. Fair condition. Medium-hard firing. No burning. Mold-made.

Type 20. "Sepphorean" Discus (cat. nos. 189–208)

The "Sepphorean" discus lamp group is characterized by a wide concave discus typically decorated with expertly carved motifs and shoulder patterns executed in a delicate linear style. Floral and faunal motifs tend to decorate the disks of this type (cat. nos. 194–95), though some examples are plain (cat. no. 190). At Sepphoris, one such discus fragment is decorated with delicately fashioned branches with leaves (cat. no. 194) and a second with a growling hyena poised for attack (cat. no. 196). The precision of the rosette depicted on the nozzle-bridge of cat. no. 189 indicates the use of a compass and is suggestive of Jewish ossuary art. Double rounded ridges surround the discus. "Sepphorean" discus lamps are generally fashioned in well-levigated chalky clay containing a high frequency of very fine sand size, gray, and well-rounded inclusions. The clay is soft fired, and the fabric color is characteristically white (2.5Y 8/2), indicative of a fabric with a high abundance of calcium carbonate.

The "Sepphorean" discus group is closely related to the Sepphorean round-bodied with large filling-hole type (originally referred to as the "Sepphoris lamp"), a mold of which was discovered east of the citadel at Sepphoris by a student during a school excursion (Adan-Bayewitz 1995: 177 n. 1). Both types are characterized by a large round body with a short, narrow spatulated nozzle. The wide and gradually sloping shoulders of both types are decorated with a distinctive and precisely executed pattern of alternating pairs of inverted *V*s and wide truncated ovules in relief (Adan-Bayewitz 1995: 177). Several "Sepphorean" discus lamp fragments included in this catalog exhibit such shoulder patterns (cat. nos. 194–95). Both types are also manufactured with a characteristic chalky white, soft fired clay. These similar characteristics point to manufacture by the same lamp-mold maker and by one or more lamp makers operating in one or more workshops with the same geographic area for both types. Despite the many features in common with the Sepphorean round-bodied with large filling-hole lamps, the "Sepphorean" discus group should be treated as a distinct and separate shape-type based on a key morphological difference: it has a concave discus, whereas the Sepphorean round-bodied with large filling-hole has a large filling-hole in place of a central discus instead.

It is plausible that both types were manufactured in the same lamp workshop. As far as I am aware, Sepphoris is the only site where examples of the "Sepphorean" discus group have been found, suggesting Sepphoris or the vicinity as the probable source of manufacture.[8] The Sepphorean round-bodied with large filling-hole type is characterized by a large filling-hole. He further notes that the filling-hole is surrounded by a raised ring and two concentric lines (Adan-Bayewitz 1995: 177). Curved parallel lines follow the sides of the nozzle.

Adan-Bayewitz (1995: 178) observes that the Sepphoris lamp resembles a significant class of lamps with wide nozzles from numerous sites in northern Palestine. As he points out (Adan-Bayewitz 1995: 179), the Sepphoris lamp differs from the latter class of lamps in that its nozzle is markedly narrower and shorter in addition to having a wider shoulder. The linear relief decoration of this lamp type is more uniform and factory-like and indicates the use of a compass and straight edge; in this respect, it stands in contrast to the freehand linear decoration and motifs of the Caesarean discus (cf. cat. nos. 177–80) and "Sepphorean" discus types. The Sepphorean, Galilean, and Caesarean discus have similar spatulate-shaped nozzles whose contours are delineated by double lines (Adan-Bayewitz 1995: 177; da Costa 2003: 1.123).

The Sepphoris mold was not used to manufacture lamps of the Galilean group, as has been suggested (da Costa 2003: 1.127), but for a different type (Adan-Bayewitz 1995: 178–79), which, in my opinion, is the Sepphorean round-bodied with large filling-hole group. The Galilean type examples found at Pella (da Costa 2003: 2. cat. nos. 195–219) correspond to the Type 22 found at Beth Shean (Hadad 2002:

8. A carved limestone lamp mold recovered at Sepphoris (reg. no. 86.1195.2X, no. 359) further substantiates the likelihood that the "Sepphorean" discus type was manufactured in Sepphoris or the area; the wide discus and splayed nozzle of the mold cavities are suggestive of this lamp type. The mold was found in an excavation area being published under the auspices of the Hebrew University team.

56–61, nos. 255–70). However, they do not correspond to the sherds of the "Sepphorean" discus or Sepphorean round-bodied with large filling-hole types found at Sepphoris. Both types exhibit a noticeably higher level of stylistic precision and sophistication, with delicately executed motifs, in comparison with the da Costa Galilean group, which is characterized by heavier and clunkier patterns, shorter and narrower nozzles, thinner walls, and lighter bodies. In the latter respect, the "Sepphorean" discus type is more akin to the ovoid with large filling-hole manufactured at Beit Nattif (see, e.g., Magness 2008: 123, figs. 5.4:9–10; 5.5:1–4). The Galilean and Sepphorean types have similar shoulder designs, nozzle shape, and lines that delineate the contour of the nozzles. In contrast to the Galilean group whose shoulders are steeply sloping and diagonal in section, the Sepphorean with round body and large filling-hole type is curved. As far as I am aware, evidence for the Sepphorean lamp type has been found only at Sepphoris and only in fragmented form. Noting that the Sepphoris lamp mold cannot be closely dated based on the present archaeological evidence, Adan-Bayewitz (1995: 180) suggests a range of manufacture sometime between the fourth and sixth centuries C.E., roughly from the Late Roman through the Byzantine periods.

Because I am unaware that this lamp type occurs at any other site, I selected the term "Sepphorean," suggesting its possible manufacture at the site. Without any molds and workshop evidence, I cannot confirm that this type was manufactured at Sepphoris and therefore enclose the place-name in quotation marks. However, in addition to its unique occurrence at Sepphoris only, the white chalky clay used to fashion this type and the type, style, and execution of the motifs and patterns of this lamp group are identical to the characteristics exhibited by the Sepphorean round-bodied type, a type for which a mold was found at Sepphoris, suggesting its manufacture there. For these reasons, I argue that the same lamp-mold workshop made this lamp type as well as the Sepphorean with round body and large filling-hole type. Their distribution is limited to Sepphoris. The Sepphoris mold and the comparatively abundant quantity of lamp fragments of this type found by the Duke team at the site (and paucity at other sites) suggests Sepphoris as the probable manufacturing place for this type.

The majority of "Sepphorean" discus lamp fragments were recovered from L85.1195, a cistern fill that contained massive amounts of whole and restorable pottery, bones, coins, and lamp fragments. The contents of the fill are dated to the Early Byzantine–Early Islamic periods. Dozens of unidentified coins and a few identified, the latest of which is a small Byzantine bronze, were found in the cistern. Taking into consideration that this lamp type's features are closely related to those of the Sepphorean with round body and large filling-hole type and the date assigned to the cistern materials, examples of the "Sepphorean" discus group were likely manufactured sometime between the fourth and sixth centuries C.E.

189. Reg. no. 85.1195.2L.1. L = 9.3 cm. W = 7.6 cm. H = 2.4 cm. Nearly complete, 90% preserved. "Sepphorean" discus. Fourth to fifth centuries C.E. Complete nozzle and base with remnant of deeply concave discus (4.3 cm diameter) and shoulder frag-

ment. Short and stumpy splayed nozzle. Wick-hole 0.8 cm diameter. On nozzle-bridge: a circle or wheel in high relief (1.4 cm diameter) with 11 inner nodules and 1 center nodule; circle inscribed in mold using a compass. On discus: remnants of rosette with highly stylized palm frond and petal composed of two concentric ovules in linear high relief. Two ridges around discus. Plain gradually sloping shoulder. Base: concave and oval-shaped with 2.0 cm × 2.1 cm diameter. No evidence for handle. Well levigated, white chalky fabric and powdery to the touch. Clay color: 2.5Y 8/2 white. Slip color: 5YR 5/6 yellowish red. Inclusions: gray quartz sand (VFS, HF). Hardness: 2.0, soft. Fair condition. Flaking fabric. Burning on nozzle. Mold-made.

190. Reg. no. 85.1195.2L.2. Preserved L = 9.9 cm; preserved W = 7.7 cm. Incomplete, 45% preserved. "Sepphorean" discus. Fourth–sixth centuries C.E. Discus, shoulder, and nozzle fragment. Incomplete, plain concave discus (5.3 cm diameter). Filling-hole 0.8 cm diameter. On discus: plain. Two concentric ridges around discus. Plain gradually sloping shoulder. Standing tongue handle with raised edges. Well-levigated, soft friable fabric. Clay color: 2.5Y 8/2 white. Slip color: 5YR 6/6 reddish yellow. Splashes of slip on nozzle-bridge, part of discus, and on interior surface. Inclusions: gray quartz sand (VFS, HF). Hardness: 2.0, soft. Good condition. Flaking. Burning. Mold-made.

191. Reg. no. 85.1195.6L.10. GD = 5.2 cm. Incomplete, 5% preserved. "Sepphorean" discus. Fourth–sixth centuries C.E. Complete nozzle fragment. Wick-hole (diameter 1.0 cm × 0.7 cm). Plain nozzle-bridge and without characteristic inverted-*V* design. Characteristic splayed lines delineating nozzle shape. Most of nozzle-underside preserved, exhibiting bloated throat. Thickness 0.5 cm, more than typical 0.2–0.3 cm of other "Sepphorean" discus examples. Soft powdery fabric. Clay color: 2.5Y 8/2 white. Slip color: 5YR 6/6 reddish yellow. Slip traces on shoulder, nozzle-bridge, and nozzle-underside. Inclusions: gray, well-rounded quartz sand (VFS, HF). Hardness: 2.0, soft. Fair condition. Burning around nozzle. Mold-made.

192. Reg. no. 84.4142b.2L.7. GD = 6.7 cm. Incomplete, 20% preserved. "Sepphorean" discus. Fourth–sixth centuries C.E. Discus, shoulder, and underside fragment. Wide concave discus (at least 5.0 cm in diameter). On discus: remnant of a rosette with five complete petals preserved in high relief; circular nodules between each petal, upper. Curved, sharply sloping shoulder. On shoulder: a single band of primitively executed guilloche. Remnant of central filling-hole. Plain nozzle-bridge and without characteristic inverted-*V* design. Characteristic splayed lines delineating nozzle shape. Most of nozzle-underside preserved, exhibiting bloated throat. Thick walled, 3.0 mm to 6.0 mm. Complete, triangular lug handle. Clay color: 2.5Y 8/2 white. Slip color: 2.5YR 6/4 light reddish brown. Slip traces on shoulder, nozzle-bridge, and nozzle-underside. Inclusions: not visible. Hardness: 2.0, soft. Fair condition. Burning around nozzle. Finger impressions. Mold-made.

193. Reg. no. 85.1195.5L.2 + 85.1195.7L.2. GD = 4.7 cm (5L.2) and GD = 2.7 cm (7L.2). Incomplete, 35% preserved. "Sepphorean" discus. Fourth–sixth centuries C.E. Mended with two fragments. Discus and shoulder fragment. Shallow concave discus. On discus: a stylized menorah with six semicircular branches and a vertical central shaft, the seventh; a crossbar extends across the top of the branches; no lamps or flames are portrayed on the branches; trace of a rectangular tripod base; to the right and flanking the menorah, a *lulav*; an oval-device at the base of the *lulav* may represent a type of hand-grip; menorah and *lulav* executed in delicate and haphazard linear style. Filling-hole: diameter 0.8 cm × 0.9 cm. Curved, relatively steeply sloping

shoulder. On shoulder: remnants of wide, consecutive radial lines. Prominent single ridge (0.3 cm wide) around discus. Soft, powdery chalky fabric. Clay color: 2.5Y 8/2 white. Slip color: 5YR 6/6 reddish yellow. Inclusions: gray, well-rounded quartz sand (VFS, HF). Hardness: 4.0. Hard fired. Fair condition. No burning. Mold-made.

194. Reg. no. 85.1195.6L.6. GD = 5.2 cm. Incomplete, 15% preserved. "Sepphorean" discus. Fourth–sixth centuries C.E. Discus and shoulder fragment. Plain, deeply concave discus. On discus: branches with leaves executed in a delicate linear style. Curved, steeply sloping shoulder. On shoulder: a pattern executed in a wide, delicate linear style: a concentric inverted *V* :: concentric ovule :: concentric inverted *V*. Double ridges around discus: outer 0.1 cm wide and inner 0.2 cm wide. Filling-hole trace. Soft, powdery, chalky fabric. Clay color: 2.5Y 8/2 white. Slip color: 5YR 6/6 reddish yellow. Underside exhibits no traces of slip. Inclusions: gray, well-rounded quartz sand (VFS, HF). Hardness: 2.0. Fair condition. Finger impressions on interior surface. No burning. Mold-made.

195. Reg. no. 85.1195.6L.7. GD = 6.5 cm. Incomplete, 15% preserved. "Sepphorean" discus. Fourth–sixth centuries C.E. Shoulder fragment. Part of shoulder underside preserved. On shoulder: a pattern executed in a broad, delicate linear style: a concentric inverted *V* :: a band of two vertical lines, inside of which are four small nodules :: a concentric ovule :: a band of two vertical lines, inside of which are four small nodules :: a concentric inverted *V*. Double ridges around discus (missing). Soft powdery fabric. Clay color: 2.5Y 8/2 white. Slip color: 5YR 6/6 reddish yellow. Underside exhibits no traces of slip. Inclusions: gray, well-rounded quartz sand (VFS, HF). Hardness: 2.0. Good condition. Finger impressions on interior surface. No burning. Mold-made.

196. Reg. no. 85.1195.5L.1. GD = 5.1 cm. Incomplete, 20% preserved. "Sepphorean" discus. Fourth–fifth centuries C.E. Discus and shoulder fragment. On discus: a growling hyena, face oriented to the right. Two front legs prancing and poised for attack; above hyena, branches with leaves. On shoulder: consecutive radials. All patterns in linear execution and in high relief. Single ridge composed of three smaller ridges around concave discus. White creamy, almost powdery fabric. Well-levigated, chalky fabric with no visible inclusions. Clay color: 2.5Y 8/2 white. Slip color: 5YR 5/6 yellowish red. Hardness: 2.0. Poor condition. Flaking fabric. No burning. Mold-made.

197. Reg. no. 85.1195.2L.2. Not shown. GD = 7.3 cm. Incomplete, 30% preserved. "Sepphorean" discus. Fourth–sixth centuries C.E. Lower half body fragment. Short, squarish splayed nozzle. Splaying indentations on each side of nozzle-underside. Base: flat, delineated by two ridges forming circle with rosette in delicately executed linear style in high relief. Well-levigated, soft powdery fabric. Clay color: 2.5Y 8/2 white. Slip color: 5YR 6/6 reddish yellow. Inclusions: gray, well-rounded quartz sand (VFS, HF). Hardness: 2.0, soft. Fair condition. Burning on underside of nozzle. Finger impressions. Mold-made.

198. Reg. no. 85.1195.6L.2. Not shown. GD = 7.6 cm. Incomplete, 30% preserved. "Sepphorean" discus. Fourth–sixth centuries C.E. Nozzle, discus, and shoulder fragment. Short, square-like splayed nozzle. On nozzle-bridge: concentric inverted-*V* pattern. Deeply concave discus (4.7 cm diameter) with central filling-hole (0.8 cm diameter). On discus: two concentric rings in high relief (3.0 cm diameter). Steeply sloping shoulder. On shoulder: consecutive radial lines in high relief but not consistently parallel. Soft powdery fabric, though somewhat gritty to the touch. Clay color: 2.5Y 8/2 white. Slip color: 5YR 6/6 reddish yellow. Slip traces on nozzle-bridge and

remnants of discus. Inclusions: gray, well-rounded quartz sand (VFS, HF). Hardness: 2.0, soft. Fair condition. Burning on nozzle-bridge. Finger impressions. Mold-made.

199. Reg. no. 85.1195.3L.1. Not shown. GD = 4.9 cm. Incomplete, 10% preserved. "Sepphorean" discus. Fourth–sixth centuries C.E. Discus and shoulder fragment. Plain discus. On shoulder: inconsistently spaced, consecutive radial lines tipped with circular nodules on the ends closest to double ridges around discus in high relief. Soft powdery fabric. Clay color: 2.5Y 8/2 white. Slip color: 5YR 6/6 reddish yellow. Inclusions: gray, well-rounded quartz sand (VFS, HF). Hardness: 2.0, soft. Fair condition. No burning. Mold-made.
200. Reg. no. 85.1195.5L.7. Not shown. GD = 3.0 cm. Incomplete, 2% preserved. "Sepphorean" discus. Fourth–sixth centuries C.E. Discus, shoulder, and nozzle-bridge fragment. Plain, deeply concave discus. Double ridges around discus. On shoulder: remnants of consecutive radial lines in high relief. On nozzle-bridge: concentric inverted-*V* pattern. Soft powdery fabric. Clay color: 2.5Y 8/2 white. Slip color: 5YR 6/6 reddish yellow. Inclusions: gray, well-rounded quartz sand (VFS, HF). Hardness: 2.0, soft. Fair condition. No burning. Mold-made.
201. Reg. no. 85.1195.5L.3. Not shown. GD = 4.0 cm. Incomplete, 5% preserved. "Sepphorean" discus, version. Fourth–sixth centuries C.E. Discus, shoulder, and handle fragment. Deeply concave discus. Double ridges around discus. On discus: remnant of the single point of a star. On shoulder: consecutive, consistently parallel radial lines in high relief. Radial lines are wider than those characteristically found on "Sepphorean" discus examples. Linear designs on this lamp fragment are thicker than the more delicate execution found on typical "Sepphorean" discus examples. Triangular handle remnant. Soft powdery fabric. Clay color: 2.5Y 8/2 white. Slip color: 5YR 6/6 reddish yellow. Slip traces on discus and shoulder. Inclusions: gray, well-rounded quartz sand (VFS, HF). Hardness: 2.0, soft. Fair condition. No burning. Mold-made.
202. Reg. no. 85.1195.6L.3. Not shown. GD = 4.0 cm. Incomplete, 2% preserved. "Sepphorean" discus. Fourth–sixth centuries C.E. Discus, shoulder, and nozzle-bridge fragment. Deeply concave discus. Double ridges around discus. On shoulder: remnants of consecutive radial lines tipped with round nodules in high relief. On nozzle-bridge: remnant of probable concentric inverted-*V* pattern. Soft powdery fabric. Clay color: 2.5Y 8/2 white. Slip color: 5YR 6/6 reddish yellow. Slip traces on nozzle-bridge. Inclusions: gray, well-rounded quartz sand (VFS, HF). Hardness: 2.0, soft. Fair condition. No burning. Finger impressions. Mold-made.
203. Reg. no. 85.1195.4L.3. Not shown. GD = 3.8 cm. Incomplete, 5% preserved. "Sepphorean" discus. Fourth–sixth centuries C.E. Nozzle with part of nozzle-bridge fragment. Wick-hole (diameter 0.9 cm × 0.8 cm). Plain nozzle-bridge. Characteristic splayed lines delineating nozzle shape. Part of nozzle-underside preserved. Soft powdery fabric. Clay color: 2.5Y 8/2 white. Slip color: 5YR 6/6 reddish yellow. Inclusions: gray, well-rounded quartz sand (VFS, HF). Hardness: 2.0. Poor condition. Flaking. Burning around wick-hole. Mold-made.
204. Reg. no. 85.1195.2L. Not shown. GD = 5.1 cm. Incomplete, 25% preserved. "Sepphorean" discus. Fourth–sixth centuries C.E. Discus and shoulder fragment. On discus: a raised single ring (2.6 cm diameter) around filling-hole (0.8 cm diameter). Plain, slightly slopping curved shoulder. Outer band decorated with wide ovule :: stylized branch :: wide ovule in delicate linear relief. On shoulder: double ridges around discus. Soft powdery fabric. Clay color: 2.5Y 8/2 white. No slip. Inclusions: gray,

well-rounded quartz sand (VFS, HF). Hardness: 2.0. Poor condition. No burning. Mold-made.

205. Reg. no. 85.1195.5L.4. Not shown. GD = 4.3 cm. Incomplete, 5% preserved. "Sepphorean" discus. Fourth–sixth centuries C.E. Discus, shoulder, and handle fragment. Plain, deeply concave discus. On shoulder: inconsistently spaced ovules closest to double ridges around discus; outer band of thick radials around band of ovules. Complete, standing triangular handle. Soft powdery fabric. Clay color: 2.5Y 8/2 white. Slip color: 5YR 6/6 reddish yellow. Slip traces on discus and shoulder. Inclusions: gray, well-rounded quartz sand (VFS, HF). Hardness: 2.0. Good condition. No burning. Mold-made.
206. Reg. no. 85.1195.7L.3. Not shown. GD = 4.6 cm. Incomplete, 10% preserved. "Sepphorean" discus. Fourth–sixth centuries C.E. Discus and shoulder fragment. On shoulder: broad in style, a concentric ovule :: concentric inverted *V* with a single miniscule nodule between each (probably to facilitate removal from mold). Double ridges around discus. Soft powdery chalky fabric. Clay color: 2.5Y 8/2 white. Slip color: 5YR 6/6 reddish yellow. Inclusions: gray, well rounded. Fair condition. Mold-made.
207. Reg. no. 85.1195.6L.9. Not shown. GD = 3.4 cm. Incomplete, 5% preserved. "Sepphorean" discus, Version 1. Fourth–sixth centuries C.E. Discus and shoulder fragment. Plain, shallow, concave discus. Flat, narrow shoulder with steeply sloping side. On shoulder: a flat band containing a spiraling line with small round nodules in delicate linear relief. Double ridges around discus. Filling-hole trace. Soft powdery chalky fabric. Clay color: 2.5Y 8/2 white. No slip. Inclusions: gray, well-rounded quartz sand (VFS, HF). Hardness: 2.0. Good condition. No burning. Mold-made.
208. Reg. no. 85.1195.4L.1. Not shown. GD = 7.6 cm. Incomplete, 45% preserved. "Sepphorean" discus, Version 1. Fourth–sixth centuries C.E. Discus and shoulder fragment. Shallow concave discus (4.6 cm diameter). On discus: a flower blossom with six connected petals fashioned in a delicate linear style; a single small circle decorates inside of each petal; two lower petals are degenerate and are poorly executed, which suggests a mistake made by the mold-carver. Flat, narrow shoulder with steeply sloping side. On shoulder: a flat band (1.0 cm wide) containing a single row of evenly spaced, small, round nodules in delicate linear relief. Double ridges around discus. Filling-hole 0.8 cm diameter. Soft powdery chalky fabric. Clay color: 2.5Y 8/2 white. Slip color: 5YR 5/6 yellowish red. Slip splashes on part of shoulder and on interior surface. Inclusions: gray, well-rounded quartz sand (VFS, HF). Hardness: 2.0. Fair condition. Friable and flaking. No burning. Finger impressions. Mold-made.

Type 21. Jerash with Zoomorphic or Tongue Handle (cat. no. 209)

The Jerash slipper lamp with zoomorphic handle is one of the most distinct types, especially the examples with a highly curved handle terminating in an animal's head and manufactured in Arabia. The head could be that of a horse, ram, or goat (Rosenthal and Sivan 1978: 139). But the handle head of a Jerash lamp found at House 119 at Umm al-Jimal in northern Jordan has another animal: it is fashioned in the shape of a bat's head. The association of bats with light makes sense: both function in darkness. A lit lamp is like a bat that "sees" in the dark (Lapp forthcoming b). While some lamps have a channel leading from the filling-hole to the wick-hole, others do not. In some examples, a cross is found on the lamp where the base of the handle joins the upper body. Each quadrant typically contains a raised circle with internal, well-centered round nodules. That several from Beth Shean (Hadad 2002: 68–71, nos. 301, 309) and Pella are decorated with crosses suggest that this lamp type was marketed to a Christian clientele in these Decapolis cities (Lapp 2007: 372–77). No trace of a cross is apparent on the Sepphoris example. All patterns and inscriptions on this lamp type are in high relief.

The Jerash lamp found at Sepphoris belongs to either the zoomorphic or tongue-handle type. In addition to the handle, other features of Jerash lamps include: a slipper-shaped body, upper shoulders usually decorated with thin consecutive parallel lines, a large filling-hole surrounded by a ridge, and a wide flat almond-shaped base. A scallop pattern with round nodules surrounds the outer filling-hole ridge of some examples, while in others two opposing, elongated curled volutes decorate the pointed nozzle end of the lamp between the filling- and wick-holes (Rosenthal and Sivan 1978: 139). Jerash lamps have varied bases; many have a small ring base with two or four pairs of volutes or a flat base that follows the outline of the lamp (Day 1942: 64–79, pls. 13–14). The flat bases of some examples are decorated with images: for example, one from Qasr Hallabat in the eastern desert of Jordan portrays an amphora containing fish (Bisheh 1982: 134–36). A ridge typically delineates the almond-shaped base.

The chief production center for the Jerash lamp type was Gerasa/Jerash in northern Jordan. Evidence for a workshop—a mold and lamp fragments—was discovered at this Decapolis city (Abu Dalu 1993: 23–34; Pierobon 1983: 95–97; Schaefer and Faulkner 1986: 431; Kehrberg 1989: 85–97, fig. 4.23). A mold used to manufacture the Jerash lamp belonging to Scholl Group II with tongue handle was found in a sounding under the narthex pavement of the Church of Bishop Marianos (Gawlikowski and Musa 1986: 145–47, 160, pl. 5.B). Impressions in the mold indicate that a cross image would have been part of the completed lamp, thus suggesting that its manufacture had been intended for a Christian clientele.

A major lamp workshop for the Jerash lamp was also discovered on the grounds of the hippodrome at Jerash (Kehrberg 1989: 85–97). Jerash bowl fine wares with distinctively painted Christian motifs were also produced at the hippodrome, and six sherds were analyzed using dcp-oes and petrography (Lapp 2004). Chemical analysis detected a high abundance of silica in the samples' fabrics, explained by the high percentage of quartz identified under the petrographic microscope (Lapp 2001a: 131–33, tables 2–3, figs. 5–10). Similarly, proto-induced X-ray emission (PIXE) analysis of Jerash lamp samples collected at the hippodrome indicate high silica abundances (da Costa 2003: 2.637–38, nos. 3021–28).

Molded Kufic inscriptions occur on some examples of the Jerash lamp on the upper shoulder and/or on the lower part of the lamp's reservoir. A unique feature of the inscriptions is that they typically identify the lamp's place of manufacture—Jerash—and the name of the lamp maker. For example, a Kufic inscription found on a Jerash lamp uncovered in Damascus identifies Jerash itself as the lamp's place of manufacture. "Yazid bin Suhayb" and "Masyad" are the artisans; and the date of manufacture is the year 94 of the Hegira, making it one of the earliest such lamps discovered (al-Khouly 2001: 194–95, *Lampe* 4, fig. 4). The majority of lamp makers' names are Arabic (with the exception of Astur and Theodore): Badur, Jayrun, Dawud, Hassan, Ibn Hassan, Busr, Ibn Hudayg, and Amar (al-Khouly 2001: 193). Four Jerash lamps decorated with inscriptions—two of which date to the early eighth and early ninth centuries C.E., respectively—are found in the al-Sabah Collection of the Kuwait National Museum (Watson 2004: 104–5, cat. Aa.14).

But Jerash may not have been the only city where this type of lamp was manufactured. A clay mold for this type was also found at Beth Shean/Scythopolis (Hadad 1997: 170, fig. 40; Hadad 2002: 127–28, nos. 500–501), suggesting that Jerash also may have been a production center. More likely, however, is that the mold is evidence for the marketing of Jerash lamp molds by Jerash mold makers to lamp manufacturers operating in Beth Shean. A comparative examination of the fabrics of examples of this lamp type from several different sites using NAA or DCP-OES with petrographic thin-section analyses would better determine whether more workshops had manufactured this lamp type.

Da Costa (2001: 1.246) observes that a deposit from Pella, with a nearly complete example from the 749 C.E. destruction, confirms the date of the Jerash lamps: the last quarter of the sixth century C.E. She points out that Jerash lamps together with the large "candlestick" type were the chief lamps used in northern Jordan in the Umayyad period. In her detailed discussion on the dating of Jerash lamps with zoomorphic handles, da Costa (2003: **vol. 1.1**, 192, 196) suggests a chronological range from the mid-(?) late sixth to the ninth centuries C.E. That the production center for the large candlestick lamp was located in the Jerusalem area, and the production of the Jerash type was in Jerash suggests interaction between the cities in the Umayyad period. Smith dates two complete Jerash lamps recovered from the North dependency (Locus 70) and the crypta of the odeum (Loci 24–25, 41) to the first quarter of the eighth century C.E. (Smith and Day 1989: 114, pls. 55.6 and 60.1).

Few Jerash lamps have been discovered in Palestine. For this reason, the fragment recovered at Sepphoris and the fragments from al-Bassa, Capernaum, Beth Shean/Scythopolis, and Hammat Gader (Iliffe 1934: 85, fig. 2; Bagatti 1963–64: 267–70, fig. 3.5; Hadad 1997: 169–72, figs. 37–40, Umayyad-period type 1; 171 n. 154) are important indicators of contact between these sites and the geographical extent of the Jerash lamp's westward distribution. The Jerash lamp sherd at Sepphoris raises the possibility that this type may be found at other Galilean sites.

In the Hauran, Jerash lamps were found at Busra and Umm al-Jimal (Wilson and Saʿd 1984: 63, figs. 22–25; Lapp 1995: 437–42, figs. 1–2, 7.1–4; Lapp forthcoming b). In Transjordan, Jerash lamps have been unearthed at Amman, al-Bassa, Gerasa/Jerash, Mount Nebo, and Pella (Lapp 1995: 441, no. 1, under "parallels"). One example has been found as far south as the legionary fortress of el-Lejjun in central Jordan (Parker 2006: 349–30, no. 374, fig. 16.74). That it was recovered from a well-stratified context from the church at the site may suggest that examples of this type were context-specific, meaning that they were used in churches and other Christian contexts. This and other examples have a cross on the shoulder where the handle attaches to the body of the lamp, indicating that makers of this type were target-marketing a Christian clientele (Lapp 2007: 372).

209. Reg. no. 84.1092.1L.1. GD = 2.6 cm. Incomplete, 5% preserved. Jerash with zoomorphic handle. Mid–(?)/late sixth to ninth centuries C.E. (da Costa 2003: 1.192, 196). Shoulder fragment with remnant of nozzle-bridge and filling-hole. On nozzle-bridge: remnant of thin and pronounced sharply curving volute. On shoulder: radials and a single band of scallop pattern around outermost filling-hole ridge. Thin, prominent double ridges around filling-hole. Clay color: 7.5YR 4/2 brown–dark brown. No slip. Inclusions: orange quartz sand (HF, VFS). Hard fired. Good condition. Mold-made. No burning. Origin: Probably Jerash, Jordan, but molds of the same type found at Beth Shean, Israel. Parallels, shape-type: Busra (Wilson and Saᶜd 1984: 63, figs. 22–25); Pella (da Costa 2003: 1. tongue handle group, nos. 521–26; zoomorphic group, nos. 528–36; probably zoomorphic, nos. 537–41); Umm al-Jimal (Lapp 1995: 437–38, 441–42, figs. 1–2, 7.1–4); el-Lejjun (Parker 2006: 349–50, no. 374, fig. 16.74).

Type 22. Slipper, Hadad Type 30 (cat. nos. 210–12)

Late Byzantine lamps served as prototypes for Early Islamic lamps, which retain many of the same characteristic features: a slipper- or almond-shaped body with a pointed nozzle; a small, circular, sunken discus with gradually sloping sides and a small filling-hole positioned in the center; and a degenerate, pyramidal or knob handle, so small that it serves a decorative purpose rather than a functional one. Some lamps have no handles. They are fashioned in a light-brown clay and are usually covered in a brown–dark brown slip (7.5YR 4/2). The base is flat and almond-shaped. No mold- or lamp maker's mark occurs. The decorative scheme employs simple patterns of floral, geometric, or zoomorphic motifs that cover the entire surface, exemplifying the aesthetic concept of horror vacuii, whereby no space is left undecorated. Human representations are seldom found on lamps of this type or, for that matter, on lamps of the Islamic period (Rosenthal and Sivan 1978: 130). Flower blossoms framed by spiraling vines decorate the shoulder of a lamp from Sepphoris (cat. no. 211; see Lapp 1996: 223, no. 119). Even the channel leading from the wick-hole to the filling-hole is ornamented with a vine tendril. Lines radiate from the central filling-hole. The channels of two lamps from Nabratein are also decorated with spiraling vines (Lapp 1996: 223, no. 120; Lapp 2009: 257, 282–83, pl. C.16–17). Lamps of this type are common only in northern Israel, specifically Galilee. Numerous examples have been recovered from Catacomb 24 in the Jewish necropolis at Beth Sheᶜarim (Hadad 2002: 72 n. 258). At Beth Shean, five fragments of this type were found (Hadad 2002: 70–72, nos. 313–16). The lamps range in date from the sixth to mid-seventh centuries C.E. (Hadad 2002: 72).

At Sepphoris, the shoulder of a lamp is decorated with intricately fashioned lions on the prowl and a flanking tree (cat. no. 212). The intricate geometrical and floral patterns decorating the shoulders of lamps of this type reflect and parallel the carved reliefs of Umayyad-period art, particularly on Walid's desert palaces, such as at Qasr Mshatta (Grabar 1987: figs. 120–23; Ali 1999: 42, pl. 16). The use of zoomorphic motifs to decorate the shoulders of the lamps reflect the Umayyads' taste for animate representations further replicated in the Indic-style statues of humans found in the desert palace at Khirbet al-Mafjar (Grabar 1987: figs. 81–83; Ali 1999:

41, pl. 14) and a lion attacking a gazelle depicted in the mosaic in the bath complex there (Ali 1999: 40, 84, color pl. 16).

210. Reg. no. 84.4106b.1L (IAA 95-3895). L = 10 cm. W = ud. H = ud. Complete. Hadad Type 30. Sixth to mid-seventh centuries C.E. Slipper-shaped body with pointed nozzle. Small round concave discus. On discus: consecutive radiating lines. Small central filling-hole. On shoulder: six complete, round flower blossoms (six petals each) framed by spiraling vines. On channel leading from filling-hole to wick-hole: spiraling vine tendril. Fabric color: 7.5YR 4/2 brown–dark brown. No slip. Inclusions: not visible. Hardness: ud. Good condition. Hard fired. Burning around nozzle. Mold-made. Published: Lapp 1996: 223, no. 119.
211. Reg. no. 84.4116b.2L. L = 9.8 cm. W = 7.2 cm. H = 3.7. Complete. Hadad Type 30. Sixth to mid-seventh centuries C.E. Slipper-shaped body. On shoulder, nozzle end: leaping lions attacking prey (missing); stylized palm with three splayed branches separates lion from prey. Plain channel leading from filling-hole to wick-hole. Flat ridge follows body contour. Fabric color: 7.5YR 7/4 pink. No slip. Inclusions: red hematite (LF, CS). Hardness: not measured. Poor condition. Substantial damage to upper half surface: filling-hole; back shoulder and base exhibit conchoidal fractures. Light salts cover surface. Hard fired. Burning around nozzle. Mold-made.
212. Reg. no. 84.5024b.2L.1. GD = 7.6 cm. Incomplete, 20% preserved. Hadad Type 30. Sixth to mid-seventh centuries C.E. Shoulder and channel fragment. Slipper-shaped body. On shoulder: five rows of consecutive horizontal notches and one row (the second) of repeated circles. Decoration in molded high relief. Plain channel leading from wick-hole to discus. Remnant of almond-shaped discus. Clay color: 7.5YR 7/4 pink; exterior surface, 7.5YR 7/6 reddish yellow. No slip. Inclusions: white (LF, VFS); red, hematite (LF, VFS). Hardness: not measured. Poor condition. Heavy burning, interior. Mold-made.

Type 23. Slipper, Hadad Type 36 (cat. no. 213)

Lamps of this type are slipper-shaped or pear-shaped. A channel runs from the wick- to filling-hole and is delineated by pronounced ridge in high relief. The ridge runs continuously around both the wick-hole and the filling-hole. At the back shoulder of the lamp is a standing conical handle. The general appearance of this lamp type is heavy patterns created by wide and prominent linear patterns in high relief. Thick-walled and heavy in weight, this was a durable lamp. The clay fabric color is variable, including brown, reddish brown, and gray. The shoulders of all lamps belonging to this type are ornately decorated with geometrical or floral patterns. Examples of geometrical patterns include: semicircles, concentric circles, circles with circular nodules, triangles, wheels, nets, and herringbone (e.g., see Hadad 2002: 82–93, nos. 356–85). Examples of floral patterns include spiraling vines and stylized branches (see, e.g., Hadad 2002: 93–94, nos. 398–401). At Beth Shean, this lamp type was the most common among the Umayyad-period lamps; 37 complete lamps, 16 nearly complete lamps, and 227 fragments were discovered at the site (Hadad 2002: 82). Two molds used for the manufacture of this type were found at Beth Shean, a city where these lamps were apparently produced. Kufic inscriptions mentioning Allāh occur on examples of this lamp type and suggest that mold makers and lamp makers intended to market them to a local Muslim clientele. At Beth Shean, a lamp with the same inscription was found (Hadad 2002: 87, 93, no. 386) and probably originated from the theater workshop at the site. Perhaps those with Kufic inscriptions mentioning Allāh were intended for use in mosques and in Islamic shrines. Other examples are decorated with Christian inscriptions and crosses, suggesting that they were meant for sale to a Christian clientele. A lamp of this type with the Greek letter *phi* (φ) decorating its channel was found at Baalbek, Lebanon (Rey-Coquais 1964: 160, no. 41). The letter *phi* should be interpreted as the Greek word *phōs* (φως), meaning "light." Menorahs portrayed on the bases of several examples found in association with the synagogues at Khirbet Shemaᶜ and Hammath Tiberias indicate that they were also manufactured for the Jewish population (see discussion in Lapp 2009: 260–61).

A distinguishing characteristic of this lamp type is an ornately decorated base, typically geometrical patterns fashioned in wide linear high relief. These include wheels, stylized branches, wheels with circular nodules, vertical lines, and one straight line flanked by two curved lines (see examples in Hadad 2002: nos. 357–60, 363, 366, 370–72, 379, 382–83, 386, 402–13, 416–418). These base-patterns may represent lamp-mold makers' marks or lamp makers' marks. If neither, perhaps they served some apotropaic purpose. The circular or almond-shaped base is typically concave and is delineated by a single, wide, rounded and pronounced ridge. The paucity of animate representations on the lamps indicates strict adherence to Islamic law forbidding the depiction of living beings. However, rare examples decorated with animals have been found at Sede Boqer in the Negev (Nevo 1985: pl. 7.1, 2, 4, 7–10).

Type 23 lamps have a widespread distribution throughout Umayyad Palestine. The molds found at Beth Shean suggest that the city's advantageous geographical location, where a number of roads intersect may explain its widespread distribution. Perhaps the molds were carved in workshops located in Beth Shean and were traded to lamp makers operating throughout Palestine. Future petrographic and trace-element analyses on numerous samples may indicate variation among the clays, suggesting different sources and origins of manufacture. At Beth Shean, Hadad dates this lamp type to the Umayyad period; they initially were manufactured at the beginning of the eighth century C.E. (Hadad 2002: 95). The inscribed lamps should not be dated earlier than the reform of ᶜAbd al-Malik in 696 C.E.

213. Reg. no. 84.4121b.1L. L = 9.1 cm. W = 6.7 cm. H = 3.2 cm. Complete. Hadad Type 36. Beginning of the eighth century C.E. Broad, heavily ornamented body. A single prominent ridge surrounds a medium-sized central filling-hole. Additional ridge defines a channel from filling-hole to wick-hole. Raised degraded chain pattern decorates channel. Broad, steeply sloping shoulder. A single band of consecutive diagonal hatches surrounds channel on shoulder. A second outer band of a single guilloche pattern decorates shoulder. Nozzle integrated into body of lamp. Plain, cone-shaped handle. Small circular wick-hole. Single ring base. Inside ring base: a spider-web pattern in high relief. Clay color: 7.5YR 7/4 dull orange. Hardness: 2.5. Inclusion(s): gray (FS, low frequency); white (FS, very low frequency). Burning on nozzle. Good condition. Mold-made. Parallels: provenance unknown (Bailey 1981:

pl. 60.Q2329); Hama (Papanicolaou Christensen, Thomsen, and Ploug 1986: 36, figs. 13.a–b; 16.h–j; and 17.e–i).

Type 24. Islamic with Large Filling-Hole and Tongue Handle (cat. no. 214)

This lamp is characterized by a large round body with a large filling-hole that differs from the slipper- or pear-shaped forms of lamps belonging to Hadad Types 36 and 37. The remnant of the lamp's nozzle suggests a wide, elongated nozzle without a channel that further distinguishes itself from the latter types. In section, however, this lamp resembles Hadad Types 36–37, and similar to Type 37, has a tongue handle. A face sporting a beard decorates the handle in molded relief. The guilloche decoration of the shoulder corresponds to similar geometrical patterns occurring on the latter Hadad types. Influenced by the Type 37 form, especially by the tongue handle, this lamp was produced locally no earlier than the end of the eighth or early ninth centuries C.E. (Hadad 2002: 105).

214. Reg. no. 85.1001.2L.1. L = 8.2 cm; W = 6.6 cm; H = 3.1 cm. Nearly complete, 90% preserved. Islamic with large filling-hole and tongue handle. Late eighth or ninth century C.E. Body fragment with complete filling-hole and part of shoulder and handle. Missing nozzle. On nozzle-bridge: remnant of a circle with a round nodule inside. On shoulder: a single band of guilloche with round nodules inside each circle in molded relief. On handle, in high relief (but not visible on drawing): a face with a beard. Wide, prominent double ridges around large filling-hole (3.1 cm diameter). Round base, slightly convex (4.4 cm diameter). Clay color: 7.5YR 5/0 gray. No slip. No visible inclusions. Well-levigated fabric. Hard fired. Poor condition. Highly friable base. Mold-made. No burning.

3. Religious Lamp Art

Jewish

Clay lamps contribute to our understanding of ancient religious art. They are one of several forms on which Jewish symbols are found at Sepphoris (fig. 2).[9] In 1996, two Jewish lamps were exhibited in the *Sepphoris: Crosscurrents of Culture Exhibition* held at the North Carolina Museum of Art in Raleigh: one belongs to a regional Syrian type (Lapp 1996: no. 117), and the second to the Caesarean discus group (Lapp 1996: no. 118). Two additional lamps with Jewish symbols build upon the repertoire of lamps and other pottery depicting Jewish symbols found at the site.

Menorah with Torah Shrine. The haphazard linear execution of the menorah and *lulav*, and large round concave discus suggest a Caesarean discus type identification for the first lamp (fourth–fifth centuries C.E.; cat. no. 180). A menorah and a

9. Menorahs have also been discovered at Sepphoris on ceramic forms (C. Meyers forthcoming), a tombstone, a bulla, an architectural element, and mosaics (Meyers and Meyers in press).

Torah Shrine are portrayed on the second lamp (cat. no. 193), which is a type made in Galilee, perhaps even at Sepphoris. The stylistic treatment of the Jewish symbols—a freehand, sketch-like graffito style—indicates a more personal Judaism, a Jewish folk art as opposed to the expertly carved, precisely rendered, almost mechanically executed menorahs decorating the disks of the mass-produced and factory-made Carthaginian form or the ornate and intricate menorahs of the Alexandrian type (Lapp in press a).

Although dissimilar in style, the combination of the menorah and Torah Shrine on the lamp reflects the Torah Shrine flanked by two menorahs depicted on the Sepphoris synagogue mosaic. The mosaic Torah Shrine with closed paneled doors closely parallels the shrine found on a Jewish lamp recovered at Miletus (Lapp 1991: 156–58, pl. 8a–d; Hachlili 1998: 364; 2000: 154 n. 82). This underscores the interweaving and consistent representation of Jewish symbols on small objects and on architectural elements throughout the ancient Mediterranean. In this respect, Jewish art at Sepphoris corresponds to Jewish art found in the Diaspora. The lamp belongs to a Late Roman–Early Byzantine round-bodied form, likely dating to the fourth–fifth century C.E. Because Sepphoris was an important center for Jewish learning from the second to fourth centuries C.E., it is not surprising that lamps with menorahs from the fourth to fifth centuries C.E. (Late Roman–Early Byzantine periods) occur at the site.

In ancient Jewish art, 23.9% of the archaeological materials on which the menorah is found are clay lamps: out of the 979 archaeological objects decorated with the menorah, 184 are clay lamps and 5 are bronze (Hachlili, 2009: 441–77). Yet, given the quantities of ancient lamps found throughout the Near East, Mediterranean, and Europe, the number decorated with the menorah is comparatively low and underscores the importance of the examples unearthed at Sepphoris.

Torah Shrine. A highly stylized Torah Shrine with closed cabinet doors flanks a menorah depicted on such a Galilean discus lamp fragment (cat. no. 184). The doors are rendered as concentric squares and the aedicula as a triangle. Similar aediculae are found on two other lamp sherds from Sepphoris (cat. nos. 182–83) and are also suggestive of Torah Shrines. Aediculae on several lamps discovered at the synagogue at Chersonesos in the Pontic region (Žuravlev 2012: 26–28, fig. 8.1–12) probably represent Torah Shrines, considering their context. Noteworthy is the highly stylized, linear, sketch-like style of the Chersonesos lamp shrines, which compare favorably with the stylized, graffito-like aediculae on the Galilean discus, Caesarean discus, and "Sepphorean" discus examples from Sepphoris. The round demi-dome gable and columns with spiraling flutes of the shrine portrayed on another Sepphoris lamp (cat. no. 185) resembles images of the Torah Shrine found in ancient Jewish art (e.g., the synagogue mosaic at Hammath Tiberias; Dothan 1983: pl. 27.1).

Christian

Christian lamp art at Sepphoris is conservative, the most common motif being a basic cross. Although the Duke University team did not discover any Christian lamps, the Hebrew University expedition at the site recovered a Jerash lamp with a cross decorating the lower handle (Gärtner 1999: no. 99, pl. 21), a lamp handle in the shape of a cross (Gärtner 1999: no. 71, pl. 16), and a cross on the shoulder of a candlestick slipper lamp commonly associated with Christian holy sites (Gärtner 1999: no. 94, pl. 20). Manufactured in the Jerusalem area, probably at Jebel Susa north of the city, the candlestick lamp was likely acquired by a Sepphorean Christian pilgrim as a souvenir while visiting a church in the Jerusalem area. Imported terra-sigillata wares stamped with crosses were recovered at Sepphoris (Vincenz 2013: 166, pl. 57.3–4).

Islamic

Islamic lamp art at Sepphoris generally follows the widespread avoidance of picturing animate images, selecting instead ornate geometrical and floral patterns (Grabar 1987: 72–98). A lion, however, is portrayed on the shoulder of a lamp (cat. no. 212) and is reminiscent of a lion flanking a tree on a mosaic in the bathhouse of the Umayyad palace of Khirbat al-Mafjar or two lions fashioned in ornately carved stone relief from the façade at Qasr Mshatta (Grabar 1987: fig. 123). Several Kufic inscriptions decorate Islamic lamps found by the Hebrew University team (Gärtner 1999: 94–95, pl. 25.111), their scripts representing calligraphic arts. A Khirbet al-Mafjar lamp found at the site (Gärtner 1999: Islamic Type 4, 94–96, pl. 24.109) belongs to a group representing a high standard of symmetrical, ornate, and detailed Islamic art.

4. Summary

Most of the lamps recovered from the residential area of the western summit came from local and regional workshops, indicating that Galilean lamp makers provided sufficiently for the lighting needs of the Sepphoris population. Lamps were manufactured in the satellite settlement of Shikhin and were brought to the Sepphoris marketplace and eventually, via the Roman road network, to other markets in Galilee and the Decapolis. Lamp molds unearthed in Sepphoris also suggest lamp production in the city. Regional lamp types represented in the Sepphoris corpus indicate interactions between Sepphoris and Judea. Too few Jebel Jofeh, Jerash zoomorphic, Darom, and Beit Nattif lamps were found on the western summit to suggest that they were traded to Sepphoreans. Rather, they represent residual finds that arrived in the city via travelers: traders and tourists. Palestinian discus lamps were popular among the Jewish residents of Sepphoris despite the mythological themes associated with this group.

CHAPTER 3

Imported Types

1. *Introduction*

Five lamp fragments excavated at Sepphoris originated from western workshops outside Palestine. Stylistic features and fabric characteristics suggest that they were manufactured in Ephesus in western Asia Minor or at Corinth and Athens in Greece. The term *import* is used here for purposes of convention. However, because it implies trade, it does not accurately describe imported lamps at Sepphoris; their low quantity indicates they were not traded or imported from abroad but represent random, isolated examples of lamps used in travel and brought by visitors to the site or acquired by Sepphoreans as souvenirs while abroad. Jewish communities are well attested in the archaeological and epigraphic record at Athens, Corinth, and Ephesus (M. H. Williams 1998: Athens, 1.21, 35, 37(?), 38–39; 2.85; 6.4; Corinth, 1.5, 21, 94; 2.145; Ephesus, 1.32, 90; 2.13, 95, 128; 4.23, 26). The discovery of lamps from a Jewish context on the western summit at Sepphoris suggests that Sepphorean Jews acquired them while engaging in trade in Greece and Asia Minor, visiting family members, or as tourists. Alternatively, the lamps may suggest that Jews or others visited Sepphoris from these Mediterranean lands.

2. *Typology*

Type 25. Ephesus, Broneer Type XIX, Howland Type 49A (cat. nos. 215–16)

A spatulated nozzle fragment (cat. no. 215) and a broken lamp body (cat. no. 216) are all that remain of an Ephesian type lamp found in the Sepphoris excavations. The elongated and splayed nozzle corresponds to the pointed form of Howland Type 49A (Howland 1958: nos. 649–64, pl. 49). A similar nozzle from Masada dates from the first century B.C.E. to the early first century C.E. (Bailey 1994: 86, no. 169). The mold-made lamp belongs to Broneer Type XIX and is characterized by a biconical-shaped oil chamber fashioned in a gray fabric covered in a black glazed slip. An inverted wheel-made collar typically surrounds a small flat discus with a large filling-hole. Three small air-holes pierce the discus. A ring handle is attached to the collar and shoulder of such lamps. The shoulder is typically decorated with alternating leaves in relief. Three leaves are preserved on the shoulder of the Sepphoris body fragment. The base is flat and oval in shape.

A lamp of the Ephesus type dated to about the second century B.C.E. was found at Samaria–Sebaste (Crowfoot, Crowfoot, and Kenyon 1957: fig. 87.5; see also Smith 1964b: 112, fig. 8). An example dating to the first century B.C.E. was found at Ashkelon (Johnson 2008: 127–28, no. 381). Three fragments from Caesarea Maritima date from the mid- to second half of the second century B.C.E. to the first century B.C.E. (125–75 B.C.E.), with some later variations in the Early Roman period (Sussman 2009: 215, nos., 5–7). Although the discovery of large quantities of lamp fragments of this type in and near Ephesus suggest a production center there, molds for this type have also been discovered in Cyprus, Pergamon, and Thasos (Bailey 1975: 90–91). Thus archaeometric analyses of the Ephesian lamp fragments found at Sepphoris, Ashkelon, Caesarea Maritima, and Samaria–Sebaste may someday reveal a local Palestinian manufacturing center as well. A body fragment of a similar Ephesus type lamp was uncovered in the Sanctuary of Demeter excavations in Corinth, Greece (Slane 1990: 11, 26, no. 11, pl. 1). Slane (1990: 26) suggests that it may date to the first century C.E. Hayes (1980: 15, nos. 55–56) dates an Ephesus type lamp in the collection of the Royal Ontario Museum from the late second century B.C.E. to the early first century C.E. The Royal Ontario Museum lamp, however, is "not in the micaceous gray fabric typical of the finds from Ephesus, but in a smoother gray ware which most closely resembles that of Knidos" (Hayes 1980: 15).

215. Reg. no. 85.3563.1L.2. GD = 3.5 cm. 10% preserved. L = 6.6 cm. Greatest preserved W = 4.9 cm. Body type: Ephesian; Broneer XIX. Nozzle type: Howland Type 49A. Origin: Ephesus, Turkey. Late second century B.C.E. to early first century C.E. (Hayes 1980: 15); First century B.C.E. to early first century C.E. (Bailey 1994: 86). Nozzle fragment. Flat, smooth upper surface. Two tapered outer edges. Wick-hole width = 1.4 cm (length not possible). Clay color: 10YR 6/1 light gray-gray. Slip color: 2.5Y 3/0 very dark gray. Very well levigated clay. Inclusions: platty muscovite mica flakes (VLF, VFS). No burning. Good condition. Mold-made. Exact nozzle shape-types: Masada (Bailey 1994: 86, no. 169); Samaria–Sebaste (Crowfoot, Crowfoot, and Kenyon 1957: 370, no. 5, fig. 87); Oboda (Sussman 2009: 66–70, 157, nos. 349–50, dated to the end of third century B.C.E. to the first century C.E.); Tel Mor (Sussman 2009: 66–70, 157, no. 351, dated from the end of third century B.C.E. to the first century C.E.); Samaria (Sussman 2009: 66–70, 157, no. 352, dated from the end of third century B.C.E. to the first century C.E.); Corinth (Slane 1990: 11, 26, no. 11, pl. 1); Royal Ontario Museum, unknown findspot (Hayes 1980: 15, no. 55, pl. 7); Adler Collection (Adler 2004: 13, nos. 69–70).

216. Reg. no. 85.1197.2L.2. Body fragment of reg. no. 85.3563.1L.2 (cat. no. 215, above). GD = 6.7 cm. Incomplete, 25% preserved. Body type: Ephesian; Broneer XIX. Origin: Ephesus, Turkey. Late second century B.C.E. to early first century C.E. (Hayes 1980: 15). Collar neck with prominent rim and filling-hole complete. Part of nozzle-bridge and upper shoulder. Collar-neck rim 4.4 cm diameter. Three air-holes (0.1 cm diameter). Remnant of channel leading to wick-hole visible. Filling-hole (1.3 cm diameter). Formed by volutes with unpreserved tips. Eight round nodules

evenly spaced (approximately 1.0 cm apart) around shoulder and close to collar join. Remnants of elongated flower petals with inner parallel striations below and/ or equal to line of nodules. Clay color: 10YR 5/1 gray. Slip color: 2.5Y 3/0 very dark gray. Very well-levigated clay. No visible inclusions. No burning. Poor condition. Severe weathering of fabric inside collar neck. Heavy salts on surface. Mold-made. Similar shape-types: Masada (Bailey 1994: 86, no. 169); Corinth (Slane 1990: 11, 26, no. 11, pl. 1); Musée de la Fondation Piérides (Oziol 1993: 29, no. 22, fig. 5); Royal Ontario Museum, unknown findspot (Hayes 1980: 15, no. 55, pl. 7:55).

Type 26. Corinthian, Broneer Type XXVIIC (cat. no. 217)

Corinthian type lamps belonging to Broneer XXVIIC are characterized by a round body with a central discus. The Sepphoris example exhibits two diagnostic features of lamps of this group: the unmistakable pale yellow, calcium carbonate rich and well-levigated fabric; and the flat pentagonal shaped nozzle. Imitations of the Corinthian vine-and-ray lamp were also manufactured in Egypt (Bailey 1988: 252, Q2070, fig. 133, pl. 44). Bailey reports that the majority of the Broneer XXVIIC lamps date to the second half of the second century and the third century C.E. (Bailey 1988: 402).

217. Reg. no. 85.3037b.3L.3. GD = 2.9 cm. Incomplete, 15% preserved. Corinthian, Broneer Type XXVIIC. Origin: Corinth, Greece. Most date to second half of the second century and the third century C.E., see Bailey 1988: 402. Part of shoulder, discus, and nozzle. On shoulder: impressed acanthus leaf band. On discus: three petal ends with round tips of a rosette. Polygonal-shaped nozzle-head. Rounded ridge (0.2 cm width) around discus. No slip. Clay color: 10YR 8/6 yellow. Very finely levigated marl clay. Soft and powdery to the touch. No visible inclusions. Good condition. Mold-made. No burning. Parallels, shape-type with impressed acanthus leaf pattern on shoulder and flat pentagonal nozzle: British Museum (Bailey 1988: 402, 405, Q3261, pl. 117, dated to first quarter of the fourth century C.E.).

Type 27. Athenian, Broneer Type XXVIII, Later Version (cat. nos. 218–19)

Two fragments belonging to two separate lamps of Athenian origin were found at Sepphoris. Both sherds belonged to lamps of the later version of Broneer XXVIII, which is characterized by oval-shaped bodies and panels. These lamps have a standing grooved handle. They are typically made of a hard, well-fired fabric of dark red or brown. The fabrics often contain mica. The Sepphoris examples of this group belong to the same type as the mid-fourth century lamp decorated with a tree of life, perhaps suggestive of a menorah, found in the Athenian agora (Perlzweig 1961: 133, no. 1065). Five fragments of this type were found at Beth Shean (Hadad 2002: Imported Type 7, 133–34, nos. 516–17). The example found at Sepphoris resembles no. 517 from Beth Shean, not no. 516. Perlzweig (1961: 17–64) dates this type of lamp to the third–fourth centuries C.E., but her chronology has been corrected and revised by Karivieri (1996: 51–52; see example on pp. 232–33, no. 236), who dates them to the late fourth to early fifth centuries C.E.

218. Reg. no. 85.1105.1L. GD = 6.4 cm. Incomplete. Athenian, Broneer Type XXVIII, Later version. Origin: Athens. Late fourth to early fifth centuries C.E. Shoulder, backside of shoulder, and complete handle fragment. Two thin grooves (5.0 mm in width) running parallel on handle. Standing, unpierced handle with tail that extends to backside. Flat shoulder. A groove runs along outline of body 7.0 mm inward, oriented toward body center. Clay color: 5YR 6/4 light reddish brown. Slip color: 5YR 5/4 reddish brown. In section, gritty fabric sharp to the touch and medium frequency of elongated voids. Brickish spongy fabric. Inclusions: white sub-rounded (VLF, VFS). Fair (–) condition. Mold-made. Parallels, shape-type only: Beth Shean (Hadad 2002: Type 7, 133–34, especially no. 517, and not no. 516 which is an earlier version). Athens (Perlzweig 1961: 133, no. 1065, pl. 22; Karivieri 1996: 206, no. 153, pl. 14, recovered at Kerameikos); and Cnidus (Bailey 1988: 410, no. Q3279, figs. 62, 144, pl. 119, dated to first half of the fifth century C.E., recovered from the Peninsula Tomb and currently in the British Museum).
219. Reg. no. 85.3100.3L. Preserved L = 4.6 cm. Preserved W = 5.3 cm. Incomplete, 15% preserved. Athenian, Broneer Type XXVIII, Later version. Origin: Athens, Greece. Late fourth to early fifth centuries C.E. Shoulder, backside, and complete handle fragment. On handle: three parallel grooves (1.5 mm in width). Standing, unpierced handle with tail that extends to backside. Flat shoulder. On shoulder: a band of horizontal notches, 4.0 mm inward from edge. Clay color: 2.5YR 6/6 light red. No slip. Inclusions: gray (LF, VFS). Poor condition. Very heavy salts on body surface. No burning. Mold-made. Parallels: Beth Shean (Hadad 2002: Type 7, 133–34, especially no. 517, and not no. 516, which is an earlier version); Athens (Perlzweig 1961: 133, no. 1065, pl. 22; Karivieri 1996: 206, no. 153, pl. 14, recovered at Kerameikos); and Cnidus (Bailey 1988: 410, no. Q3279, figs. 62, 144, pl. 119, dated to first half of the fifth century C.E., recovered from the Peninsula Tomb and currently in the British Museum).

3. Summary

Few imported lamps were excavated at Sepphoris. Only five lamp fragments of foreign manufacture were unearthed by the Duke team (cat. nos. 215–19), and an additional six were found by the Hebrew University excavations (Gärtner 1999: nos. 115–20, pl. 26). Similarly, a small number of imported lamps were found at the seaport of Ashkelon; they came from Italy, Asia Minor, Egypt, and North Africa (Johnson 2008: 127–31, nos. 380–89). The paucity of imported lamps at Sepphoris indicates that they were not marketed to the local inhabitants but, rather, represent random isolated finds. Their occurrence at the site suggests that traders, tourists, and other visitors from Turkey and Greece brought them to use while on their journey to Sepphoris or perhaps as gifts to colleagues, friends, or family. The lamps may also be souvenirs purchased by Sepphorean travelers to Athens, Corinth, and Ephesus. Archaeological, epigraphic, and literary evidence indicate the presence of Jewish communities in Athens, Corinth, and Ephesus (M. H. Williams 1998: Athens, 1.21, 35, 37(?), 38–39; 2.85; 4.4; Corinth, 1.5, 21, 94; 2.145; Ephesus, 1.32, 90; 2.13,

128; 4.23, 26). Thus the occurrence of these lamps at Sepphoris may represent connections between members of the Jewish community at Sepphoris and the other cities. However, their paucity indicates that links between Sepphoris and those cities were neither consistent nor strong. The presence of foreign lamps from Greece and Asia Minor at Sepphoris, however, further complements the recovery of significant numbers of Late Roman and Byzantine fine-ware sherds from Italy, Cyprus, Egypt, and North Africa (Balouka 2013: 50–51; Vincenz 2013: 142–45, 162–67). Together, the imported lamps and fine wares support the perspective that Sepphoreans interacted with the greater Mediterranean world of late antiquity.

The general absence of Italian imports is noteworthy considering the presence of Roman soldiers of the Sixth Legion stationed at nearby Leggio. At Masada, for example, 31 Italian imports were found (Bailey 1994: 79–85, nos. 138–68), and their occurrence there can be attributed to the Roman soldiers who had sieged the cliff-side fortress. The distinctive and enormously popular Carthaginian lamps of North Africa are also generally absent at Sepphoris, although a fragment was found by excavations conducted by the Hebrew University of Jerusalem (Gärtner 1999: 106–7, Imported Type 5, no. 120, pl. 26; see also Hadad 2002: 135 n. 32). Their paucity at Sepphoris is comparable to their overall rarity at sites in Palestine: at Beth Shean, for example, only two sherds were found (Hadad 2002: imported type 8, 133–35, nos. 518–19). They tend to be found in more-significant quantities in the south, such as at Gaza and in the Red Sea port city of Roman Aila (Lapp forthcoming a).

CHAPTER 4

Petrographic and DCP-OES Analysis of Lamps from Sepphoris and the Decapolis

1. Introduction

Ancient potters provided light sources by forming clay into oil-containing, wick-holding lamps. The manipulation of light also plays an important role in modern times for determining the origin of ancient oil lamps: in order to find out where lamp makers collected their clays and manufactured their lamps, archaeologists rely on the optical properties of light. The polarizing petrographic microscope and the optical spectrometer, for example, are instruments used to identify clay sources. The arrangement of atoms within any given mineral is unique to that mineral type and will exhibit differences in plain or polarized light. The petrographic microscope manipulates light by refracting it. As light passes through the planes of the mineral, one can identify the types of nonplastic mineral inclusions found in a clay lamp fabric and relate them to their geological environments. Micromorphological features such as cracks, voids, and vapor veins can also be observed.

Different clay sources located in proximity to one another can exhibit similar petrological characteristics in thin-section. As a result, the polarizing microscope alone cannot establish the exact provenance of ceramic wares. Thus a more effective means for identifying the provenance of lamps is necessary: major- and trace-element analyses. Neutron activation analysis (NAA), inductively coupled plasma emission spectrometry (ICPS), and particle-induced X-ray emission or proton-induced X-ray emission (PIXE) are commonly used for trace-element characterization of pottery. Trace-element analysis is based on the premise that the chemical composition of one clay source can be distinguished from others. In order to determine the provenance of clay lamps, one must find a compositional match with reference material (e.g., pottery, wasters, or raw clay) recovered from a site where a strong case for local manufacture can be made. The matching of the compositions of clay lamps and reference pottery is based on the analysis of as many major and trace elements as possible, for the abundance of a single element, or a few, may be the same in completely separate clays.

2. *Short History of Research*

The archaeometric analysis of clay lamps is a relatively recent development. When I initially embarked on my archaeometric analysis of clay lamps in the early 1990s for my 1997 dissertation, few petrographic and chemical analyses had been conducted on lamp fabrics of any chronological period in Israel and Jordan. Similarly, few characterization studies had examined lamps recovered from sites elsewhere in the ancient world (Allen 1993: 326; Hughes 1988: 461–85). Trace-element analysis was the first method used to establish clay sources of vessels in Israel and Jordan. In the early 1980s, Asaro conducted NAA on a single lamp-nozzle fragment among other ceramic wares recovered at Meiron in Israel (Asaro 1981: 141–43, samples "Meir–9," "Meir–11," table 4). The lamp sherd belonged to the Herodian type with a black slip. Later on, Gunneweg and Perlman (1984–85: 79–83) analyzed the chemical compositions of 6 lamp fabrics of the Herodian type from the Jerusalem area, including 3 from Jerusalem and 3 from Avdat (Oboda). ʿAmr (ʿAmr 1987: 29–37, table B.4, pls. 10–21, sample nos. PL1–60) conducted NAA on 60 lamp samples unearthed in Petra. Yellin (1994: 107–24) analyzed 27 samples belonging to the Judean radial, Herodian, and molded lamps with floral decorations types from Masada, the Cave of Horror, and Aroer. It was not until 1987 that a lamp was included in a petrographic thin-section analysis: an Iron Age example from Tel Michal (Goldberg, Singer-Avitz, and Horowitz 1987: 265, no. 11, fig. 9.11.8; without photomicrographs). Another petrographic thin-section analysis on an Iron Age sample from Ḥorvat Teiman appeared almost a decade later (Goren 1995: 206–7, no. 3; without photomicrographs). This lack of interest can be explained in part by a preference for coarser-grained wares and the assumption that finer-grained fabrics such as sigillata and lamps would yield little in terms of inclusions and micromorphological features (see discussion in D. F. Williams 1979).

3. *Purpose*

The purpose of this chapter is to compare the petrographic (micromorphological) and chemical compositions of clay oil-lamp fabrics from Sepphoris and from other sites in Galilee, the coastal plain, and the Decapolis. Forty-three lamp samples belonging to the Palestinian discus, bilanceolate, and Beit Nattif types were examined. Versions of the Palestinian discus type were examined. A sample of a lamp belonging to the Tarsus group, presumably an import, was analyzed. One sample of the Gilead group was also examined. Whereas previous characterization studies had analyzed lamps in the corpus of finds from a single site located in one region, the present analysis represents an interregional survey of lamp fabrics collected from multiple sites located in several geographical areas in Israel and Jordan: the coastal plain, Galilee, the northern Jordan Valley, and Transjordan.

The purpose of this study thus is to determine whether variation exists among the lamp fabrics and whether distinct fabrics correspond to specific lamp types. Were tempering agents added to the fabrics by the lamp makers, and if so, can these nonplastic constituents be linked to a specific geological area? What types of clays were used? No clay lamps of local Roman Palestinian and Arabian provenance had yet been examined petrographically. Recognizing the wide variety of different lamp fabrics found in the region merely by observing the broken sections of lamp sherds with a hand lens, I had good reason to suspect that different petrological and textural features regarding the fine-grained lamp fabrics could be distinguished using the petrographic microscope. DCP analysis of the lamp fabrics was conducted in order to determine the chemical compositions of the lamp fabrics and to determine whether the samples cluster into distinct chemical groups, suggesting the same clay sources and, if so, which fragments match each another. When compared with the petrographic findings, these chemical "clusters" would hopefully aid in the determination of the samples' provenances.

4. *Research Problem*

The most significant shortcoming of this study was the absence of reference material sampled directly from workshops that manufactured Palestinian discus, bilanceolate, and Beit Nattif lamps. No such workshops have been discovered. Reference materials (e.g., lamp wasters, sherds, and clays from settling pools) could then have been analyzed and compared to the lamp samples of this study to help determine their origin. I was aware of this problem from the outset of my project. However, the main purpose was to determine the petrographic characteristics and chemical compositions of the sampled lamps and to compare these findings on a preliminary level.

In essence, the dcp-oes component of this study is a prototype that helps lay the groundwork for future, more extensive analyses. The Jerash bowls were analyzed in part to determine whether any of the lamp samples were chemically similar to the bowls, suggesting perhaps that the lamps had been manufactured at the Jerash hippodrome pottery workshop. If so, future studies with a significantly larger sampling would be necessary to corroborate the preliminary results.

Sample size is another problem. A substantially higher quantity of lamp samples and reference materials must be chemically analyzed to corroborate the initial dcp-chemical composition groups of this study. The dcp-cluster groups of this study (Group A–D) are preliminary, a fact that is reflected by the hatched lines in fig. 13 (see p. 144). Only Group D is represented as a solid line, because only the Jerash bowl samples are reference material collected from a known workshop. An adequate quantity of Palestinian discus and bilanceolate lamp samples have been examined in this study to provide us with an accurate understanding of the general petrographic and micromorphological characteristics associated with each group. The petrographic findings provide solid data on the types of clays and tempers used for the manufacture of the Palestinian discus, bilanceolate, Beit Nattif, Gilead, and Tarsus lamps in additional to micromorphological features such as cracks, voids, and vapor veins. Petrographic examination of additional lamp samples is always encouraged to detect any further variation within each shape-type group.

5. *Methods and Procedures*

Comparative petrography and direct current plasma-optical emission spectrometry (dcp-oes) were adopted as methods for characterizing the lamp fabrics. Direct current plasma-optical emission spectrometry had not yet been used to characterize artifacts. It was therefore necessary for me first to test this archaeometric method by analyzing four Jerash bowl fragments. The DCP analysis of the bowl samples was successful, and DCP was adopted as the analytical technique of this study.

Because petrographic thin-section and DCP analyses are destructive processes requiring approximately 500 mg, or about a 1 cm × 1 cm sample, the selected samples were first drawn and photographed for documentary purposes. Additionally, their respective macroscopic characteristics were noted. All samples were then thin-sectioned for petrographic analysis. Before preparing select samples for DCP analysis, we grouped lamp fragments on the basis of their respective clay fabric types to aid further in the final selection of samples for chemical analysis. The following features were considered in selecting the lamp fragments for DCP characterization: lamp class, decoration and patterning, iconographic elements, lamp makers' marks, clay fabric type, archaeological findspot, and chronology. Shape-type, typological distribution, petrographic, and chemical data received equal consideration when we interpreted the evidence.

Macroscopic Analysis

Macroscopic examination of all samples preceded their preparation for petrographic thin-section and DCP analyses. The term *macroscopic* refers to features visible without the use of a microscope: shape, decoration, iconographic elements, motifs, slip, grits, fabric color, and carbon traces. The lamp fragments were then grouped under their corresponding types based on shape, decoration, and fabric. Descriptions of each sample were recorded (see appendix B). Although not a visual characteristic, the hardness of each sample was also determined using the Mohs scale. Hardness is not a well-defined property, and its limitations as a criterion of classification are well known; but it was nonetheless tested as one more helpful comparative feature (for a discussion concerning hardness and its limitations with respect to pottery, see Shepard 1957: 113–17). Using the international standard Munsell Soil Color Charts (Munsell 1971), we determined the clay fabric color of each sample from a broken edge.[1] If any traces of slip were evident, the color also was noted. The grit (or types of inclusions appearing in the fabric according to color, size, and frequency) was noted. The sorting of inclusions follows the scale defined by Barraclough (in Orten, Tyers, and Vince 1993: 139, 240). Traces of carbon were noted because they indicate burning and use of the lamp.

Petrographic Thin-Section Analysis

Once photographed and drawn, thin sections were prepared from slices of each sample fixed to glass slides with epoxy and abraded to a standard thickness (0.03 mm; Humphries 1992: 53–56). The sections were carefully polished on plates of glass with water and iron grit of grades 200, 400, and 1000.

DCP-OES Analysis

Direct current plasma-optical emission spectrometry (dcp-oes) centers upon the analysis of samples prepared as solutions (fig. 7). It is an effective, cost-efficient, and non-nuclear method for the characterization of archaeological material culture. Following the method outlined by Emily M. Klein (Klein, Langmuir, and Staudigel 1991: appendix B), the samples were ground using a steel tool-alloy set and the resulting grains were sifted. The grains were then cleansed with deionized water and desiccated in an oven at 40°C for 24 hours. Under a binocular microscope, grains contaminated by oxides, slip, ink, and glue (from inventory labeling) or other impurities were removed and saved in a glass vial marked "picked out." The uncontaminated grains were selected and placed in a second glass vial marked "picked." The "picked" grains were then carefully powdered using an agate mortar and pestle. Because it is not known whether or to what extent the ancient lamp makers had levigated their clays, I did not subject the clay samples to any treatment except for standard DCP preparatory procedures. Approximately 0.25 g of powdered sample was weighed into an alumina crucible and heated in a muffle furnace at 800–900°C for 30 minutes, after which the total weight-change due to the loss on ignition (l.o.i.) was determined. This step is necessary to burn off any atmospheric water absorbed by the sample (otherwise, one would be weighing water in addition to the sample itself) and to oxidize iron to Fe+3 for proper standard calibration.

Next, the sample was brought into solution using a lithium metaborate fusion technique. Ultrapure lithium metaborate ($LiBO_2$) flux (0.4+/–0.0004 g) and a 0.1+/–0.0004 g powdered sample of standard were weighed into preignited graphite crucibles (Spex), and the mixture was thoroughly homogenized by stirring. The ultrapure lithium metaborate flux was purchased from Spex, Inc. The mixture was then fused at approximately 1030–1060°C for 13 minutes. The resulting pebble-shaped melt was removed from the oven and immediately poured into a plastic bottle containing 24.0 mL of nitric acid (1 N HNO_3) and agitated until all fused glass had dissolved. This concentrated solution (1/250 dilution) was then ready to be analyzed

1. For a discussion on the use of Munsell charts in measuring color, see Rice 1987: 339–43.

for trace elements as described below. In order to make a more dilute (1/6250) solution for analysis of major elements, one milliliter of this solution was added to each of 13 smaller plastic bottles; 4.0 mL of Red Pot Standard A and "blank" were pipetted into their respective bottles. The weights of these solutions taken were recorded. Twenty-four milliliters of 1 N HNO_3 + 3500ppm Li were added to each of the 13 containers; 96.0 mL of the same were added to the smaller bottles containing Red Pot Standard A and "blank" dilutions, respectively. These solutions were then analyzed for major elements as described below (see tables 8–9, pp. 157, 160).

Once in solution, the samples were ready for DCP analysis. Each solution was aspirated and its aerosol directed into the plasma. Plasma refers to the high temperature region used to excite the electrons in the sample, thus giving off light for spectrometric analysis. As described by Potts (1995: 192–94), the direct current plasma is created by the initiation of an electrical discharge within a stream of argon gas flowing between two electrodes. Once the solution has been nebulized, its aerosol particles experience desolution, vaporization, and atomization. This occurs in the "excitation zone" where the sample temperature ranges between 5700 and 6000° K (fig. 7). At this temperature, the elements' atoms undergo transitions to lower electronic states and emit their excess energy as quanta of light. Because these wavelengths of light are

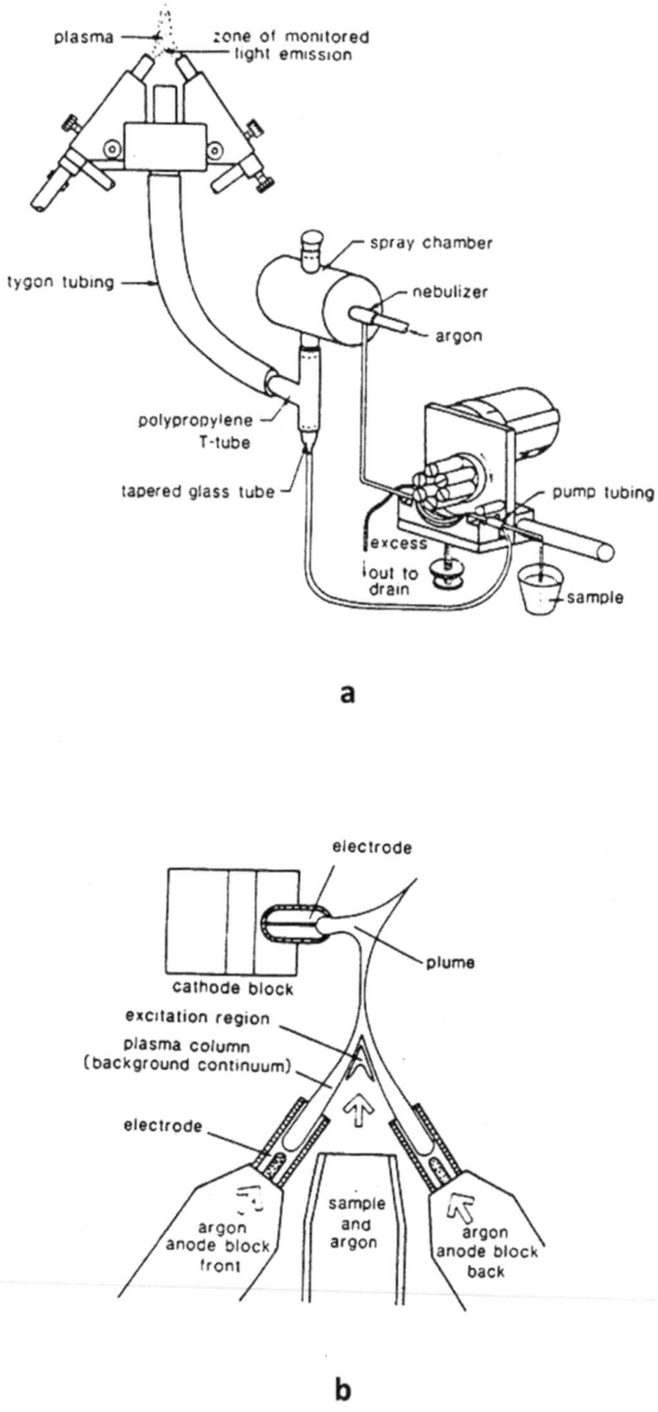

Fig. 7. Diagram showing: (a) principal components of the dcp-oes, and (b) detail of the plasma column and excitation region. After Potts 1995: 193.

specific to the emitting element, the identification of a tested element can be discerned by a spectrometer able to resolve different wavelengths of light. The intensities of emission are measured by the spectrometer, the measurements of which indicate the elemental abundances in any given sample. Samples atomized in a direct current plasma experience significantly lower atomization temperatures than those in the inductively coupled argon plasma of ICP analysis.

Each "run" included 10 lamp samples, 3 rock standards, a blank of fused $LiBO_2$, and a "High Standard" (especially, Red Pot Standard A) used to monitor machine drift throughout the run. The rock standards used for calibration during the study were K1919, IOBC, and IORC. The Red Pot Standard A originates from a crushed flower pot made of Georgia (U.S.A.) red clay. Typically, 0.4000 +/– 0.0004 g of Red Pot Standard A was mixed and fused with 1.6000 g $LiBO_2$ in 96.0 mL of 1 N HNO_3. A "blank" containing only $LiBO_2$ was also fused.

Eighteen elements were determined: 10 major elements as the oxides SiO_2, Al_2O_3, Fe_2O_3, MgO, CaO, Na_2O, K_2O, TiO_2, MnO, and P_2O_5, and 8 trace elements (Cr, Ni, Y, V, Ba, Cu, Sr, and Sc). Because of detection problems encountered early on in the analyses, the values for zircon (Zr) were not determined. The concentrations of elements in each sample are presented in table 5 (see p. 131). For the major elements, the concentration is given as a weight percent (wt%) oxide; for the trace elements, in parts per million (ppm).

6. *Sample Selection*

This study provides a special opportunity to investigate and compare fabric interrelationships between cities and some villages in multiple regions, including the coast, Galilee, and the Decapolis (fig. 8, p. 132; tables 3–4). Forty-three lamp samples (11 shown on fig. 9, p. 133) belonging to the following types were selected for analysis: Palestinian discus (N = 17); Palestinian discus (Abila version) (N = 1); Palestinian discus (Scythopolis version) (N = 1); Tarsus (N = 1); Beit Nattif (northern version) (N = 6), Gilead (N = 2); and bilanceolate (N = 15; pl. 2, tables 3–4, appendix B). The lamp fragments were collected from multiple sites located in a variety of geographical regions (with the findspots of the lamp samples indicated in parentheses): the coastal plain (Ashkelon and Caesarea Maritima); the Galilee (Ḥorvat Ḥazon, Meiron, and Sepphoris); the northern Jordan Valley (Scythopolis); and Transjordan (Abila, Gadara, Gerasa, and Pella). Nine clay samples were also taken from the following sites: Gadara (N = 1), Pella (N = 5), Scythopolis (N = 2), and Sepphoris (N = 1) (table 4). In addition, 4 Jerash bowl fragments recovered from the hippodrome lamp workshop at Gerasa were also sampled and characterized.

Samples exhibiting a wide variety of diagnostic characteristics were selected to ensure diversity: specific types and subtypes (e.g., Palestinian discus, Palestinian discus versions, bilanceolate, Beit Nattif, etc.), unique motif types (e.g., bi-associative and erotic), lamp makers' marks (e.g., Greek initials), and different slipping techniques (e.g., red, brown, and bichrome). Such variation is important in that it in-

Table 3. Findspots and Registration Information of Samples Selected for Analysis

Sample	*Archaeological Context*	*Registration Information*
DAB4	Unavailable	
DA2	Street-front room of house (fill)	Combined find grids
DA3	Street-front room of house (fill)	FG43 (sherd base with Greek *eta*)
DA9	Street-front room of house (fill)	FG43
DC6	Possibly synagogue	A118/3 (sherd base with Greek *eta*)
DHH2	Pit	180.A/20
DM10	Cistern 4 (C-7), lower city, Area MVII	R5236/Spec. 80-385
DP3	Room of house, east cut, Area III, Plot P (pit)	L735
DP11	Room of house, east cut, Area III, Plot P (pit)	L436
DSCY6	Shop no. 11, western stoa (foundation trench)	L2768 (7733/2)
DSCY7	Surface of Area GK (theater and Roman temples)	L40600/B511232
DSCY8	West bathhouse, dressing room (fill on bedrock)	L51168/B511232
DS4	Cistern (residential quarter)	84.1068.13L
DS9	Cistern (residential quarter)	84.1068.10L
DS10	Cistern (residential quarter)	84.1068.(?).11
DS21	Cistern (residential quarter)	84.1068.10L
DS22	Cistern (residential quarter)	84.1068.7L.3
DS24	Cistern (residential quarter)	84.1068.3L.38
DS29	Cistern (residential quarter)	84.1068.15L.6
DS30	Cistern (residential quarter)	84.1068.7L
BAB1	Unavailable	A92/H32/007/1114; removed from sherd A92/H32/010/110
BG1	South tower of Tiberias Gate (Area XII)	UQ
BG5	Underground mausoleum entrance hall (Area XI)	UQ87XI.E.16
BG8	Underground mausoleum entrance hall (Area XI)	UQXI.E.457
BG10	Underground mausoleum entrance hall (Area XI)	UQXI.339
BP1	Room of house, east cut, Area III, Plot P (pit)	L373
BP2	Room of house, east cut, Area III, Plot P (pit)	L313
BP3	Room of house, east cut, Area III, Plot P (pit)	L373
BP7	Room of house, east cut, Area III, Plot P (pit)	L368
BSCY1	Lower market, outside row of shops (fill)	L100131 (B1001347)
BSCY4	Lower market, shop no. 9 (cleaning of floor)	L90181 (B901822N/E44)
BSCY5	Lower market, outside row of shops (fill)	L90107 (B901055)
BS3	Northwest of villa lavatory chamber (fill)	85.2041.1X
BS4	Fill (residential area)	84.2232.3L?
BS9	Fill (residential area)	86.1021.3X.2
BNG1	Underground mausoleum entrance hall (Area XI)	UQXI.E.433
BNG8	Underground mausoleum entrance hall (Area XI)	UQXI.289
BNG9	Underground mausoleum entrance hall (Area XI)	UQ88XI.E.112
BNG10	South tower of Tiberias Gate–Area XII (fill)	UQXII.77

Table 3. Findspots and Registration Information of Samples Selected for Analysis (cont.)

Sample	*Archaeological Context*	*Registration Information*
BNJ2	Unavailable	Locus 1/11.XII-19.XII.8
BNJ4	Unavailable	Locus 1/11.XII-19.XII.82
BNS1	Cistern (residential area)	84.2077.4L.5
BNS3	Cistern (residential area)	84.2077.4L.6
JB1	Jerash hippodrome pottery kiln (Chamber W2)	Hippodrome W24
JB2	Jerash hippodrome pottery kiln (Chamber W2)	Hippodrome W24
JB3	Jerash hippodrome pottery kiln (Chamber W2)	Hippodrome W24
JB5	Jerash hippodrome pottery kiln (Chamber W2)	Hippodrome W24

creases the number of features that can be compared and thus aids in the determination of provenance. For example, if lamp fabrics sharing the same diagnostic features (such as bichrome slipping or the same lamp maker's mark) belong to the same petrographic group and "match" chemically, this suggests the same possible clay source or technique of manufacture unique to a specific workshop.

Caution was exercised to ensure that a representative sampling of lamps from a number of sites was analyzed. Freestone (1995) suggests petrographic examination of at least 10 samples when one is characterizing pottery. The sampling of the Beit Nattif type for this study fell short of the suggested minimum: only 6 Beit Nattif lamp samples were selected for analysis–2 from Sepphoris and 4 from Gadara. In my dissertation analysis, I identified Jerash samples BNJ2 and BNJ4 as bow-shaped nozzle examples; but I agree with da Costa's later observation (2003: 1.71–72) that they are probably Gilead sherds. Originally, I also considered the Palestinian discus versions from Abila and Scythopolis to be strictly Palestinian discus, but in this revision I distinguish them from the major group. Although the quantity of samples from Tarsus, Gilead, and Beit Nattif types is low, I nevertheless believe that these samples serve sufficiently for comparing with the larger sampling of Palestinian discus and bilanceolate groups.

Palestinian Discus

Clay lamps of the Palestinian discus type were sampled because they were widely popular throughout Roman Palestine, including among both Jewish and non-Jewish Sepphoreans. The popularity of this group can be explained by this type's concave discus, which enabled the representation of a vast repertoire of mythological, religious, floral, geometric, faunal, and erotic scenes and motifs. The central discs of some examples were intentionally broken, perhaps for ritual purposes or for the creation of photo amulets (Lapp in press a; Lapp in press b). Seventeen Palestinian discus lamp fragments recovered from 7 sites were selected for petrographic and chemical analysis. They include samples DA2 and DA3 from Ashkelon; sample DC6 from Caesarea Maritima; sample DHH2 from Ḥorvat Ḥazon;

Table 4. Types and Diagnostic Characteristics of Samples Selected for Analysis

Sample	*Site*	*Type*	*Diagnostic Characteristics*
DAB4	Abila	Palestinian Discus	Lamp maker's mark: double-axe; distinctive local red fabric
DA3	Ashkelon	Palestinian Discus	Lamp maker's mark: *eta*; red fabric
DC6	Caesarea	Palestinian Discus	Lamp maker's mark: *eta* with tri-bud floral motif; orange fabric
DM10	Meiron	Palestinian Discus	Lamp maker's mark: *eta*; pale yellow fabric
DP11	Pella	Palestinian Discus	Lamp maker's mark: *nu, alpha*; red fabric
DS24	Sepphoris	Palestinian Discus	Lamp maker's mark: *eta*; orange fabric
DS30	Sepphoris	Palestinian Discus	Lamp maker's mark: *eta*; pale yellow fabric
DP3	Pella	Palestinian Discus	Bi-associative motif: ibis and crab; pale yellow fabric
DS10	Sepphoris	Palestinian Discus	Bi-associative motif: Bacchus and chalice; pale yellow fabric
DS9	Sepphoris	Palestinian Discus	Erotic scene; pale yellow fabric
DA2	Ashkelon	Palestinian Discus	Bichrome; pale yellow fabric
DHH2	Ḥorvat Hazon	Palestinian Discus	Bichrome; pale yellow fabric
DS4	Sepphoris	Palestinian Discus	Bichrome; pale yellow fabric
DSCY6	Scythopolis	Palestinian Discus	Unfired; possibly local; lamp maker's mark: *nu, eta* (?)
DSCY7	Scythopolis	Palestinian Discus	Lamp maker's mark: *šin* design; unique local (?) variant of Palestinian discus lamp (three vertical ridges decorate nozzle-underside); coarse sand grits; brown fabric
DSCY8	Scythopolis	Palestinian Discus	Lamp maker's mark: *nu*
DS21	Sepphoris	Palestinian Discus	Lamp maker's mark: tri-bud floral motif; pale yellow fabric
DS22	Sepphoris	Palestinian Discus	Lamp maker's mark: *alpha*; yellow fabric
DS29	Sepphoris	Palestinian Discus	Lamp maker's mark: *gamma, alpha* (?); yellow fabric
DA9	Ashkelon	Tarsus	Distinctive well-levigated pale yellow fabric
BAB1	Abila	Bilanceolate	On shoulder: deeply impressed acanthus leaf pattern
BG1	Gadara	Bilanceolate	On shoulder: impressed circles; pinched handle
BG5	Gadara	Bilanceolate	On shoulder: deeply impressed acanthus leaf pattern; narrow elongated body
BG8	Gadara	Bilanceolate	On shoulder: impressed circles
BG10	Gadara	Bilanceolate	On shoulder: deeply impressed acanthus leaf pattern
BP1	Pella	Bilanceolate	On shoulder: deeply impressed acanthus leaf pattern
BP2	Pella	Bilanceolate	On shoulder: deeply impressed acanthus leaf pattern
BP3	Pella	Bilanceolate	On shoulder: deeply impressed acanthus leaf pattern
BP7	Pella	Bilanceolate	On shoulder: impressed circles
BSCY1	Scythopolis	Bilanceolate	On shoulder: deeply impressed acanthus leaf pattern
BSCY4	Scythopolis	Bilanceolate	On shoulder: deeply impressed acanthus leaf pattern
BSCY5	Scythopolis	Bilanceolate	On shoulder: impressed circles
BS3	Sepphoris	Bilanceolate	On shoulder: deeply impressed acanthus leaf pattern; rope pattern around filling-hole

Table 4. Types and Diagnostic Characteristics of Samples Selected for Analysis (cont.)

Sample	*Site*	*Type*	*Diagnostic Characteristics*
BS4	Sepphoris	Bilanceolate	On shoulder: deeply impressed acanthus leaf pattern; rope pattern around filling-hole
BS9	Sepphoris	Bilanceolate	On shoulder: deeply impressed acanthus leaf pattern; rope pattern around filling-hole
BNG1	Gadara	Beit Nattif	On nozzle-bridge and shoulder: ornate jewel pattern in high relief; unslipped with hard reddish fabric
BNG8	Gadara	Beit Nattif	On shoulder: generally plain except for two circles in high relief
BNG9	Gadara	Beit Nattif	Generally plain shoulder
BNG10	Gadara	Beit Nattif	On shoulder: small circles in high relief
BNS1	Sepphoris	Beit Nattif	On shoulder: consecutive circles in high relief; friable fabric with red slip
BNS3	Sepphoris	Beit Nattif	On shoulder: ornate jewel pattern in high relief
BNJ2	Jerash	Gilead	On shoulder: impressed, narrow acanthus leaf pattern
BNJ4	Jerash	Gilead	On shoulder: impressed, narrow acanthus leaf pattern
JB1	Jerash	Jerash Bowl	Confirmed source of origin: hippodrome pottery workshop at Jerash; rim fragment
JB2	Jerash	Jerash Bowl	Confirmed source of origin: hippodrome pottery workshop at Jerash; body fragment
JB3	Jerash	Jerash Bowl	Confirmed source of origin: hippodrome pottery workshop at Jerash; base fragment
JB5	Jerash	Jerash Bowl	Confirmed source of origin: hippodrome pottery workshop at Jerash; unfired body fragment

sample DM10 from Meiron; samples DP3 and DP11 from Pella; samples DSCY6 and DSCY8 from Scythopolis; and samples DS4, DS9, DS10, DS21, DS22, DS24, DS29, and DS30 from Sepphoris. Two separate versions of the Palestinian discus group were selected from Abila (sample DAB4) and Scythopolis (sample DSCY7). Special attention was given to the selection of Palestinian discus lamp fragments taken from, or once belonging to oil lamps whose bases bear lamp makers' marks. I hoped that these distinct features would aid in the determination of provenance. For example, the clustering together of all discus lamps with bichrome slipping into a single, distinct chemical group would suggest the same clay source and possibly the same workshop. The discus lamp fragments were selected for DCP analysis on the basis of 5 macroscopic diagnostic characteristics: (1) bichrome slipping; (2) the lamp maker's mark of the Greek initial *eta* (H) depicted either alone or in combination with another initial or mark; (3) bi-associative motifs (especially, ibis-and-crab;

Table 5. Clay Samples Collected in the Field for DCP-OES Analysis

Sample	*Site*	*Clay Type*	*Field Observations*
L1	Pella	Marl	Poor plasticity
L2	Pella	Alluvial marl	High plasticity
L3	Pella	Chalky marl	Poor plasticity
L4	Pella	*Terra rossa*	Poor plasticity
L5	Pella	*Terra rossa*	High plasticity
BS1	Scythopolis	Chalky marl	Poor plasticity
BS2	Scythopolis	Chalky marl	Poor plasticity
G2	Gadara	*Terra rossa*	High plasticity
S1	Sepphoris	Chalky marl	Poor plasticity

Bacchus-and-chalice); (4) a variety of different lamp makers' marks (especially, a double axe, *gamma alpha*, *alpha*); and (5) an erotic scene.

The purpose of characterizing (i.e., identifying the petrological constituents and chemical composition of the ceramic sample) the discus lamp fragments is to determine whether their respective fabrics exhibit similar or entirely different petrological and micromorphological characteristics. Chemical analyses of their fabrics should reveal whether the samples cluster into one or more distinct chemical composition groups and whether the composition groups reflect these petrographic similarities or differences. Depending on the results, this would suggest the same possible clay sources and provenance. The discus motifs, lamp makers' marks, iconographic elements, and any other diagnostic characteristics occurring on the lamps that compose each of these groups would be compared to determine whether any theme is a distinctive of each cluster. For example, do all bichrome lamps group together? Do lamps with the same lamp makers' mark of *eta* share similar chemical compositions? Do lamps with pale yellow fabrics cluster into a distinct chemical group while lamps with red fabrics cluster into another? And do lamps with different diagnostic features that one would not anticipate as being related share similar compositions? Finally, the discus lamps are analyzed to examine whether they group with samples from other lamp groups–namely, the Beit Nattif and bilanceolate forms. If they do cluster with lamps belonging to different types, this would suggest the same clay source or sources quarried and used by lamp workshops to manufacture different lamp types. And given the fact that each lamp type is chronologically specific (e.g., Palestinian discus [late first–third centuries C.E.], Beit Nattif [fourth century C.E.], and bilanceolate [late fourth–fifth centuries C.E.]), matching chemical profiles of Middle Roman discus lamps with Late Roman–Byzantine period bilanceolate samples would suggest the possible continuous use of the same clay source or sources over a substantial period of time. The findspots of the Palestinian discus lamp samples are four different geographical regions: the coastal plain, Lower Galilee, the northern Jordan Valley, and Transjordan.

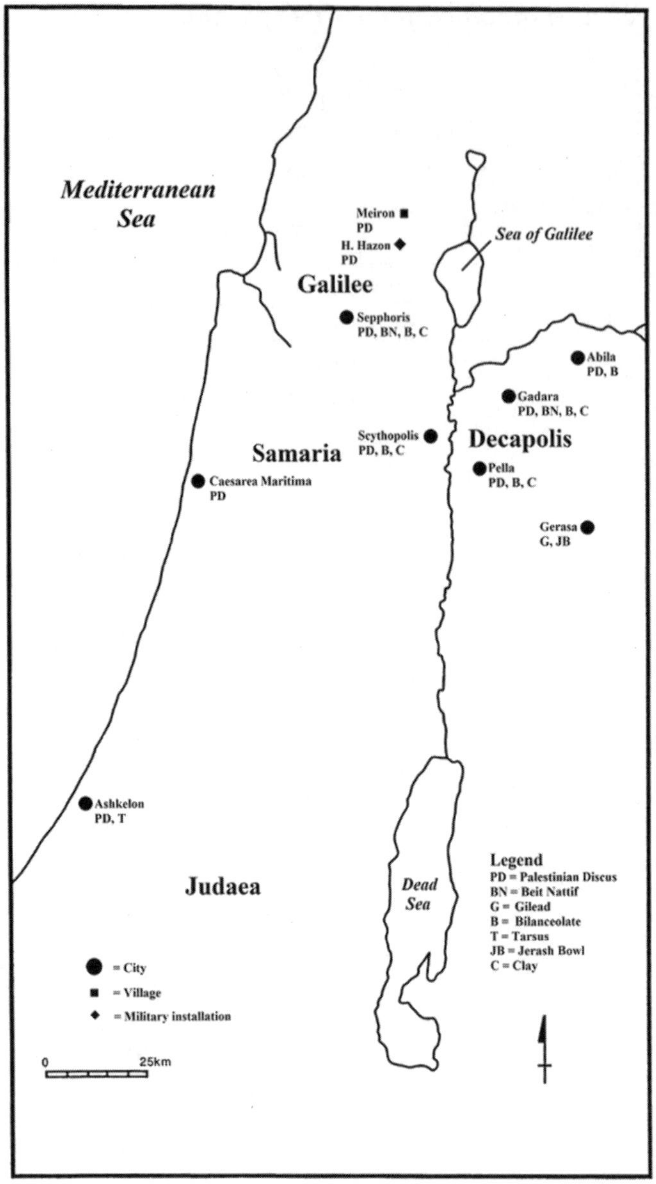

Fig. 8. Map showing sites of lamp sherds collected for analysis. Map drawn by E. C. Lapp.

Bichromes. Three Palestinian discus lamp fragments were selected for petrographic and chemical analysis because they exhibit bichrome slipping: these include sample DA2 from Ashkelon, sample DHH2 from Ḥorvat Ḥazon, and sample DS4 from Sepphoris (Lapp 1997a: figs. 96–97). The term *bichrome* refers to the two-color slipping technique used to decorate certain discus lamps: the central discus is slipped red and the shoulders and body a dark blue or black. The purpose of analyzing these samples was to determine whether their fabrics belong to the same petrographic fabric group and whether they share similar chemical compositions. This would suggest the same clay source and possibly the same workshop. The bichrome slipping technique was likely used by a specific lamp maker (or workshop) to distinguish his products from others; or the red–dark blue combination was intended to serve an apotropaic purpose (Lapp in press b). The findspots of the samples are two geographical regions: the southern coastal plain and Lower Galilee.

Lamp Maker's Mark–Greek Eta (H). Six Palestinian discus lamp sherds were chosen for analysis because their bases are incised with a lamp maker's mark—the Greek initial *eta* (H). These samples include DA3 from Ashkelon; DC6 from Caesarea Maritima; DM10 from Meiron; DSCY6 from Scythopolis; and DS24 and DS30 from Sepphoris. The purpose of the analysis of these fragments was to determine whether their respective fabrics belong to the same petrographic fabric group and share the same chemical fingerprint. If so, this would imply a similar clay source, workshop, or both. These samples come from three different geographical regions: the coastal plain, Lower Galilee, and the northern Jordan Valley.

Other Lamp Makers' Marks. In addition to the special group of lamp fragments with the makers' mark of the Greek initial *eta* (H), other lamp sherds with bases

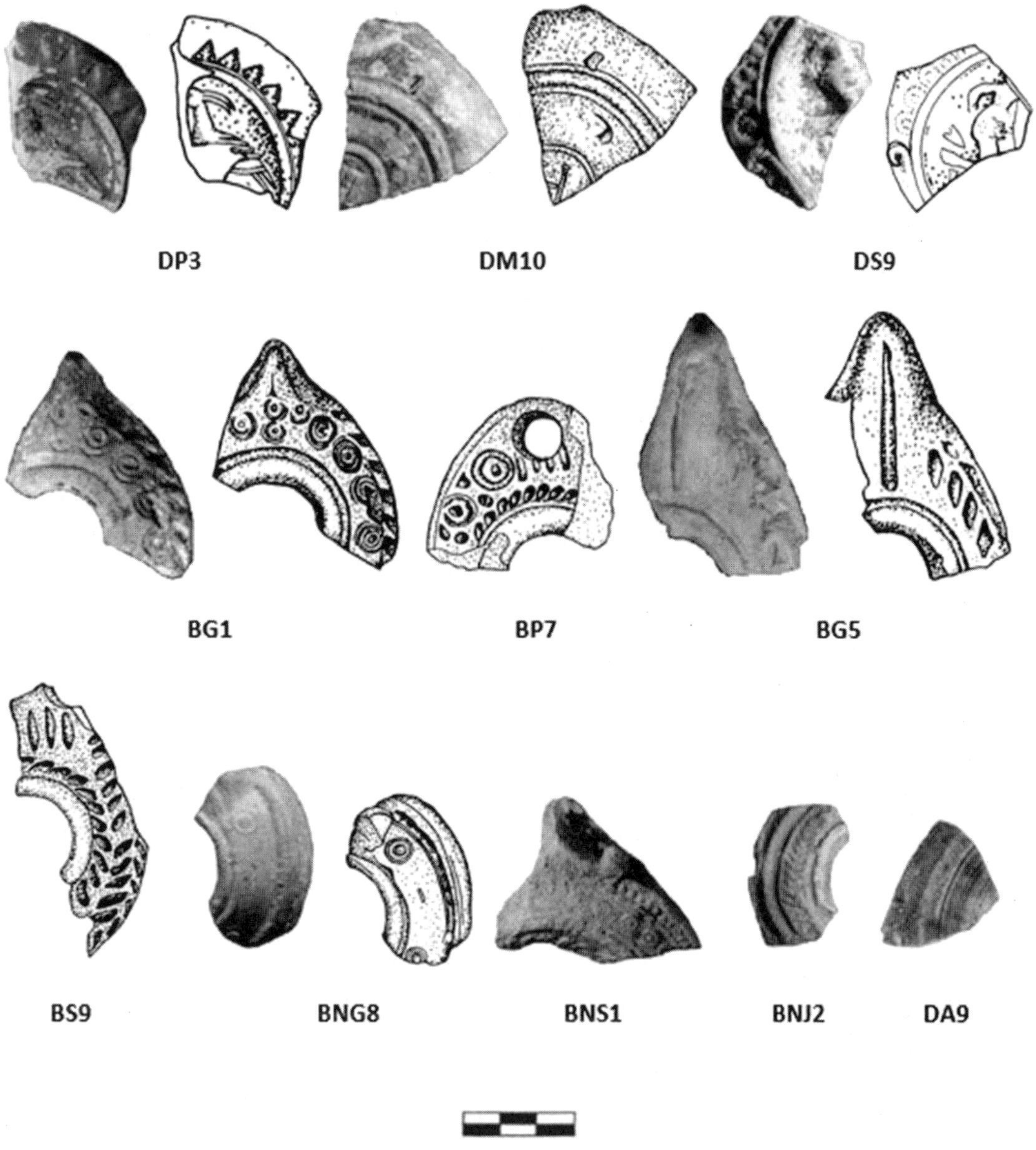

Fig. 9. Drawings and photographs of select lamp samples chosen for petrographic and dcp-oes analyses. Drawings and photographs by E. C. Lapp.

bearing lamp makers' marks were examined: Abila (sample DAB4), Pella (sample DP11), Scythopolis (samples DSCY7 and DSCY8), and Sepphoris (samples DS21 and DS22). Their lamp makers' marks include: (1) a tri-bud floral mark with curved branch (DS21); (2) the Greek initial *alpha* (A) (DS22); (3) a mark that resembles a Hebrew *šin* (?) (DSCY7); (4) the Greek initial *nu* (N) (DSCY8); (5) the Greek initials *nu* and *alpha* (NA) (DP11); and (6) a double-axe mark (DAB4). A variety of makers' marks on lamps from findspots of wide geographical range were chosen for petrographic and chemical characterization to determine whether they are compositionally similar. Samples DSCY7 and DSCY8 are particularly noteworthy in that the unique characteristics of their respective fabrics, styles, and lamp makers' marks suggest local manufacture either at Scythopolis or in the vicinity of the site. Similarly, sample DAB4 appears to have been manufactured locally at Abila (or in the vicinity) because: (1) its red clay fabric resembles other lamps from the site and is a distinct red in comparison with other samples, (2) the treatment and execution of the body shape and base rings are different, and (3) an incised double axe found on the lamp's base is a rare lamp maker's mark. If the fabrics of samples DSCY7, DSCY8, and DAB4 matched the compositions of any of the other lamp fabrics, they would act as an anchor and suggest a local provenance at these respective sites. The findspots of these discus lamp samples are located in three different geographical regions: Lower Galilee, the northern Jordan Valley, and Transjordan.

Bi-associative Motifs (Ibis-and-Crab and Bacchus-and-Chalice). Two Palestinian discus lamp fragments were selected because of the bi-associative motifs portrayed on their respective discs. These include sample DP3 unearthed at Pella and DS10 from Sepphoris. The orientation and style of the ibis depicted on the inner right curve of the central discus of the Pella sample DP3 (shown on fig. 9, top left) compares favorably with those represented on complete specimens depicting a second motif of a crab to the left of the ibis. Thus it is likely that this Pella fragment belongs to the same bi-associative motif type: ibis-and-crab. Similar Palestinian discus lamps with the ibis-and-crab theme have been recovered at Caesarea Maritima, Samaria, and Beth Guvrin (Sussman 2012: 255–57, nos. 453, 456, 464). Discus fragment DS10 depicts a single drinking vessel with handles (a chalice), and its orientation on the inner left curve of the central discus parallels complete lamps whose discuses include a bust of Bacchus in profile. Thus it is likely that the Sepphoris lamp sherd belongs to the same bi-associative motif type of Bacchus-and-chalice. The purpose of analyzing these fragments is to compare their respective petrographic fabric groups and to determine whether they match chemically. A close correlation between the petrographic and chemical groups might suggest that the same clay source was being exploited by one or more workshop. It is also important to determine whether the bi-associative lamp fabrics match any other discus lamps, and if so, which ones. Such a match would suggest the same clay source. The findspots of these discus lamp samples are two different geographical zones: Lower Galilee and Transjordan.

Erotic Scene. A Palestinian discus lamp fragment was selected for analysis because a heterosexual erotic scene is portrayed on its discus. The sample (DS9) was recovered from L84.1068, the fill in a cistern (L84.1066) in Area 84 of the western summit. Sample DS9 closely resembles sample DS10 with respect to the slipping treatment, the fabric color and type, and the execution of the central motif. Taking these similarities into consideration, I analyzed the fabric of this fragment to determine whether its chemical composition matched that of DS10 or any other discus lamps with pale yellow fabrics. If so, this would suggest the same clay source and possibly the same workshop manufacturing two different types of motifs: the Erotic Type 1 scene and the bi-associative motif of Bacchus-and-chalice.

Tarsus

A lamp fragment from Ashkelon (sample DA9; see also Lapp 1997a: fig. 97, pl. 1) with a single curved pseudo-handle decorating its shoulder and incised concentric rings decorating its central discus was petrographically and chemically analyzed. Ashkelon Sample DA9 is unique in that: (1) it exhibits a nearly white clay fabric that is noticeably different from lamps typically found in the region and could represent an import; (2) its occurrence is rare at Roman-period Palestinian and Arabian sites; and (3) it exhibits a high quality of execution and unique design (especially: incised lines and pseudo-handle). The purpose of analyzing sample DA9 is to determine whether it is petrographically and chemically similar to any of the known indigenous types examined in this study (e.g., the bilanceolate or Beit Nattif lamps). If it clusters with the indigenous Palestinian lamps, this would suggest that lamps of this group are not necessarily imports but, rather, local products. Because its clay in thin-section is pale yellow and lacks inclusions, it is important to determine how it will compare chemically with the other samples with the same fabric characteristics (especially, the pale yellow discus lamp fragments, such as DS9 and DS10). The sample was found in a room with hundreds of other lamp fragments at the southern coastal city of Ashkelon.

Bilanceolate

Fifteen lamp fragments belonging to the bilanceolate type were selected for petrographic and DCP analysis. They include samples from Sepphoris (BS3, BS4, and BS9) and from the Decapolis cities of Abila (BAB1), Gadara (BG1, BG5, BG8, and BG10), Pella (BP1, BP2, BP3, and BP7), and Scythopolis (BSCY1, BSCY4, and BSCY5). Given the limited geographical distribution of bilanceolate lamps and the petrological similarity of the samples' fabrics, I characterized select fragments to determine whether they belong to the same petrographic fabric group and whether their chemical compositions match. This would suggest that the same clay source was used by one or more lamp workshops. If the samples from these five sites do match chemically, this would further indicate cultural contact between the predominantly Jewish population of Sepphoris and their co-religionists and non-Jewish contemporaries in the cities of the Decapolis. Given the high concentration and widespread distribution of this lamp type at sites located in or close to the northern Jordan Valley and Galilee, bilanceolate lamps were probably manufactured somewhere in this area. The distribution of the bilanceolate lamp is most concentrated at Scythopolis and Pella. Their fabrics are consistently foraminifera-rich marly clays. One region where foraminiferal marls occur is central and northern Israel (Goren 1995: 206–7, nos. 5, 6, 12). Bilanceolate lamp fragments belonging to Types 1–3 were examined to determine whether their respective clays had originated from the same or different sources. The latter would suggest different workshops. Special attention has also been paid to determine whether the samples cluster according to their respective find sites, thus generating five different and distinct chemical groupings. If such a scenario materializes, the similar petrographic and micromorphological characteristics of their fabrics could be explained by the fact that the same type of marly clay was being used at each site. It is highly likely then that five separate workshops manufacturing bilanceolate lamps were located at these urban centers. The findspots of these discus lamp samples are three different geographical regions: Lower Galilee, the northern Jordan Valley, and Transjordan.

Beit Nattif

The chemical compositions of six lamp fragments belonging to the Beit Nattif (northern versions) type and collected from two sites were characterized: Gadara (samples BNG1, BNG8, BNG9, and BNG10) and Sepphoris (samples BNS1 and BNS3; Lapp 1997a: fig. 104; pl. 4). Originally in my dissertation, I identified these samples as bow-shaped nozzle rather than Beit Nattif lamp types, as two other lamp types were found in the cisterns at Beit Nattif (da Costa 2003: 1.76). But the term "bow-shaped nozzle" is equally misleading because several regional lamp types—such as the Herodian, Darom, Sepphorean spatulate, and "Sepphorean" discus forms—also have bow-shaped nozzles. The samples selected for petrographic and dcp-oes analyses belong to northern versions of the Beit Nattif group with bow-shaped nozzles and ornately decorated shoulders and nozzle-bridges. Because all eight samples belong to the same lamp type, examining them microscopically would determine whether their fabrics belong to the same petrographic and micromorphological group. If they were found to cluster into the same chemical group as determined by DCP analysis, this would suggest the same clay source used by one or more workshops. If the two samples selected from each site clustered into three distinct petrographic and chemical groups, this would indicate three different clay sources and centers of manufacture specific to each site. The samples were collected from sites of two geographical areas: Lower Galilee and northern Transjordan.

Gilead

Two samples belonging to the Gilead type were selected for petrographic and chemical composition analyses. They include samples BNJ2 and BNJ4 excavated at Gerasa. In my original study, they were identified as bow-shaped nozzle type lamp fragments (Lapp 1997a: 134). However, I agree with da Costa's suggestion (2003: vol. 1.i, 71–72) that they may in fact belong to the Gilead group.

7. *Reference Materials*

The ceramic reference materials analyzed in this survey include four fragments belonging to the Jerash bowl pottery group (samples JB1, JB2, JB3, and JB5) and an unfired Palestinian discus lamp fragment from Scythopolis (sample DSCY6). The Jerash bowl sherds date to the Late Byzantine to Early Islamic periods. The samples were recovered from the excavations of the hippodrome kilns at Gerasa. Additional reference materials include raw clays collected from several locations among the cities of the Decapolis and Sepphoris: Gadara (sample G2), Pella (samples L1, L2, L3, L4, and L5), Scythopolis (samples BS1 and BS2), and Sepphoris (sample S1).

Jerash Bowls

For comparative purposes, four samples of non-lamp pottery were selected for petrographic and chemical examination: three fired Jerash bowl fragments (samples JB1, JB2, and JB3) and one unfired (sample JB5). The samples were collected in situ from the hippodrome kilns at Jerash (see table 3, p. 127). The reasons for examining these fragments are: (1) the exact provenance of the Jerash bowl is known, and therefore it is important to determine whether any of the samples chemically match any of the lamp fabric compositions; (2) to compare the micromorphological features indicative of different manufacturing techniques for mold-made lamps as opposed to wheel-thrown fine wares (e.g., tempering and parallel orientation of cracks); and (3) to compare the type of clay used for lamps to that used for fine wares. Because the origin of the Jerash bowl samples is known, a chemical match with any of the lamp samples would link them to Jerash as their place of manufacture. Petrographic examination of the Jerash bowl sherds shows that their fabrics differ from those belonging to the lamps; hence, I did not expect a chemical match. Nonetheless, given the importance of Jerash as a major pottery-producing center, it is important to include these fragments on the outside chance that a match might occur. Furthermore, because these fragments belong to bowls rather than lamps, petrographic and chemical analysis enables us to compare the fabrics belonging to two functionally different pottery vessels and to discern whether the potters used different clays for different vessel types.

Unfired Palestinian Discus Fragment, Scythopolis Version

Upon thin-sectioning sample DSCY6, I realized that it was unfired for, while I proceeded to cut the specimen, the non-epoxied portion began to dissolve rapidly from the water of the thin-section, saw as did the unfired Jerash bowl sample JB5. Nonetheless, I succeeded in making a thin-section from this sample (Lapp 1997a: pl. 9.2). This unfired fragment is important because: (1) it points to Scythopolis as a possible manufacturing center for Palestinian discus lamps, and (2) if other lamp fragments match its chemical composition, sample DSCY6 would help determine their respective provenances. The fact that the discus lamp fragment is unfired suggests the existence of a kiln located in or near Scythopolis: it is unlikely that unfired lamps were brought in from distant workshops because of the potential for breakage during transport.

Raw Materials (Clays)

Nine clay samples were collected from four sites in Israel and Jordan for compositional analysis (the number of clay samples collected from each site is enclosed within brackets): Gadara [1], Pella [5], Scythopolis [2], and Sepphoris [1]. Previous studies have shown that the success rate of identifying ancient clay beds suitable for pottery-making is low (Adan-Bayewitz 1993: 45–46). It was necessary, therefore, to decide whether to analyze local clays. Numerous clay deposits exist throughout the region, and many ancient clay sources suitable for lamp and pottery manufacture may have been exhausted in antiquity. Also, taking into account the paucity of actual lamp kilns operating throughout the region in the Roman period, one finds it difficult to determine the actual clay sources. Lamp kilns are necessary to provide wasters and reference material of known provenance to match chemically with the clays and other lamp samples.

Abundant clay sources exist throughout the relatively widespread geographical range covered in this survey. For example, while surveying the Pella area for clay sources, I observed a number of different clay sources in roughly a 1,000-m radius centered around the Odeon of Pella. Thus lamps produced at several different workshops using clays from a variety of sources at a site such as Pella would appear noticeably different chemically as well as petrologically and would lead one to expect several separate origins when, in fact, all were manufactured at a single site. Until lamp kilns containing important reference material are found, it is a nearly impossible task to determine the exact provenance of manufacture.

However, I decided to sample and analyze select raw clays in the hope that some connection could be made between the lamp fabrics and the actual raw clay material to determine how they compare with one another chemically. Additionally, there is always the outside possibility of identifying a potential ancient clay source. Several varieties of clay were collected from each site in order to obtain some sort of "feel" for the chemical compositions of the clays in these ancient cities and the range of clays. While sampling clays, I intentionally allowed some degree of randomness because we do not know exactly what governed clay selection among ancient lamp makers. I did not sample only "good" clays, which exhibited plasticity and probable suitability for lamp manufacture. For example, due to the risk of spalling caused by the release of carbon dioxide, clays high in calcium carbonate are considered "poor" clays by modern standards for pottery manufacture; but the archaeological record indicates that such clays were commonly used for the making of pottery wares as well as lamps. Additionally, although the addition of temper to pottery served a practical function, it remains unclear whether or not it was also introduced into the clay body in some instances for apotropaic purposes. Out in the field, I conducted a simple plasticity test of adding water to the collected clay and rolling it to determine whether it adhered (Hamer and Hamer 1993: 346).

Taking into consideration that wadis offer an ideal source for the quarrying of clay for lamp and pottery manufacture, I concentrated my clay sampling at these localities. Wadi cliff sequences often comprised clay-rich stratified layers. The fact that these clay deposits have already been exposed by water processes and mechanical weathering would have greatly facilitated their extraction by ancient lamp makers, as opposed to ground-level surface quarrying, which would have required considerable labor to expose the desired deposits. One or more wadis often run through or are situated in proximity to urban centers (e.g., Abila, Ashkelon, Gerasa, Pella, and

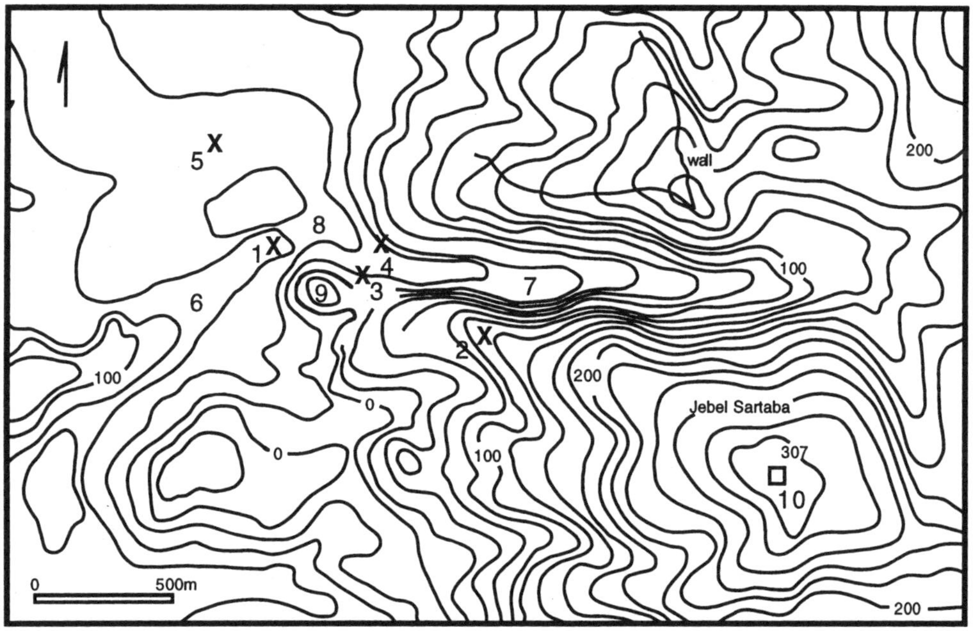

Fig. 10. Map showing clay sample collection sites at Pella (1 = clay sample L1; 2 = L2; 3 = L3; 4 = L4; 5 = L5; 6 = Wadi Jirm; 7 = Wadi Malawi; 8 = acropolis; 9 = Tell el-Husn; 10 = Hellenistic fortification). After Edwards 1992: 290, fig. 41.

Scythopolis), where lamps and pottery were typically manufactured. Because copious amounts of water are necessary in lamp and pottery manufacture, one would expect the location of kilns near a water source, such as a wadi bed or a perennial spring. The identification of pottery kilns situated in a wadi bed in the vicinity of Ashkelon suggests this practice (da Costa 2003: 1.387). For these reasons, I focused on sampling clays from the walls of wadis. Special attention was paid to the sampling of clays from several sites where evidence for lamp manufacture (e.g., lamp wasters, molds, and/or kilns) has been found and/or from sites where lamp fragments in this study were collected for DCP and petrographic analyses.

Lamp distribution maps were weighed heavily with respect to clay sampling. For example, with respect to the bilanceolate lamp type, findspot distribution patterns and the heavy concentration of samples in this group from Scythopolis and Pella point to their possible manufacture in the Jordan Valley. Thus I collected seven clay samples in this area: five from Pella and two from Scythopolis. Using the Hebrew University Clay Standard of Perlman and Asaro (1969: 21–52), I calibrated the chemical compositions of clays collected in Jerusalem and Lower Galilee to the Duke University Clay Standard.

Pella

In 1982, Ian Edwards sampled clays from Pella to assess their suitability for pottery making (Edwards 1992: 286–98). Based on the fact that the clays of Pella and vicinity have been the most extensively sampled and examined for their suitability as potters' clay, I concentrated on the collection of clays from this important site. Using Edwards's study as a guide, I resampled select clays he described and added two collection sites in the Wadi Malawi (fig. 10). After firing his clay samples at several temperatures to determine their workability, Edwards assigned a rating to each clay sample's potential as potters' clays: Grade "A" clays had very good potential as a potter's clay, "B" good potential, and "C" possible potential. The clays that were probably used by Chalcolithic and Bronze Age potters are deeply buried beneath the sediments of the Wadi Jirm (personal communication and site visitation with Stephen Bourke in April 1993). The clays most accessible to the Pella potters during the Late Roman and Byzantine periods, however, lie above these sediments.

Because of the nature of the clay deposits and topography at Pella, I posit that the potters there probably collected their clays from surface clay outcrops and wadi walls. The walls of the Wadi Malawi in particular would have been a superb source of clay in terms of accessibility and raw material supply potential. Outcrops of clay such as those found on the surface of Jebel Sartaba and the slopes of Wadi Malawi would also have been easily accessible.

Clay Sample L1 (= Edwards's S12). Clay sample L1 was collected in a dry fissure traversing the northernmost point of the Wadi Jirm. The fissure is oriented in a north–south direction. Edwards observes that the alluvial clays found in shallow depressions along the bed of the Wadi Jirm are suitable for pottery-making, despite their poor development and high lime content. For the following information regarding the suitability of clays for pottery-making at Pella, I rely on Edwards 1992: 290–93, appendix 10. He notes that the optimum firing temperature for these clays falls in the range of 850–950°C and writes: "The discontinuous nature and limited amounts of this material available throughout the period of settlement at Pella seem to indicate limits on its use. Seasonally heavy rains in winter would presumably have replenished the supply of this clay-rich material" (Edwards 1992: 292). Taking into account the A rating (especially good potential as a potter's clay) of S2 as determined by Edwards and the fact that the Umayyad biscuit ware CN206 that he analyzed matches the Wadi Jirm clay (except for the potassium component), I considered the sampling and composition analysis of clay from this locus as important comparative reference material. Given the similarity of the pale yellow and fine grained fabrics of the yellow Palestinian discus lamps (e.g., samples DM10, DS30, DP3, DS9, DS10, DS4, DS21, DS22, DS29, DA9, DHH2, and DA2) to the Umayyad biscuit ware, I was curious about whether clay sample L1 would match the composition of these lamp fabrics.

Clay Sample L2. Sample L2 was collected in the lower south wall of the Wadi Malawi, located at the first southerly bend of the wadi's mouth. The clay itself is a reddish brown, very fine, and compact. It was sampled near the surface bed of the wadi.

Clay Sample L3 (= Edwards's S5). Clay sample L3 is off-white and very compact. It was collected at two locations on the high ridge of Jebel Sartaba. The first sampling site (Location A) was located on the surface of the hill as an outcrop of clay.

Several centimeters of the surface were removed first in order to obtain "fresh" clay, understanding that the clay found on the surface may contain contaminants such as fertilizers and other chemicals resulting from water runoff and soil leaching. The second sampling site (Location B) was approximately 500 m down the slope of the Jebel Sartaba, below Location A. It was collected in a shaft facing the southeast side of Tell el-Husn. Because the clay sampled from the shaft of Location B was much deeper than that collected on the surface at Location A and thus more free from potential surface contamination, I chose to analyze it instead.

Among the clays sampled at Pella, sample L3 was petrographically and chemically analyzed to determine whether it chemically matched the clay fabrics of several select lamp sherds from Pella that have been analyzed in this study. Edwards notes that his clay sample S5 (Sample L3 of this study) was suitable for making pottery at temperatures of ca. 1000°C and that the firing color of the resulting fired clay fabric was a distinctive pale greenish yellow and matched one of the Umayyad wares in his study (Edwards 1992: 292). This pale greenish yellow is also similar to the fabric of Palestinian discus lamps from Pella and other sites included in this study. With this in mind, I chose to conduct DCP analyses on this clay sample to determine whether these lamp fabrics were similar to the clay itself. It was also important to determine whether the similar clay fabrics of the bilanceolate and imported Tarsus type lamp fragment from Ashkelon (sample DA9) matched L3 chemically. A close match would further strengthen the petrographic findings and give us a better idea of the exact provenance. Edwards observes that the pale yellowish-green color suggests a high lime content (Edwards 1992: 293).

Clay Sample L4. Sample L4 is reddish brown and very compact. It was collected on the north cliff face of the Wadi Malawi. A thin veneer of white phosphate or oxide stained the outer surface. To avoid contamination, I first scraped away this residue and removed several centimeters of the hard outer surface of the clay deposit. Compact clay from approximately 25 cm beneath was then collected. Cobble-to-gravel chert layers were along and above this clay deposit. The sampling site of L4 was approximately 25 m south of, and in view of the sampling site of L2 across the wadi: both clays belong to the same unit. Similar clay was located on the surface of the slope east of sampling site L4. Sample L4 was collected to determine its similarity to clay L2 and the other Pella samples.

Clay Sample L5 (= Edwards's S12). Sample L5, a *terra rossa* clay, was collected from the south face of the North Wadi of Tabaqat Fahl, approximately 40 m east of Edwards's wheat-fields track. Samples of L5 were collected at three sites, Locations A, B, and C. Sample L5 collected at Location B was analyzed. As Edwards reports, this clay has ferrous oxide and stony inclusions and has considerable potential as potters' clay; it is rated A at 850°C and B at 950°C and 1050°C. Edwards's samples from the North Wadi (S9, S11, and S12) are representative of the clays still used by local villagers for building clay ovens. If thoroughly levigated and/or ground, all three would produce serviceable pottery. Their optimum firing temperature would be in the range of 850–950°C.

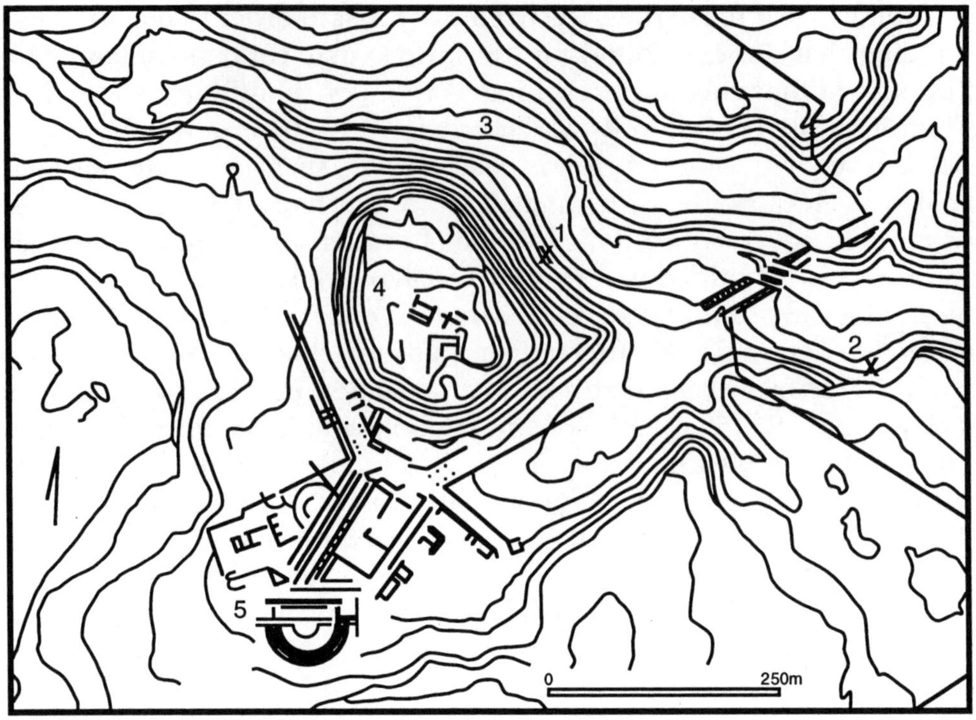

Fig. 11. Map showing clay sample collection sites at Beth Shean/Scythopolis (1 = clay sample BS1; 2 = BS2; 3 = Wadi Beth Shean; 4 = Tel Beth Shean; 5 = acropolis). After A. Mazar 1994: 66.

Scythopolis

Clay Samples BS1 and BS2. Clays were collected at two sampling sites located at Beth Shean/Scythopolis (fig. 11). The first sample (BS1) was collected from a cut located on the north-northeast side of Tel Beth Shean and the second sample (BS2) from an exposed cut along the south bank of Nahal Harod, approximately 175m east of the Roman–Byzantine period bridge. I sampled clays from Scythopolis because lamp molds were found at the site and suggest lamp manufacture there (Hadad 2002: 127–29, nos. 500–506). Given the proximity of Scythopolis to Pella, I was curious about how the clay samples of these sites would compare with one another and about whether clay samples BS1 and BS2 would cluster with any of the bilanceolate lamp fabrics. The Scythopolis samples are pale green in color and contain rounded pebble- to gravel-size limestone.

Gadara

Clay Sample G1. Sample G1 was collected on the southeastern cliff face of the wadi, a place contemporary Gadarans call the Ira el-Shomar. This clay is referred to as *hazeirah* and is located in the topsoil formation just above the chert and lower limestone formations. It was used to make ovens as well as pottery. A second clay was collected from the chert formation located above the limestone formation of the cliff that local Gadarans also refer to as *hazeirah*. This formation is characterized by gravel- to cobble-size chert. The clay itself occurs in rock form and is found intertwined among the chert inclusions. Considerable grinding and processing is required to extract sufficient amounts of workable clay. I was informed that only ovens can be made from this clay, which is too hard and thick for making pottery.

Clay Sample G2. Sample G2 was collected approximately 5 m directly in front of the Western Gate at Gadara (fig. 12). It is a *terra rossa* and very compact.[2] The ferrous clay itself appears to have percolated to ground surface where it formed small mounds, similar to those produced by burrowing moles. The sampling site was located in a depression in front of the gate. Water filled the westernmost portion of the depression. The local name for this area of Gadara (Umm Qeis) describes this phenomenon of occasional flooding of the basin. A poorly preserved wall composed of limestone blocks supported a portion of the westernmost edge of the depression. A plowed field is located just beyond the depression. If pesticides or fertilizer are sprayed on this field and thus enter the groundwater here, they could adversely affect the chemistry of these clay samples. I was unable to determine whether pesticides and fertilizers are used on this field. A handful of clay G2 can be readily formed and shaped. It will be important to determine whether this clay is chemically similar to any lamp fragments in this study, particularly the Beit Nattif samples from Gadara (BNG1, BNG8, BNG9, and BNG10) and, if so, whether they should be extensively sampled for future chemical analyses.

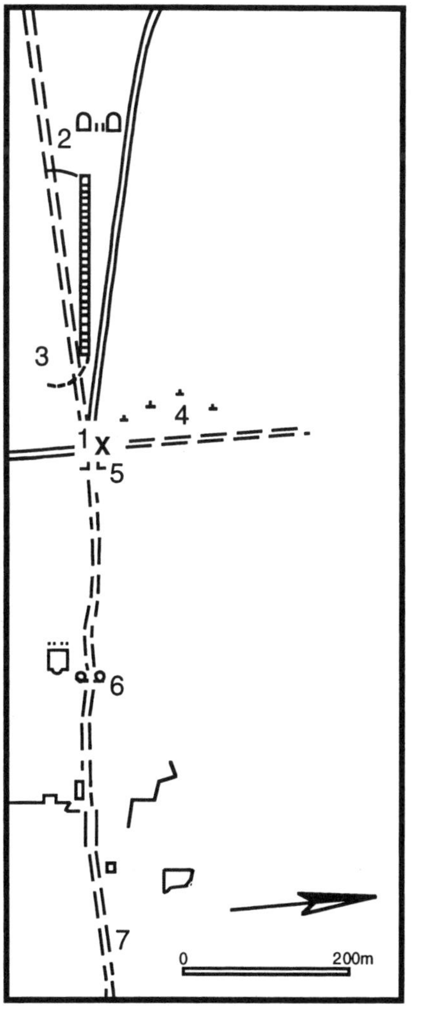

Fig. 12. Map showing the clay sample collection sites at Gadara (1 = clay sample G2; 2 = monumental arch; 3 = hippodrome; 4 = Muslim cemetery; 5 = Western Gate; 6 = Tiberias Gate; 7 = *decumanus maximus*). After Weber and Hoffman 1990: 322, fig. 1.

Sepphoris

Clay Sample S1. A yellow chalky clay similar to Sample L3 from Pella was sampled at a road section located approximately 2 km south of Tel ʿEin Zippori in Lower Galilee (fig. 13). The sample was analyzed to determine whether it chemically matched any of the pale yellow-green fabrics of the discus lamps from Sepphoris or any other sites. I also was interested in how it would compare with similar clays collected at Pella and Scythopolis.

2. For a detailed discussion on the distribution, formation, morphology, and chemical and mineralogical characteristics of *terra rossa* soil, see Singer 2007: 91–106.

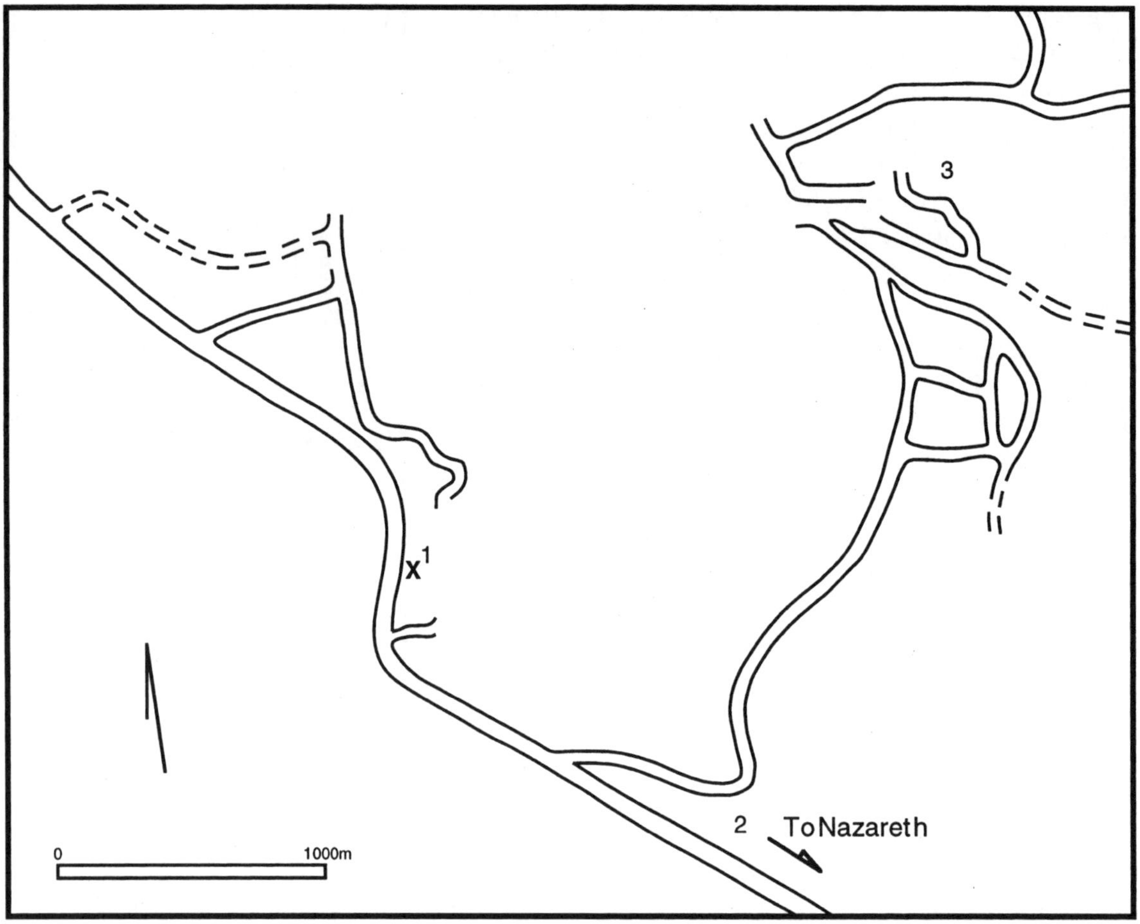

Fig. 13. Map showing clay sample collection sites at Sepphoris (1 = clay sample S2; 2 = Tel ʿEin Zippori; 3 = western summit). Map drawn by E. C. Lapp.

8. *Results and Discussion*

Petrographic Findings

The results of the petrographic thin-section analysis indicate that the nonplastic inclusions occurring in the lamp fabrics are consistent with the predominant, sedimentary geologic lithology of the region: quartz sand, dolomite, calcite, limestone, hematite, and microfossils. Rhombohedral dolomite crystals were observed in several clay fabrics (e.g., Sample BP7). Dolomite rocks are widespread in Cretaceous formations in Israel and are exposed, for example, in the Moẓa and ʿAmminadav formations (Cohen-Weinberger and Goren 2011: 219). These were intensively exploited for clay used for diachronological pottery manufacture (Goren, Finkelstein, and Naʾaman 2004: 263–64). Typically, the quartz sand is angular and poorly sorted. The silt size of the quartz inclusions suggests that it is naturally occurring in the clay. As observed by Adan-Bayewitz and Wieder, the large amount of silt-size quartz grains–as in the Kefar Hananiah common ware–resulted from the dissolution of

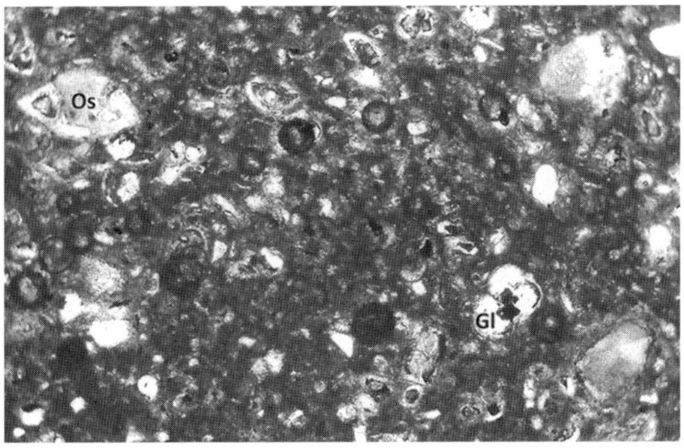

Photo 3. Photomicrograph of Sample DS10, a Palestinian discus lamp fragment with a bi-associative image of chalice and Bacchus. Poorly preserved and mottled microfossils exhibiting distinct boundaries (center left). Concave ostracod valve (Os) and globeriginid foraminifera (Gl). Poorly sorted, sub-angular to sub-rounded quartz sand inclusions. Opaque grains. *Terra rossa* nodules (darker than matrix) added as temper. Epoxy air bubbles (spheres). Foraminiferous marl. Plain polarized light (×10). Sepphoris, Lower Galilee. Photomicrograph by E. C. Lapp.

hard limestone or dolomitic limestone mixed with aeolian dust material originating in the Sinai and Sahara deserts (Adan-Bayewitz and Wieder 1992: 201; Yaalon and Ganor 1973). Leached *terra rossa* soil nodules were observed in several lamp fabrics of this study (e.g., samples DAB4, BS9, and BNG1) and were added as temper by the lamp maker (see discussion in Adan-Bayewitz et al. 2008: 61, 68).

Calcite typically occurs as a secondary replacement mineral of microfossils comprising many of the clay fabrics. Lamps belonging to the bilanceolate variety are particularly rich in foraminifera. The presence of planktonic foraminifera in these clays indicates that the lamp makers used foraminiferous marls in the manufacture of bilanceolate lamps. Foraminiferous marls are found in central or northern Israel (Goren 1995: 206–7, appendix A, nos. 5–6, 12), but also occur in Jordan (Lapp 1997a: 163–64; da Costa 2003: 1.63). The foraminifera in most lamp fabrics tend to be disarticulated, suggesting a highly agitated depositional environment. Another explanation for the poor preservation of the microfossils' internal structures is that foraminiferous-rich limestone was ground by the lamp maker and then added to the lamp fabrics as a tempering agent. Relatively high firing temperatures were responsible for the decomposed skeletal tests and mottling exhibited by some microfossils in several samples (photo 3, Sample DS10).

No diagnostic tempers such as basalt rock fragments were observed in the lamps' fabrics: only a single basalt-derived rock fragment was found in the fabric of sample DSCY8 from Scythopolis (Lapp 1997a: pl. 5.2). In parts of Lower Galilee near Chorazin north of the Sea of Galilee and Hammath Tiberias to the west, basalt is a common source rock. The occurrence of basalt temper in any of the lamp fabrics would have suggested this region as a likely source at least for this tempering agent. Based on the sedimentary nature of the mineralogy and presence of calcareous microfossils, the petrology indicates that the lamps were manufactured with calcareous clays. With the addition of dilute HCl acid, the samples effervesced vigorously, further exhibiting their high concentrations of calcium carbonate in their respective clay fabrics.

Petrographic and micromorphological variation occurs among the sampled lamp fabrics (table 6). The majority of Palestinian discus lamp fabrics are characterized by

Table 6. Petrographic Descriptions of Representative Samples by Type: Palestinian Discus, Bilanceolate, Beit Nattif, Gilead, and Tarsus

Palestinian Discus, Group 1, Sample DS10	
Characteristics	Pale rendzina soil material
Coarse material	Coarse silt to medium sand-size quartz grains: 40–340 µm, low sphericity, sub-angular, diffuse, moderately sorted, 15% of matrix
Fine material	Calcareous groundmass
Pedofeatures	*Terra rossa* clay nodules: medium sand-size, 350–540 µm, rounded, < 1% of matrix
Biorelicts	Foraminifera grains: very fine to fine sand-size, 80–240 µm, well rounded, diffuse, poorly sorted, hematite staining, poorly preserved, replaced by secondary calcite, ca. 3% of matrix
Pores, cracks, veins	None
Palestinian Discus, Group 2, Sample DM10	
Characteristics	Reddish-brown rendzina soil material
Coarse material	Medium silt-size quartz grains; 20–30 µm, sub-angular, distinct, low sphericity, poorly sorted, 3% of matrix; medium silt to very fine sand-size quartz: 80–270 µm, angular, low sphericiticity, distinct, poorly sorted, 10% of matrix; travertine: medium silt, 30–220 µm, sub-rounded, low and high sphericity, 30% of matrix; chlorite (?): medium silt-size grain, 30 µm, sub-angular, < 1% of matrix
Fine material	Calcareous groundmass
Pedofeatures	*Terra rossa* nodules containing no carbonates: medium silt to fine sand-size, 20–230 µm, rounded, low sphericity, poorly sorted, diffuse, ca. 5% of matrix, naturally occurring temper (note: a single grain: coarse sand-size, 700 µm)
Biorelicts	Globeriginid grains: very fine to medium sand-size, 70–300 µm, well rounded, diffuse, poorly sorted, hematite staining, ca. 5% of matrix
Pores, cracks, veins	None
Palestinian Discus, Scythopolis Version, Sample DSCY8	
Characteristics	Reddish-brown rendzina soil material; abundant travertine
Coarse material	Coarse silt to medium sand-size quartz grains: 50–450 µm, sub-angular low sphericity, well-sorted, distinct, ca. 25% of matrix; travertine grains: medium silt to coarse sand size, 25–1625 µm, well rounded, mostly high sphericity, distinct, moderately sorted, ca. 20% of matrix
Fine material	Calcareous, Fe-rich groundmass
Pedofeatures	*Terra rossa* clay nodules: coarse silt to very fine sand size, 50–100 µm, low sphericity, round, 3% of matrix
Biorelicts	Terrestrial mollusk shells, rounded, low sphericity, poorly sorted, distinct, < 1% of matrix, coarse sand size, 650 µm
Pores, cracks, veins	Elongated pores: very fine sand size to very coarse sand size, 75–2625 µm, sub-angular, 5% of matrix; vapor veins: 1% of matrix
Palestinian Discus, Abila Version, Sample DAB4	
Characteristics	Reddish-brown rendzina soil material
Coarse material	Quartz grains: medium silt to find sand, 20–190 µm, sub-angular, low and high sphericity, poorly sorted, distinct, ca. 7% of matrix; rhombohedral dolomite grains: coarse silt size, 40–60 µm, poorly sorted, distinct, ca. 7% of matrix
Fine material	Calcareous groundmass
Pedofeatures	*Terra rossa* clay nodules containing no carbonates, coarse to fine sand size, 20–590 µm, high sphericity, well rounded, moderately poorly sorted, added as temper, ca. 25% of matrix

Table 6. Petrographic Descriptions of Representative Samples by Type: Palestinian Discus, Bilanceolate, Beit Nattif, Gilead, and Tarsus (cont.)

Biorelicts	None
Pores, cracks, veins	Cracks (replaced by secondary calcite) follow contours of *terra rossa* clay nodules; elongated and round pores: very fine sand to medium sand size, 110–380 μm, ca. 3% of matrix
Tarsus, Sample DA9	
Characteristics	Oolitic limestone parent-source rock
Coarse material	Coarse silt to very fine sand-size quartz grains: 50–100 μm, low sphericity, well rounded, poorly sorted, diffuse, 5% of matrix
Fine material	Calcareous, argillaceous groundmass
Pedofeatures	Clay nodules: medium silt to fine sand size, 30–170 μm, low sphericity, well rounded, poorly sorted, residual, ca. 1% of matrix
Biorelicts	Oolites: coarse silt to fine sand size, 50–100 μm, well-rounded, diffuse, poorly sorted, hematite staining, ca. 3% of matrix
Pores, cracks, veins	Elongated pores: very fine sand to coarse sand size, 70–610 μm, ca. 3% of matrix; rounded pores (destroyed oolites): coarse silt to very fine sand size, 50–100 μm, 3% of matrix
Bilanceolate, Sample BG1	
Characteristics	Pale rendzina soil material
Coarse material	Medium to coarse silt-size quartz grains: 20–50 μm, sub-rounded, low sphericity, 2% of matrix, poorly sorted; fine sand size, rhombohedral dolomite grains: medium to coarse silt size, 20–40 μm, angular, < 1% of matrix
Fine material	Highly calcareous groundmass
Pedofeatures	Residual hematite (black) clay nodules: medium silt to medium sand size, 10–400 μm, sub-angular, low and high sphericity, diffuse, ca. 7% of matrix
Biorelicts	Foraminifera grains: medium silt to fine sand size, 10–230 μm, well to moderately sorted, distinct, very well preserved, secondary calcitic replacement, some hematite staining, ca. 80% of matrix
Pores, cracks, veins	Elongated pores: very fine sand to coarse sand size, 840–1150 μm, ca. 10% of matrix
Beit Nattif, Group 1: Sample BNG8	
Characteristics	Reddish-brown rendzina soil material; abundant travertine
Coarse material	Coarse silt to fine sand-size quartz grains: 40–170 μm, angular, low sphericity, well sorted, distinct, 10% of matrix; travertine grains: medium silt to coarse sand size, 20–950 μm, rounded, mostly high sphericity, distinct, moderately sorted, 30% of matrix; plagioclase feldspar: coarse silt-size grains, 40 μm, angular, low sphericity, residual, < 1% of matrix; high birefringence minerals: medium to coarse silt-size grains, 10–40 μm, angular, low sphericity, 1% of matrix
Fine material	Calcareous, Fe-rich groundmass
Pedofeatures	*Terra rossa* clay nodules: medium silt to very fine sand size, 10–80 μm, round, low sphericity, 10% of matrix
Biorelicts	Terrestrial mollusk shells: medium to coarse sand size, 280–710 μm, low sphericity, poorly sorted, veins, distinct, ca. 3% of matrix
Pores, cracks, veins	Elongated and round pores: coarse silt to medium sand size, 50–400 μm, 5% of matrix; vapor veins, < 1% of matrix
Beit Nattif, Group 2, Sample BNG1	
Characteristics	Reddish-brown rendzina soil material

Table 6. Petrographic Descriptions of Representative Samples by Type: Palestinian Discus, Bilanceolate, Beit Nattif, Gilead, and Tarsus (cont.)

Coarse material	Travertine: medium silt, 30–220 μm, sub-rounded, low and high sphericity, 30% of matrix; chlorite (?): medium silt-size grains, 30 μm, sub-angular, < 1% of matrix
Fine material	Calcareous groundmass
Pedofeatures	*Terra rossa* nodules containing no carbonates: medium silt to fine sand size, 20–230 μm, rounded, low sphericity, poorly sorted, diffuse, ca. 5% of matrix, naturally occurring temper
Biorelicts	Globeriginid grains: very fine to medium sand size, 70–300 μm, well rounded, diffuse, poorly sorted, hematite staining, ca. 5% of matrix
Pores, cracks, veins	None
Beit Nattif, Group 3, Sample BNS1	
Characteristics	Reddish-brown rendzina soil material
Coarse material	Coarse silt to fine sand-size quartz grains: 40–160 μm, sub-angular, low sphericity, well sorted, distinct, ca. 10% of matrix; travertine grains: fine to medium sand size, 160–490 μm, well rounded, low sphericity, distinct, poorly sorted, ca. 2% of matrix; high birefringence minerals: coarse silt size, 50–60 μm, low sphericity, sub-angular, < 1% of matrix
Fine material	Calcareous groundmass
Pedofeatures	*Terra rossa* clay nodules: medium silt to very coarse sand size, 25–1500 μm, rounded, low sphericity, 10% of matrix, poorly sorted
Biorelicts	Oolites: medium silt to very fine sand size, 80–240 μm, rounded, high sphericity, poorly sorted, 1% of matrix
Pores, cracks, veins	Elongated pores: fine to very fine sand size, 150–960 μm, 3% of matrix; rounded pores: very fine to medium sand size, 60–400 μm, < 1% of matrix; vapor veins, 1% of matrix
Gilead, Sample BNJ2	
Characteristics	Groundmass of well-sorted quartz grains exhibiting high sphericity
Coarse material	Coarse silt to fine sand-size quartz grains: 60–160 μm, sub-angular, low sphericity, poorly sorted, 5% of matrix
Fine material	Medium silt-size quartz sand: 10–20 μm, high sphericity, well sorted, round, ca. 80% of groundmass
Pedofeatures	*Terra rossa* clay nodules: medium silt to medium sand size, 20–480 μm, rounded, low sphericity, diffuse, 5% of matrix; secondary calcitic replacement of cracks around grains
Biorelicts	None
Pores, cracks, veins	Elongated pores: fine to coarse sand size, 170–260 μm, sub-angular, 1% of matrix; cracks: < 1% of matrix
Jerash Bowl, Sample JB1	
Characteristics	Lower Cretaceous sandstone
Coarse material	Silt to medium sand-size quartz grains: 40–320 μm, low sphericity, very angular, well sorted, 30% of matrix
Fine material	Very ferruginous groundmass
Pedofeatures	Hematite clay nodules: silt to medium sand size, 40–320 μm, well rounded, high sphericity, poorly sorted, ca. 2% of matrix
Biorelicts	None
Pores, cracks, veins	None (note: ca. 30% desiccation cracks occur in fabric of JB2

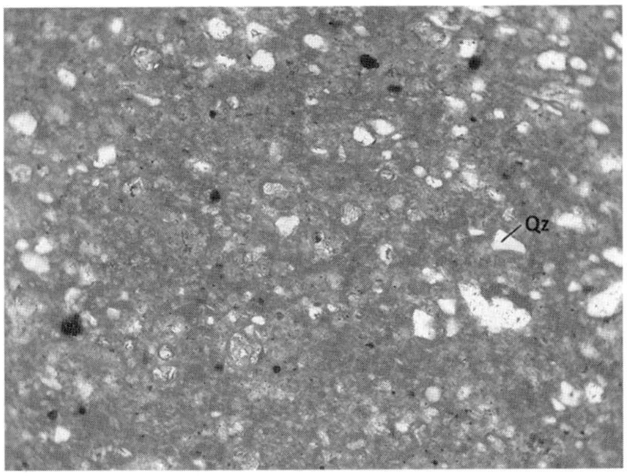

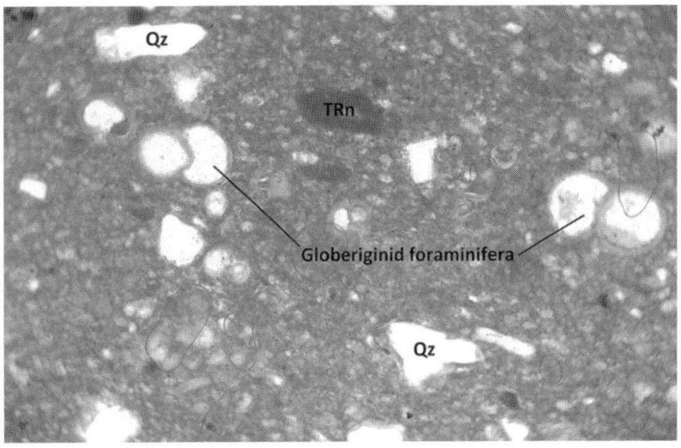

Photo 4 (left). Photomicrograph of Sample DP3, a Palestinian discus lamp fragment with a bi-associative image of ibis and crab. Moderately sorted, sub-angular quartz sand inclusions (Qz). *Terra rossa* nodules. Plain polarized light (×10). Pella of the Decapolis, Transjordan. Photomicrograph by E. C. Lapp.

Photo 5 (right). Photomicrograph of Sample DM10, a Palestinian discus lamp fragment with a lamp maker's mark on the base, the Greek initial *eta*. Globeriginid foraminifera. Poorly sorted, sub-angular quartz sand inclusions (Qz). *Terra rossa* nodules (TRn) added as temper. Plain polarized light (×10). Meiron, Upper Galilee. Photomicrograph by E. C. Lapp.

a high percentage of poorly preserved and mottled bioclastic inclusions. The bilanceolate lamp fabrics are characterized by a very high percentage of well-preserved foraminifera, ostracod valves, and shell fragments. Several Palestinian discus and Beit Nattif lamp samples also exhibit similar characteristics. The Beit Nattif lamp fabrics are characterized by either a high percentage of travertine inclusions, a high concentration of quartz sand exhibiting high sphericity, or a comparatively low percentage of nonplastic inclusions.

Petrographic Fabric Groups and Micromorphological Characteristics

Group Rz (Rendzina Group)

Palestinian Discus. Pale rendzina or brown rendzina soil material was selected by lamp makers for manufacturing Palestinian discus lamps (tables 6–7). Palestinian discus lamps made with pale rendzina soil material exhibit a highly calcareous micromass and a low abundance (< 10%) of poorly sorted quartz sand grains. Pale rendzina soils occur in the hill and mountain regions of Galilee, Samaria, and Judea. The chief diagnostic characteristic of lamp fabrics belonging to this petrographic group is a relatively high abundance of poorly preserved microfossils (photos 3–7). These bioclastic inclusions (globerginids, shell fragments, and disarticulated ostracod valves) exhibit a high degree of mottling, suggestive of higher firing temperatures (approximately in the range of 700°C) than those used for the samples comprising the other fabric groups. The vitrified appearance of the fabrics also suggests higher firing temperatures. Lamp fabrics belonging to this group have a high percentage of

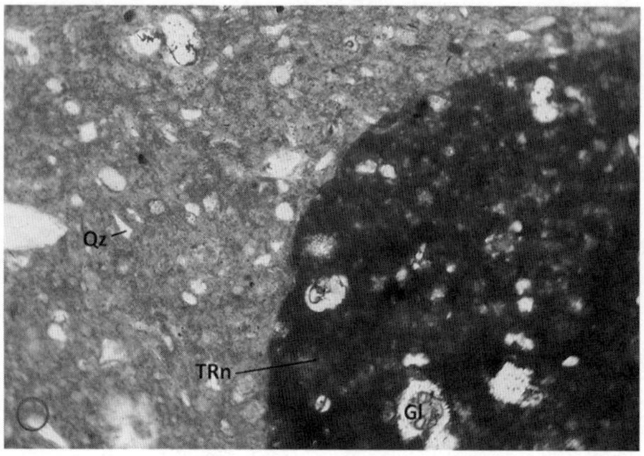

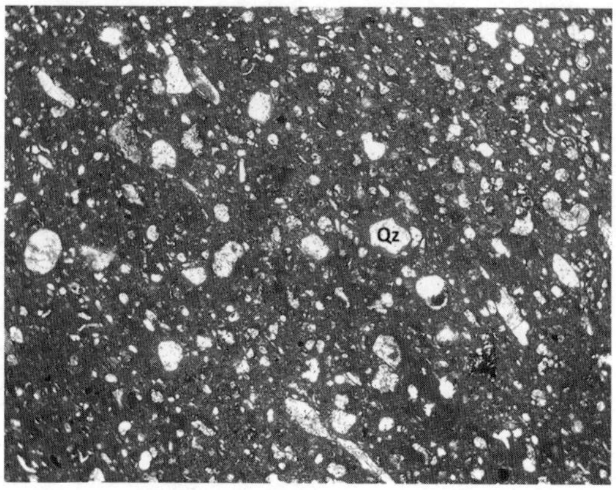

Photo 6 (left). Photomicrograph of Sample DHH2, a Palestinian discus lamp fragment with a bichrome slip (red discus and dark blue body). Poorly preserved globeriginid foraminifera (Gl). Poorly sorted, sub-angular quartz sand inclusions (Qz). *Terra rossa* nodules (TRn) added as temper. Plain polarized light (×10). Ḥorvat Ḥazon, Upper Galilee. Photomicrograph by E. C. Lapp.

Photo 7 (right). Photomicrograph of Sample DS9, a Palestinian discus lamp fragment with an erotic scene. Poorly preserved and highly disarticulated bioclasts. Well-sorted, angular quartz sand inclusions (Qz). Plain polarized light (×10). Sepphoris, Lower Galilee. Photomicrograph by E. C. Lapp.

sub-rounded quartz sand exhibiting low sphericity. Dark green material also occurs in these fabrics and may represent organic material. The nonplastic inclusions are poorly sorted. Twelve lamp samples belong to this fabric group: DA2, DA9, DHH2, DM10, DP3, DS4, DS9, DS10, DS21, DS22, DS29, and DS30.

The Abila version of the Palestinian discus group (sample DAB4) exhibits a relatively high concentration of soil nodules of leached *terra rossa* soil material added as temper (photo 8). Evidence for the use of soil nodules of leached *terra rossa* soil material as a tempering agent has also been observed petrographically in Herodian lamp samples recovered at Sepphoris, Dora, and Scythopolis (Adan-Bayewitz et al. 2008: 60–61, table 3, figs. 15.2, 16.1–2, 18.1). Fabrics of this group are further characterized by a high percentage of fine quartz sand inclusions that exhibit fair to poor sorting and a relatively low percentage of microfossils. The relatively high abundance of hematite and quartz sand found in sample DAB4 explains the percentages of Fe2O3 (7.01%) and SiO2 (50.83%) occurring in the fabric as determined by DCP analysis. Only sample DSCY7 exhibits as high an Fe2O3 (7.84%) and SiO2 (50.83%) content, but it does not exhibit the same petrographic and micromorphological fabric characteristics. The fabric of sample DSCY6 is unfired and contains disarticulated microfossils (photo 9).

Tarsus. The lamp maker of the Tarsus type (sample DA9) used a pale yellow calcareous clay with a very low abundance of poorly sorted, angular to sub-rounded quartz grains (photo 18, tables 6–7). Very few microfossils (< 1%) occur in the fabric. This creates a generally "clean" appearance. It is essential that additional samples

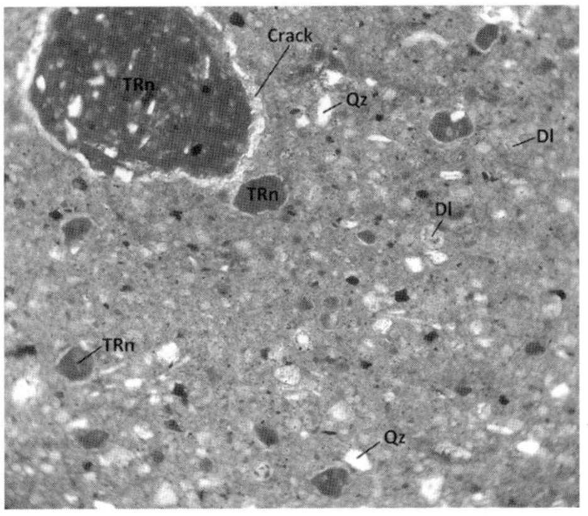

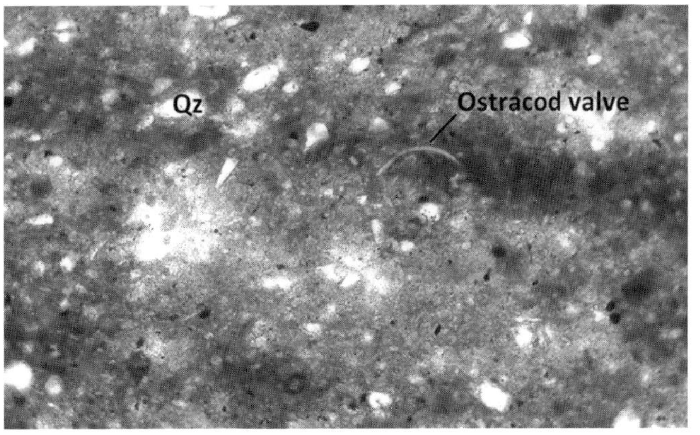

Photo 8 (left). Photomicrograph of Sample DAB4, a Palestinian discus (Abila version) lamp fragment with a double-axe lamp maker's mark on the base. Reddish-brown rendzina soil material. Poorly sorted, sub-angular quartz sand inclusions (Qz). Rhombohedral dolomite grains (Dl). *Terra rossa* nodules added as temper (TRn). Plain polarized light (×10). Abila of the Decapolis, Transjordan. Photomicrograph by E. C. Lapp.

Photo 9 (right). Photomicrograph of Sample DSCY6, an unfired Palestinian discus lamp fragment with a lamp maker's mark on the base, the Greek initials *nu eta*. Moderately sorted, sub-angular to sub-rounded quartz sand inclusions (Qz). Plain polarized light (×10). Scythopolis of the Decapolis, Transjordan. Photomicrograph by E. C. Lapp.

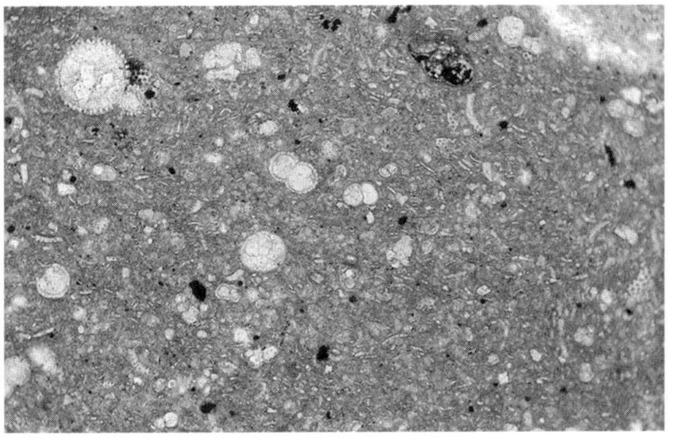

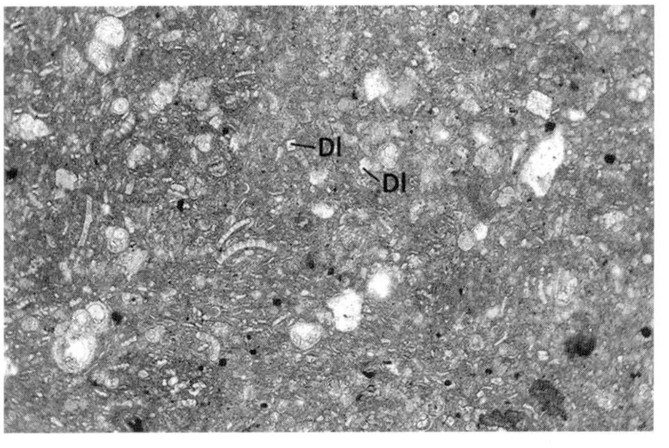

Photo 10 (left). Photomicrograph of Sample BG1, a bilanceolate lamp fragment. Pale rendzina soil material. Extremely high abundance of well-preserved and well-sorted foraminifera. Poorly sorted, sub-angular quartz sand inclusions. Rhombohedral dolomite grains. Plain polarized light (×10). Gadara of the Decapolis, Transjordan. Photomicrograph by E. C. Lapp.

Photo 11 (right). Photomicrograph of Sample BP7, a bilanceolate lamp fragment. Pale rendzina soil material. Extremely high abundance of well-preserved and well-sorted foraminifera. Poorly sorted, sub-angular quartz sand inclusions. Rhombohedral dolomite grains (Dl). Plain polarized light (×10). Pella of the Decapolis, Transjordan. Photomicrograph by E. C. Lapp.

of the Tarsus group from Ashkelon and Tarsus itself are examined petrographically and compared to determine the general micromorphological characteristics specific to the group and whether fabric variation exists.

Table 7. Petrographic Groups

Petrographic Group	*Diagnostics*	*Suggested Provenance*
Group Rz–Rendzina Group	*Sample DS3:* Pale rendzina soil material with highly calcareous groundmass of low abundance (< 10%) of poorly sorted quartz sand grains	Local and regional: hill and mountain regions, Galilee and eastern Samaria
	Sample DS10: Pale rendzina soil material with highly calcareous groundmass and high abundance of poorly preserved and mottled foraminifera (ca. 40%); very low abundance of poorly sorted, silt-size quartz sand grains (< 5%)	
	Sample DAB4: Brown rendzina soil material with *terra rossa* nodules added as temper; low abundance (< 5%) of poorly sorted rhombohedral dolomite sand grains; low abundance (< 10%) of poorly sorted, silt-size quartz sand grains	
Group Fm–Foraminifera-Rich Marl Group	*Samples BS9, BP1, BP2,BP7, BG1, BG5, BSCY1, BSCY5:* Very high abundance of Tertiary foraminifera, ca. 80% of matrix; well-preserved bioclastic tests; low abundance (< 10%) of poorly sorted, silt-size quartz sand grains; low abundance (< 5%) of poorly sorted rhombohedral dolomite sand grains	Local, unidentified foraminiferal marl source
Group TRm–Travertine and Mollusk Shell Group	*Samples BNG8, DSCY8, DSCY7:* Ferruginous clay with travertine and terrestrial mollusk shells	Local to Beth Shean Valley
	Samples BNS1, BNS2: Carbonaceous clay with travertine and terrestrial mollusk shells; some chlorite (?)	
Group SQz–Spherical Quartz Group	*Samples BNJ2, BNJ3:* Groundmass of well-sorted (?), silt-size quartz grains exhibiting high sphericity	Jerash area, local wadi (?) sand
Group FQz–Ferruginous Clay with Quartz Temper	*Samples JB1, JB2, JB3, JB4, JB5, and JB6:* Very high abundance of well-sorted, angular, mature quartz grains (ca. 80% of matrix) added as temper; and set in a ferruginous groundmass; corresponds to Beth Shean Petrographic Group Qz2 (Cohen-Weinberger and Goran 2011: 219, 222, 224)	Lower Cretaceous sandstones exposed in Jerash area and confirmed Jerash origin at hippodrome pottery workshop

Group Fm (Foraminifera-Rich Marl Group): Bilanceolate

Bilanceolate lamp fabrics are characterized by a very high abundance of well-preserved Tertiary foraminifera (about 80%) and a low percentage or absence of quartz inclusions (photos 10–13, tables 6–7). Lamp makers typically used reddish-brown marly clays rich in foraminifera to manufacture bilanceolate lamps. The range of microfossils found in the fabrics of this group include: planktonic foraminifera, highly disarticulated ostracod valves, terrestrial mollusk shell fragments, and coral fragments. The internal structure of the tests of the globigerina comprise bulbous

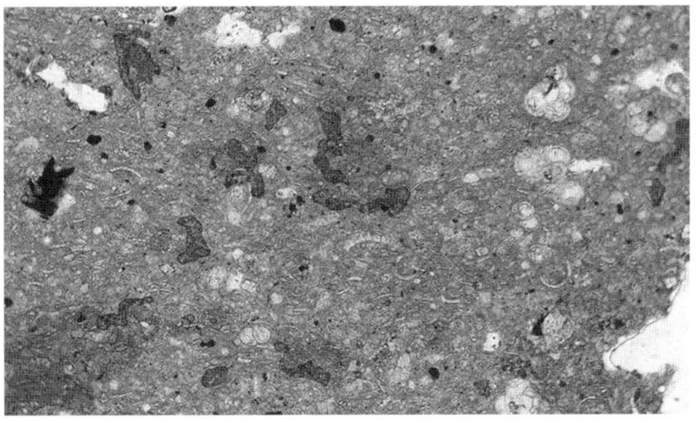

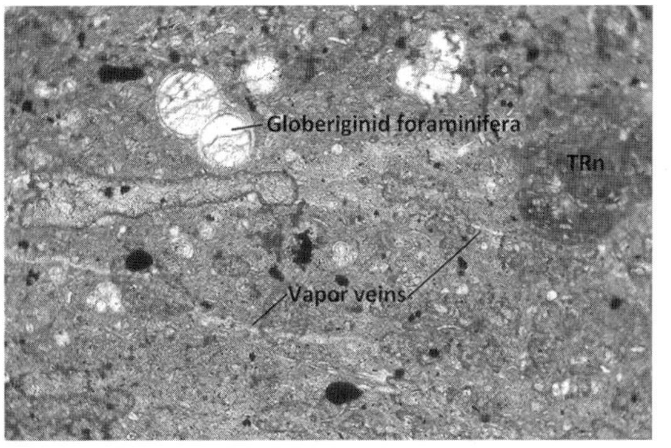

Photo 12 (left). Photomicrograph of Sample BG5, a bilanceolate lamp fragment. Pale rendzina soil material. Extremely high abundance of well-preserved and well-sorted foraminifera. Poorly sorted, sub-angular quartz sand inclusions. Rhombohedral dolomite grains. Plain polarized light (×10). Gadara of the Decapolis, Transjordan. Photomicrograph by E. C. Lapp.

Photo 13 (right). Photomicrograph of Sample BS9, a bilanceolate lamp fragment. Pale rendzina soil material. Extremely high abundance of well-preserved and well-sorted foraminifera. Poorly sorted, sub-angular quartz sand inclusions. Rhombohedral dolomite grains. *Terra rossa* nodules (TRn). Vapor veins. Plain polarized light (×10). Sepphoris, Lower Galilee. Photomicrograph by E. C. Lapp.

chambers replaced by secondary platty calcite. The good state of preservation of the bioclasts can be explained by a low firing temperature (approximately 400°C) that did not destroy the internal structure of the microfossils, as is evident with respect to the poorly preserved bioclasts of the Palestinian discus fabrics. The bioclastic nonplastic material exhibits a very low degree of merging with the fabric, and the boundaries are well defined. The concentration of foraminifera is significantly greater than that observed in any other fabric group of this study. The micromass is highly calcareous. Rhombohedral crystals of dolomite and opaque matter occur in the fabrics. A low abundance or complete absence of silt-size quartz sand was found in the bilanceolate fabrics and tended to be sub-angular to sub-rounded, and exhibit low sphericity. Samples BS9, BG5, and BP1 are representative examples of this group. The abundant foraminifera content of this fabric group indicates that the source or sources were a marly clay. Ten lamp samples belong to this fabric group: BG1, BG8, BP2, BP3, BP7, BSCY1, BSCY5, BS3, BS4, and BS9 (photos 10–13). Although there is variation in the petrographic composition of bilanceolate lamps, the consistent appearance of most components, specifically the well-preserved foraminifera indicates a homogenous petrographic group.

Group TRm (Travertine and Mollusk Shell Group): Beit Nattif

The Beit Nattif samples BNG1, BNG8, BNG9, and BNG10 recovered from Gadara are characterized by a high percentage of well-rounded travertine grains and contain terrestrial mollusk shells (photos 14–15, tables 6–7). They were manufactured with a reddish-brown rendzina soil material and exhibit an iron-rich groundmass. The fabrics contain well-sorted, angular coarse silt to fine sand-size

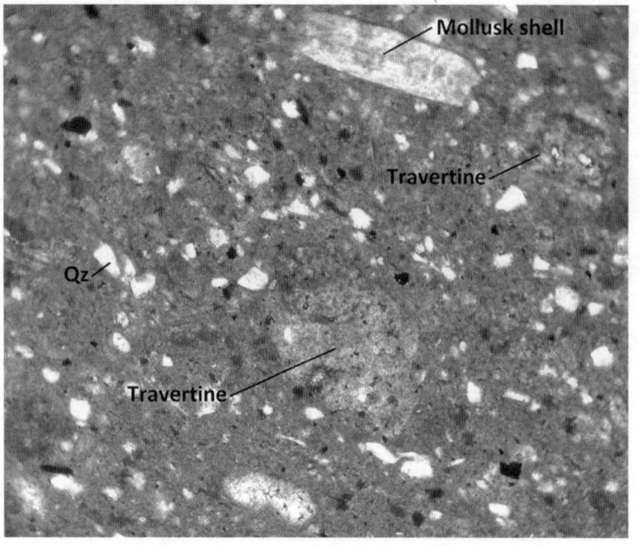

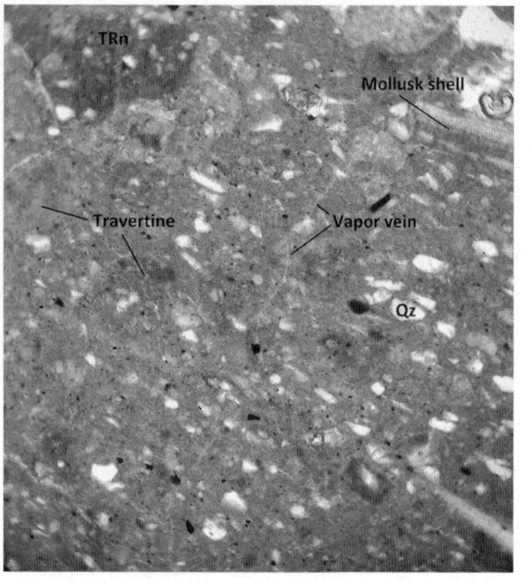

Photo 14 (left). Photomicrograph of Sample BNG8, a Beit Nattif lamp fragment. Reddish-brown rendzina soil material. Calcareous Fe-rich groundmass. Abundant travertine. Well-sorted, sub-angular quartz sand inclusions (Qz). *Terra rossa* nodules (TRn). Terrestrial mollusk shells. Elongated pores and vapor veins. Plain polarized light (×10). Gadara of the Decapolis, Transjordan. Photomicrograph by E. C. Lapp.

Photo 15 (right). Photomicrograph of Sample DSCY8, a Palestinian discus lamp fragment. Reddish-brown rendzina soil material. Calcareous Fe-rich groundmass. Abundant travertine. Well-sorted, sub-angular quartz sand inclusions (Qz). *Terra rossa* nodules (TRn). Terrestrial mollusk shells. Elongated pores and vapor veins. Plain polarized light (×10). Scythopolis of the Decapolis, Transjordan. Photomicrograph by E. C. Lapp.

quartz grains and *terra rossa* clay nodules. Elongated pores and vapor veins are also present. The petrographic characteristics of these samples suggest local manufacture and closely resemble the Travertine Group (Group TR) local to Beth Shean (Cohen-Weinberger and Goren 2011: 216–17, table 10.1).

Two distinct grain-size groups of inclusions can be distinguished: Group A grains are very fine and Group B, fine to medium. Quartz belongs to both grain-size groups and limestone to only the latter. The limestone fragments exhibit a high degree of merging while the Group B quartz, low merging. Some of these inclusions are poorly preserved microfossils that have lost their primary internal structures. They are a distinctive dark-gray color. The Gadaran fabrics contain a greater travertine abundance than any other of the samples in this survey. Three lamp samples belong to this fabric group: DSCY7, DSCY8, and BNG8. This is a particularly important petrographic group in that lamp samples belonging to two entirely different types—the Palestinian discus (samples DSCY7 and DSCY8) and Beit Nattif (BNG8)—share the same petrographic characteristics and belong to the same fabric group. These samples demonstrate that lamps belonging to different typological classes may in fact belong to the same petrographic fabric group. This suggests the use of the same or a similar clay source for the manufacture of lamps belonging to these

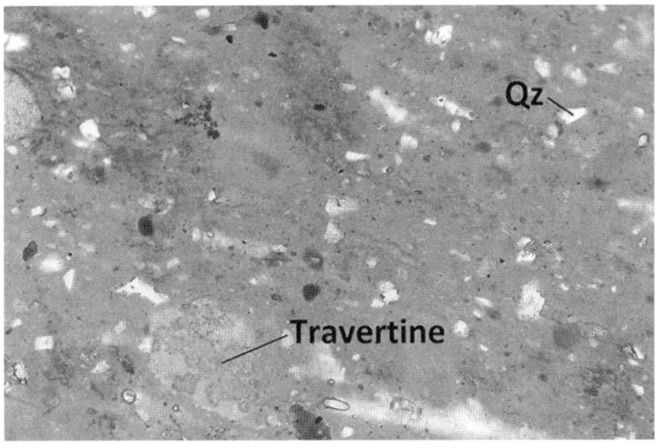

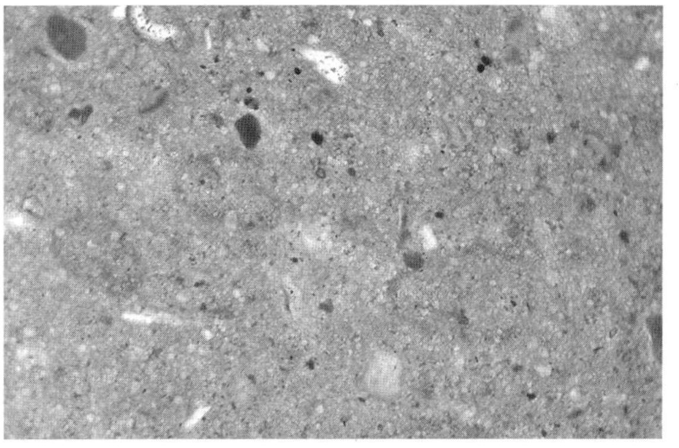

Photo 16 (left). Photomicrograph of Sample BNS1, a Beit Nattif lamp fragment. Pale rendzina soil material. Calcareous groundmass. Travertine grains. Well-sorted, sub-angular quartz sand inclusions (Qz). Oolites. *Terra rossa* clay nodules (TRn). Vapor veins. Plain polarized light (×10). Sepphoris, Lower Galilee. Photomicrograph by E. C. Lapp.

Photo 17 (right). Photomicrograph of Sample BNJ2, a Gilead lamp fragment. Groundmass consisting of a high abundance of well-sorted quartz sand inclusions exhibiting high sphericity. No biorelicts. *Terra rossa* clay nodules (TRn). Elongated pores and cracks replaced by secondary calcite. Plain polarized light (×10). Gerasa of the Decapolis, Transjordan. Photomicrograph by E. C. Lapp.

different groups. They will be analyzed to determine whether they share the same chemical fingerprint. If so, it would suggest that the same raw clay source had been exploited for their manufacture. The presence of travertine fragments and quartz sand inclusions in the unfired fabric of sample DSCY6 suggests it belongs to this fabric group.

The Beit Nattif samples found at Sepphoris (BNS1 and BNS3) are characterized by a very fine micritic, calcareous groundmass and a very low percentage of microfossils that include planktonic foraminifera: a low abundance of poorly preserved oolites and disarticulated ostracods (photo 16, tables 6–7). The sorting of the allochems ranges from poor to fair. Similar to the Gadaran samples, the Sepphorean fabrics contain well-sorted, angular coarse silt to fine sand-size quartz grains and *terra rossa* clay nodules. Only two lamp samples belong to this group: BNS1 and BNS3.

Group SQz (Spherical Quartz Group): Gilead

Gilead lamp fabrics (samples BNJ2 and BNJ4) are characterized by a very high concentration of well-sorted and well-rounded quartz sand inclusions that exhibit a high degree of sphericity (photo 17, tables 6–7). The quartz grains are very fine to fine in size. The petrographic and micromorphological features of their fabrics are distinct from all others in the sampling of the study, including the Beit Nattif samples. This supports their identification as a separate lamp type—the Gilead group—as observed by da Costa (2003: 1.71–72). Archaeological evidence (molds and numerous such lamps) suggests the manufacture of Gilead lamps in Gerasa or in the vicinity of the Decapolis city. Only two lamp samples belong to this group:

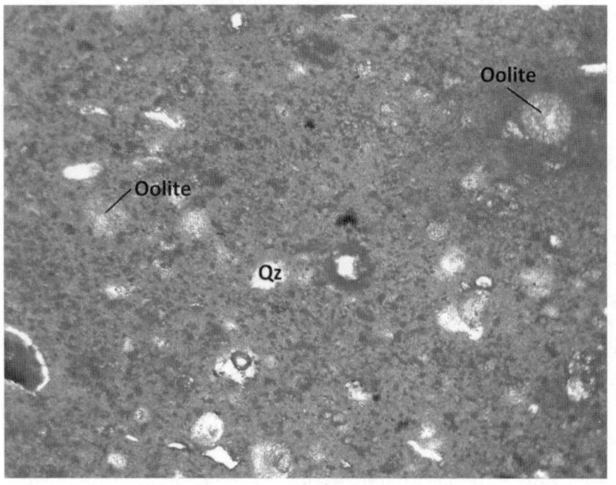

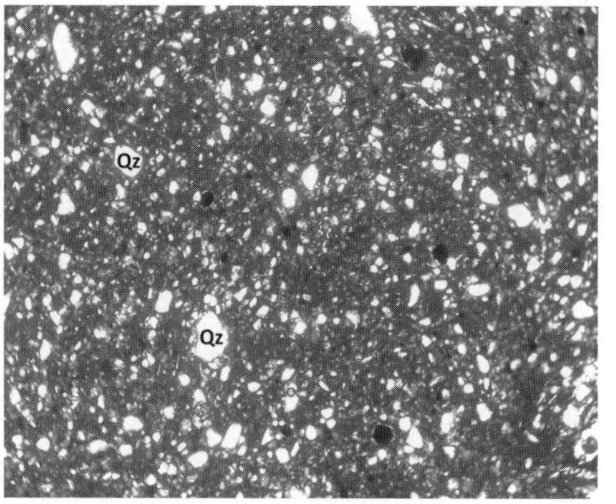

Photo 18 (left). Photomicrograph of Sample DA4, a Tarsus lamp fragment. Calcareous argillaceous groundmass. Poorly sorted oolites. Poorly sorted, well-rounded quartz sand inclusions (Qz). Elongated and rounded pores. Plain polarized light (×10). Ashkelon, coastal plain. Photomicrograph by E. C. Lapp.

Photo 19 (right). Photomicrograph of Sample JB1, a Jerash bowl base fragment. Lower Cretaceous sandstone. Highly ferruginous groundmass. High abundance of well-sorted angular quartz grains (Qz). Biorelicts absent. Plain polarized light (×10). Gerasa of the Decapolis, Transjordan. Photomicrograph by E. C. Lapp.

BNJ2 and BNJ4. Both samples originally belonged to lamps of the Gilead type. That the findspot of the samples comprising this group is Gerasa, and that this site is located near the Nubian sandstone geological formation show that these lamps could be derived from Nubian sandstone clays from the vicinity of Gerasa (Bender 1968: 40–41). Gilead lamp fabrics have a low percentage of microfossils. Inclusions include mainly quartz and limestone. The *terra rossa* clay nodules were likely added as temper. Gilead fabrics also contain disarticulated ostracod valves.

Group FQz (Ferruginous Clay with Quartz Temper): Jerash Bowls

All six Jerash bowl fabrics (JB1–JB6) are characterized by a high abundance of well-sorted, angular quartz sand (photo 19, tables 6–7; see also Lapp 2001a: 131–33, tables 1–3). No lamp fabrics of this study are similar to this group. The quartz grains account for the high abundance of silica (SiO2) on average 71.66% determined by the dcp-oes analysis. The quartz grains range in size from 0.04 mm to 0.48 mm (silt to medium sand). Overall the percentage of quartz sand ranges between approximately 20–30% of the samples' fabrics. Limestone inclusions only occur in sample JB3 and represent less than 1.0% of the fabric. *Terra rossa* nodules comprise approximately 1.0–2.0% of the fired fabrics JB1, JB2, and JB3. The quartz-rich Lower Cretaceous Jerash bowl fabric group corresponds to "Group FQz–Ferruginous Clay with Quartz Temper" analyzed from the theater pottery workshop at Beth Shean (Cohen-Weinberger and Goren 2011: 219). The potters operating in the hippodrome workshop at Jerash had easy access to locally exposed Lower Cretaceous clay sources (Cohen-Weinberger and Goren 2011: 223).

Table 8. Chemical Compositions of Samples as Determined by DCP-OES

	DA2	*DA3*	*DA9*	*DAB4*	*DC6*	*DHH2*	*DM10*	*DP3*
SiO_2	44.89	46.15	45.55	50.83	45.04	48.08	48.77	44.32
Al_2O_3	14.91	13.7	11.13	15.1	16.7	17.11	15.12	15.21
Fe_2O_3	5.36	4.91	5.00	7.01	5.47	5.69	5.15	6.46
MgO	1.84	1.64	2.94	2.34	1.9	2.57	2.79	3.75
CaO	29.8	29.5	30.84	18.8	25.5	22.6	24.45	27.08
Na_2O	0.34	0.39	0.42	0.3	0.33	0.38	0.24	0.57
K_2O	2.31	2.59	1.35	1.34	2.98	2.59	2.39	2.32
TiO_2	0.76	0.73	0.46	0.9	0.79	0.83	0.79	1.03
P_2O_5	0.5	0.00	0.10	0.2	0.00	0.51	0.58	0.80
MnO	0.02	0.405	1.12	2.72	0.502	0.03	0.027	0.066
Sum	100.73	100.0	98.94	99.6	99.3	100.43	100.29	101.61
LOI[a]	-0.667	-0.859	-0.637	-0.378	-0.474	-0.418	-0.53	-0.174
Ba	576.7	521	4284	2502	653	457.7	514	380
Sr	762.04	712	1012	899	732	565.28	518	729
Y	43.6	46.6	80.38	195	48.7	42.4	50	59
V	93.04	133	107.85	173	142	82.95	125	129
Sc	18.9	22.8	32.86	25.8	23.8	18.8	19	24
Cu	38.36	156	236.46	79.5	60.8	37.41	31	34
Cr	114	128	195.6	238	149	124	113	215
Ni	59.6	74	151.9	238	76.7	58.5	69	95

	DP11	*DSCY6*	*DSCY7*	*DSCY8*	*DS4*	*DS9*	*DS10*	*DS21*
SiO_2	37.66	35.39	50.83	43.31	45.37	42.82	40.02	47.55
Al_2O_3	11.43	9.25	11.80	11.36	15.62	14.69	14.59	15.52
Fe_2O_3	5.63	5.37	7.84	6.44	5.66	5.27	5.46	5.89
MgO	2.15	3.07	4.64	3.59	2.47	2.11	2.34	1.92
CaO	36.01	44.16	21.93	32.5	25.9	31.65	34.37	26.08
Na_2O	0.6	0.44	0.62	0.69	0.43	0.63	0.69	1.12
K_2O	3.19	1.14	1.25	1.20	3.67	2.67	2.0	1.89
TiO_2	0.76	1.05	1.17	1.05	0.79	0.75	0.75	0.83
P_2O_5	0.047	0.21	0.23	0.09	0.46	0.02	0.02	0.68
MnO	0.89	0.08	0.085	0.36	0.03	0.71	0.72	0.03
Sum	98.38	100.15	100.39	100.63	100.39	101.37	101.0	101.5
LOI[a]	-1.016	-1.449	-0.948	nd	-0.594	-0.542	-0.427	-0.429
Ba	354.6	357	1212	773.5	341.9	826.8	823.9	558
Sr	1136	676	542	655.5	527.42	742	790.8	640
Y	52.69	46	44	46.68	48.2	45.29	48.75	51
V	179.93	102	106	116.1	121.3	98.49	94.08	137
Sc	33.44	20	18	24.24	21.3	33.06	36.15	22
Cu	66.91	141	46	141.93	45.06	77.90	55.63	65
Cr	189.4	119	99	117.64	146	131.1	135.6	143
Ni	130.2	98	91	78.88	70.1	60.87	63.6	67

Table 8. Chemical Compositions of Samples as Determined by DCP-OES (cont.)

	DS22	*DS24*	*DS29*	*DS30*	*BAB1*	*BG1*	*BG5*	*BG8*
SiO_2	43.15	45.69	47.35	44.68	37.03	30.44	35.13	30.48
Al_2O_3	13.53	16.1	16.25	16.2	16.36	12.90	14.99	13.03
Fe_2O_3	5.05	5.7	5.56	5.43	5.28	4.5	4.76	4.59
MgO	2.20	1.73	2.17	2.25	1.45	2.0	1.46	1.79
CaO	32.01	24.6	24.89	28.4	37.99	45.5	40.67	47.82
Na_2O	0.74	0.4	0.93	0.92	0.34	0.17	0.26	0.18
K_2O	2.91	3.53	2.17	2.26	1.05	1.12	1.45	1.16
TiO_2	0.70	0.79	0.81	0.79	0.533	0.45	0.49	0.46
P_2O_5	0.91	0.509	0.02	0.544	1.75	1.94	2.11	1.93
MnO	0.03	0.00	0.69	0.00	0.1	0.09	0.098	0.085
Sum	101.23	99.1	100.88	101.00	101.9	99.07	101.43	101.53
LOI[a]	-0.664	-0.738	-0.33	-0.509	-0.978	-1.143	-1.096	-1.56
Ba	694	601	661.1	387	2983	2214	1930	2448
Sr	632	524	588.3	627	971	974.98	967	1098
Y	50	47.8	46.94	47.6	89.6	100.0	90	97
V	100	153	134.07	135	169	179.7	161	151
Sc	20	23.9	29.62	23.2	27.5	30.1	25	26
Cu	70	59.3	44.23	53.7	55.9	68.64	46	46
Cr	118	134	137.4	144	271	329	276	321
Ni	58	78.6	60.21	69.3	137	122	120	113
	BG10	*BP1*	*BP2*	*BP3*	*BP7*	*BSCY1*	*BSCY4*	*BSCY5*
SiO_2	31.68	32.22	33.70	34.49	34.07	35.45	30.50	35.86
Al_2O_3	13.80	13.35	14.68	15.86	15.41	16.40	14.68	16.47
Fe_2O_3	4.48	4.69	4.77	5.02	4.77	5.59	5.48	5.62
MgO	1.54	1.68	1.86	1.85	1.83	1.71	1.80	1.92
CaO	43.6	43.06	40.21	39.1	41.27	37.43	43.92	36.96
Na_2O	0.19	0.33	0.42	0.45	0.44	0.54	0.34	0.42
K_2O	1.52	2.30	1.72	1.78	1.67	0.90	0.64	0.82
TiO_2	0.47	0.44	0.48	0.53	0.51	0.56	0.48	0.56
P_2O_5	2.37	2.04	2.19	2.05	1.77	2.32	2.10	1.89
MnO	0.09	0.074	0.09	0.12	0.075	0.07	0.08	0.07
Sum	99.71	100.19	100.17	101.29	101.82	101.01	100.04	100.60
LOI[a]	-1.133	-1.293	nd	nd	-1.384	nd	nd	-1.626
Ba	560	1119	2135	1665	1439	2148	1040	3612.28
Sr	1027.4	951	939	1013	947	1104	1055	1062.97
Y	103	102	99.72	108.9	88	99.83	101.8	91.20
V	182.6	158	178.5	177.8	171	130.7	152.9	131.53
Sc	30.5	24	29.21	30.88	25	28.07	30.85	35.21
Cu	57.25	41	62.52	49.10	49	91.86	123.08	149.58
Cr	331	293	280.37	267.45	262	271.8	302.19	255.83
Ni	112	105	119.02	143.31	115	141.05	221.98	187.08

Table 8. Chemical Compositions of Samples as Determined by DCP-OES (cont.)

	BS3	*BS4*	*BS9*	*BNG1*	*BNG8*	*BNG9*	*BNG10*	*BNJ2*
SiO_2	36.06	33.86	30.58	32.06	42.31	37.32	36.67	23.09
Al_2O_3	15.37	14.89	12.25	12.94	10.96	14.83	14.85	7.05
Fe_2O_3	4.68	4.85	4.61	4.77	6.31	5.56	5.14	2.77
MgO	1.38	1.9	2.07	0.94	3.28	0.95	0.86	1.80
CaO	39.6	40.62	46.44	44.28	31.47	36.01	36.45	61.22
Na_2O	0.39	0.39	0.36	0.23	0.59	0.19	0.22	0.31
K_2O	1.27	1.43	0.94	1.43	2.16	1.57	1.66	1.97
TiO_2	0.51	0.49	0.41	0.58	1.03	0.69	0.65	0.45
P_2O_5	0.09	0.10	0.10	3.48	0.20	3.05	0.08	0.40
MnO	2.32	1.99	2.31	0.07	0.09	0.09	3.10	0.034
Sum	101.71	100.46	100.10	100.79	98.40	100.27	99.67	99.10
LOI[a]	nd	nd	nd	–0.984	–0.941	–0.707	–0.91	-1.40
Ba	798.2	923.4	650.6	874	374	1088	793	125
Sr	1011	909.8	1042	898	613	848	835	331
Y	102.4	105.2	119.9	148	39	133	129	42
V	147.1	146	161.6	171	116	173	176	103
Sc	28.90	29.70	31.96	30	20	30	28	20
Cu	51.56	58.32	46.58	51	50	48	51	30
Cr	259.54	277.86	306.42	290	106	259	248	84
Ni	116.31	116.57	106.52	94	63	113	107	40
	BNJ4	*BNS1*	*BNS3*	*JB1*	*JB2*	*JB3*	*JB5*	*HU*[b]
SiO_2	27.78	36.18	33.34	70.09	75.90	71.07	69.61	62.74
Al_2O_3	8.78	9.19	9.98	17.27	12.74	16.25	15.38	30.67
Fe_2O_3	3.67	3.94	4.31	5.23	4.53	4.99	4.81	1.51
MgO	2.08	2.48	1.86	1.32	1.17	1.16	1.32	0.528
CaO	55.3	45.7	48.21	1.92	1.92	1.99	3.91	0.49
Na_2O	0.21	0.36	0.39	0.29	0.24	0.32	0.22	0.39
K_2O	2.41	1.87	1.88	1.88	1.70	1.99	1.92	1.70
TiO_2	0.51	0.52	0.54	1.57	1.31	1.52	1.45	1.37
P_2O_5	0.242	0.752	0.65	0.10	0.11	0.10	0.14	0.072
MnO	0.1	0.1	0.07	0.06	0.06	0.06	0.04	0.009
Sum	101.00	101.00	101.22	99.75	99.67	99.45	98.81	99.47
LOI[a]	-1.24	-0.011	-2.35	-0.095	-0.04	0.77	-0.424	-0.189
Ba	108	1318	2633	338	326	299	301	697.72
Sr	517	927	993	136	144	117	155	126.25
Y	43.7	90.7	75	39	30	35	35	21.75
V	126	158	147	114	93	98	127	175.09
Sc	25.3	25.8	22	14	12	12	14	16.42
Cu	41.9	58.4	42	37	33	38	32	63.14
Cr	91.5	187	169	106	83	98	95	111.2
Ni	58.9	107	83	45	38	43	37	267.9

a. LOI = loss on ignition
b. HU = Hebrew University Clay Standard

Table 9. Averages and Standard Deviations for Four Replicate Analyses of Red Pot Standard A

Element/ Oxide	*Average Measured*	*Standard Deviation*
SiO_2	67.49	0.4049
Al_2O_3	22.23	0.49512
Fe_2O_3	5.62	0.06759
MgO	0.64	0.03167
CaO	0.82	0.0221
Na_2O	0.62	0.00885
K_2O	1.34	0.02329
TiO_2	0.95	0.01407
P_2O_5	0.05	0.00222
MnO	0.07	0.00295
Sr	56.74	1.58815
Ba	448.46	3.22525
Y	28.59	0.5302
V	132.83	3.39276
Sc	15.37	0.18695
Cu	33.62	3.09088
Cr	65.34	2.18501
Ni	31.55	3.60123

DCP Findings

Chemical Compositions

DCP-OES analysis corroborates the petrographic findings, indicating highly calcareous clay fabrics for Roman Palestinian lamps (tables 8–11).[3] Although the use of calcareous clays in pottery-making is not common, the high calcium values among the samples of this study indicate that lamps from the sampled sites are predominantly of this kind. Ariel et al. (1985: 148–49, samples DOR 8, 9, 10, 11, table 3) point out that, in addition to the Hellenistic storage-jar vessels from Dor, sites lo-

3. High calcium abundances in clays can complicate the establishment of provenance for pottery. Ariel et al. observe:

> Calcite weighs 2.5 times as much as its calcium component; therefore a vessel containing 20% calcium was made from clay with 50% calcite. Since the elements used for fingerprinting are almost exclusively in the clay minerals, their abundances are strongly depressed by the dilution of the clay component by calcite. If the amount of calcite is variable, so are the dilution factors. It is not possible to calculate the amount of dilution from the calcium alone because, depending on the kiln temperature, varying amounts of calcite decompose with the emission of carbon dioxide. The test for a dilution effect is that a single arithmetic factor applied to one group, or piece of pottery, will bring all of the diverse elements into line with another group. (Ariel et al. 1985: 149; see also Adan-Bayewitz 1993: 255–57; Widemann et al. 1975: 50–57, table 7.)

Table 10. Chemical Compositions of Clay Powders as Determined by DCP-OES

	L1	*L2*	*L3*	*L4*	*L5*	*G2*	*BS1*	*BS2*	*S1*
SiO_2	29.26	5.91	8.47	7.12	26.73	50.14	16.23	41.39	8.77
Al_2O_3	5.93	1.12	1.65	1.18	10.25	17.29	3.72	10.19	2.84
Fe_2O_3	3.22	1.11	0.56	1.17	4.48	9.39	3.62	6.03	1.70
MgO	2.62	0.62	0.46	1.88	1.33	3.65	1.96	2.68	1.00
CaO	55.23	84.1	78.51	83.42	52.53	16.37	71.45	36.49	84.81
Na_2O	0.22	0.15	0.42	2.17	0.19	0.26	1.09	1.11	0.15
K_2O	0.50	0.09	0.25	0.20	0.57	1.78	1.17	1.75	0.07
TiO_2	0.52	0.13	0.13	0.12	0.50	1.38	0.64	1.10	0.16
P_2O_5	0.05	0.006	10.12	3.00	3.75	0.69	0.12	0.12	0.47
MnO	2.60	6.91	0.01	0.01	0.03	0.14	0.07	0.05	0.03
Sum	100.18	100.17	100.57	100.25	100.37	101.08	100.06	100.90	100.00
LOI[a]	nd	nd	-0.019	-0.03	-1.411	-0.856	-2.367	-1.919	-2.349
Ba	216.6	69	118.72	48.31	431.86	643.60	201.25	218.20	63.44
Sr	857.8	1990	2469.55	1826.98	1336.82	228.84	599.46	368.00	1086.06
Y	71.45	162.2	263.83	114.01	116.13	45.75	35.04	39.83	85.39
V	142.7	162.3	115.27	138.70	218.37	163.84	100.51	108.67	90.09
Sc	23.82	29.25	51.56	48.40	42.11	25.99	48.31	31.29	53.37
Cu	64.08	87.01	170.79	87.91	149.84	59.62	34.74	39.83	44.90
Cr	256.42	446.51	609.99	520.64	652.05	156.76	93.67	119.14	140.89
Ni	119.78	166.4	190.40	102.10	280.33	117.67	61.17	88.75	66.57

a. LOI = loss on ignition

Table 11. Comparison of Select Major Elemental Average Concentrations of Samples (in %)

	Palestinian Discus[a]	*Bilanceolate*	*Beit Nattif*	*Jerash Bowls*	*Clays*
SiO_2	44.87	33.44	33.60	71.67	21.56
Al_2O_3	14.27	14.70	11.07	15.41	6.02
Fe_2O_3	5.72	4.91	4.56	4.89	3.48
CaO	28.5	41.61	44.83	2.44	62.55

a. Samples DSCY7 and DAB4 are not included because they are versions of the Palestinian discus type, and sample DA9 was also excluded because it belongs to the Tarsus group.

cated in the northern coastal region of Israel extending into Lebanon yield ancient pottery exclusively made of calcareous clays. The following samples contain comparatively high calcium abundances: Beit Nattif group (CaO: 44.83%) and bilanceolate (41.61%; table 11). By marked contrast, the Jerash bowl fragments exhibit very low calcium carbonate values (2.44%; table 11). High calcium abundances also occur

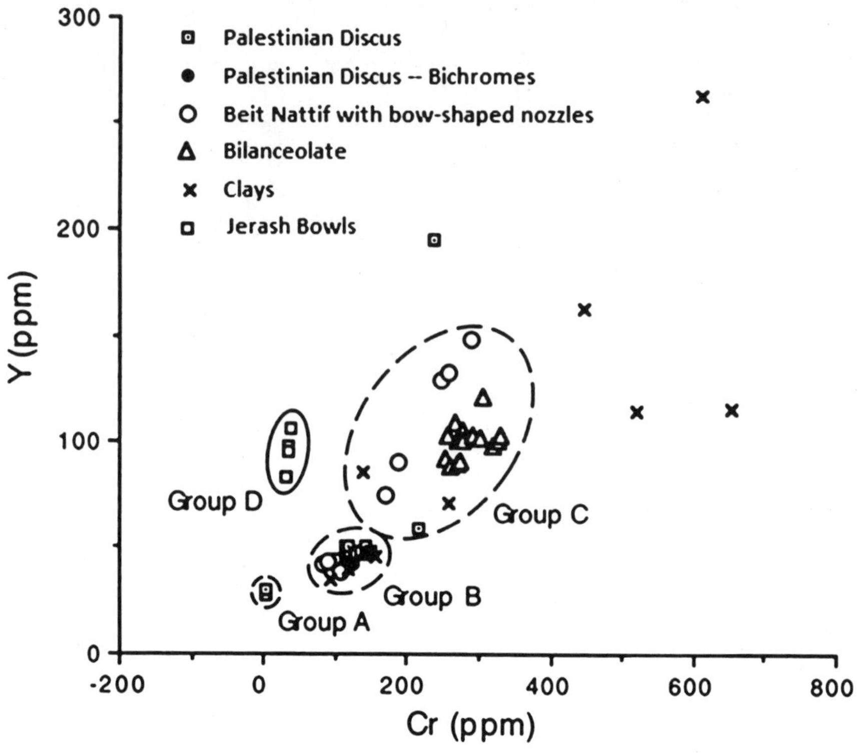

Fig. 14. Graph showing Y/Cr (in ppm) chemical composition results of lamp, Jerash bowl, and raw clay samples analyzed by dcp-oes.

among the local clays analyzed: with the exception of sample G2 exhibiting the lowest calcium value among the clays (CaO: 16.37%), the Pella clays on average contain 70.76% CaO, and the Sepphoris sample (S1) contains 84.81% CaO (table 10). The Pella clay samples with good potential for pottery-making (especially L1 and L5) contain on average 53.88% CaO (Edwards 1992: 290–93). Ariel et al. (1985: 148) observe that the amount of the element calcium present in clay minerals is typically one or two percent; hence, by comparison, the high calcium values found in the lamp fabrics indicate a large component of fine chalky calcite. As petrographic thin-sectioning indicates, calcium carbonate microfossils replaced by secondary calcite and limestone rock fragments account for the high percentage of CaO found in the lamp fabrics.

The Palestinian discus lamp samples on average have a higher percentage of SiO_2 (44.87%) than the bilanceolate and Beit Nattif lamp fabrics (table 11). By comparison, the average chemical compositions of the bilanceolate and Beit Nattif lamp fabrics more closely resemble each other and have higher abundances of CaO than the Palestinian discus group (table 11). The SiO_2 content of the Jerash bowl fragments (71.67%) is demonstratively higher than that found in the lamp and clay samples (table 11). This can be explained by the high percentage of quartz sand found in their respective fabrics, as determined by petrographic thin-sectioning. The CaO content of the clay samples on average is high (62.55%). By comparison, the SiO_2, Al_2O_3, and Fe_2O_3 percentages are low (table 11). The higher abundances of phosphates in Pella clay samples L3 (P_2O_5: 10.12%), L4 (P_2O_5: 3.00%), and L5 (P_2O_5: 3.75%) probably reflect the effects of human and animal occupation at the site,

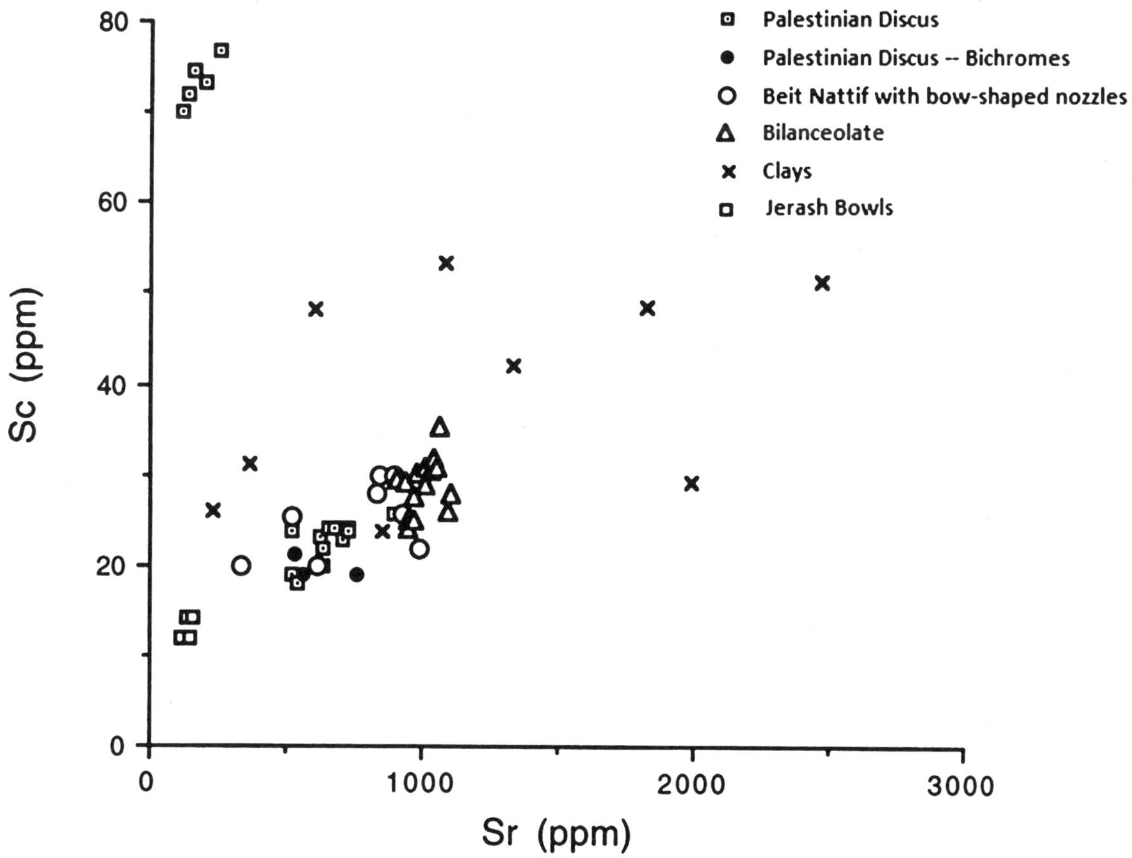

Fig. 15. Graph showing Sc/Sr (in ppm) chemical composition results of lamp, Jerash bowl, and raw clay samples analyzed by dcp-oes.

which increases the phosphates in the soils; the remaining clays averaged a much lower P_2O_5 value (0.17%; table 8). The lamp, bowl, and clay sample fabrics similarly reveal a low percentage of Fe_2O_3. Quartz sand found in the fabrics account in part for the high SiO_2 content recorded by the DCP analyses. As petrography has shown, the presence of hematite in the fabric of many of the samples accounts in part for the Fe_2O_3 abundance as determined by the chemical analysis.

Chemical Composition Groups A–D

The lamp samples cluster into three distinct chemical composition groups when compared on a Y/Cr plot (fig. 14; table 8). Group A consists of Palestinian discus lamp samples only. The majority of Palestinian discus lamp fragments and several Beit Nattif samples compose Group B. Group C consists of samples belonging to all three types: Palestinian discus, bilanceolate, and Beit Nattif forms. All 15 bilanceolate lamp samples cluster into Group C only. As anticipated, the Jerash bowl samples clustered together in Group D because they were collected directly from the Jerash hippodrome workshop. The Y/Cr plot groupings are further replicated in a Sc/Sr plot (fig. 15).

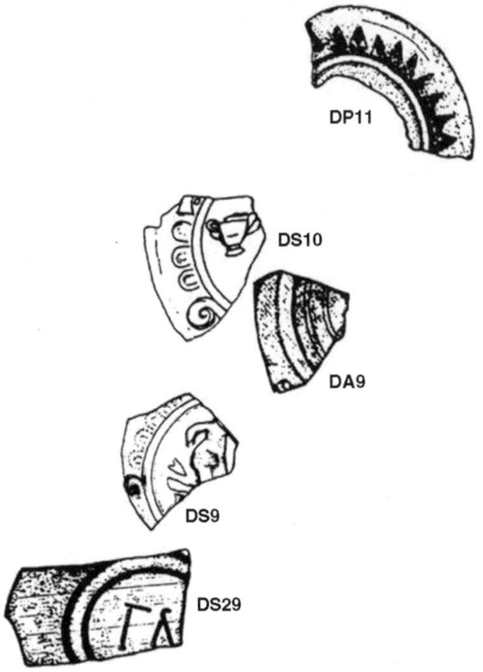

Fig. 16. One-to-one plot of each lamp sample in Group A in relation to its determined chemical position on a trace-element plot of Sc/Sr (in ppm). Drawing by E. C. Lapp.

The DCP findings suggest that at least three clay sources were used for the manufacture of the lamp samples and an additional source for the production of the Jerash bowl samples. Different chemical groups, however, do not necessarily indicate different provenances (Gunneweg and Perlman 1984–85: 232). For example, absolute reference material composed of terra-sigillata kiln wasters recovered from excavations of a pottery workshop at Lyons, France, clustered into three distinct chemical groups (Widemann et al. 1975: 50–57, table 7). Thus as at Lyons, the chemical groups that derive from the lamp fragments selected for this study suggest the exploitation of three clay sources, not necessarily at three separate provenances, but possibly only at one place.

Chemical Composition Group A. Group A consists of samples DS9, DS10, DS29, DP11, and DA9 (fig. 16). That these samples share a similar chemical fingerprint and cluster into a distinct chemical composition group may indicate the use of a chemically similar clay source by one or more lamp workshops. The diagnostic characteristics associated with these fragments include an erotic motif (DS9), the bi-associate chalice-and-Bacchus theme (DS10), and lamp makers' marks or initials *gamma alpha* (DS29) and *nu alpha* (DP11). The Tarsus imported lamp fragment exhibits radial lines and a curled decorative device (DA9). Taking into account the chemical similarity, molding execution, slipping, and fabric exhibited by the erotic and bi-associative motifs (the Bacchus-and-chalice theme), it is plausible that samples DS9 and DS10 originated from the same workshop. If this were the case, the two different discus images suggest that a single workshop produced lamps with more than one theme and did not necessarily mold a single motif only (e.g., only erotic or bi-associative). Notably, the two Palestinian discus lamp fragments belonging to the same bi-associative motif group (DS10, Bacchus-and-chalice and DP3, ibis-and-crab) have different chemical compositions. The fact that DP3 does not cluster with DS10 suggests that different clay sources were used to manufacture these lamps. Taking into account the similar fabric and slipping technique of sample DS29 with DS9 and DS10, it is possible, too, that this sample was made at the same workshop. Although belonging to the same DCP Group A, the molding style and execution of sample DP11 suggest a different mold and possibly a separate workshop.

It is possible, however, that the same workshop could have used a variety of different lamp molds carved by different mold makers. It is noteworthy that sample DA9 clusters with the presumably locally and regionally made Palestinian discus

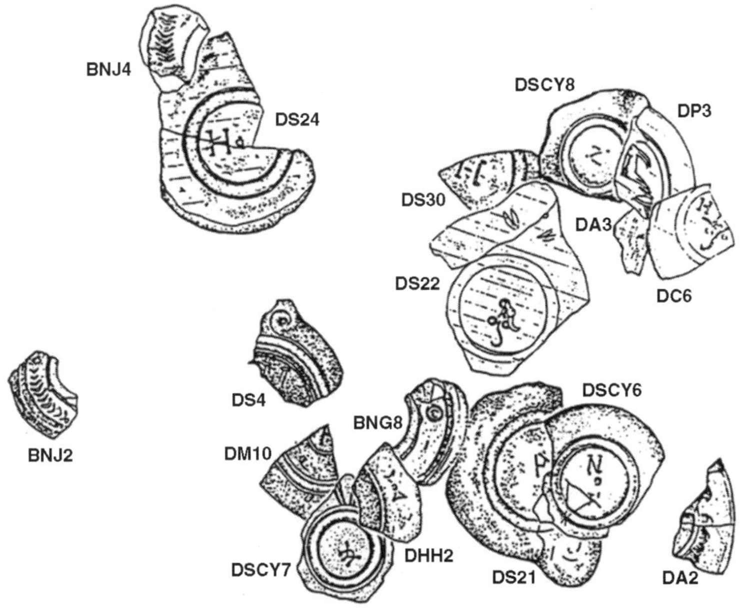

Fig. 17. One-to-one plot of each lamp sample in Group B in relation to its determined chemical position on a trace-element plot of Sc/Sr (in ppm). Drawing by E. C. Lapp.

lamps of Group A, especially because it is an import belonging to the Cilician Tarsus lamp type. That this lamp from Ashkelon has a chemical composition similar to that of sample DP11 suggests perhaps that it and others like it found in Ashkelon were manufactured locally or regionally using lamp molds acquired from a mold maker operating in Tarsus. Also, Cilician-style lamp molds carved in Tarsus may have been exported to and sold in the marketplace of Ashkelon. The Group A samples were recovered from different geographical regions: three from Sepphoris (Lower Galilee), one from Pella (Transjordan), and one from Ashkelon (southern coastal plain). The distribution of the findspots of the Group A lamps reflects cultural interactions between the cities.

Chemical Composition Group B. The results of the dcp-oes analysis indicate that a second clay source was likely exploited for the manufacture of the Palestinian discus lamps. Group B (fig. 17) consists of Palestinian discus samples from Ashkelon (DA2 and DA3), Caesarea Maritima (DC6), Ḥorvat Ḥazon (DHH2), Meiron (DM10), Sepphoris (DS4, DS21, DS22, DS24, and DS30), Scythopolis (DSCY6, DSCY7, and DSCY8), and Pella (DP3). The majority of Palestinian discus lamp samples occur in this chemical composition group. Samples DSCY6, DSCY7, and DSCY8 were recovered from deposits at Scythopolis and are chemically similar to the Palestinian discus lamps found in the group. Notably, sample DSCY6 is an unfired Palestinian discus lamp and, as such, presumably did not travel far from its place of manufacture in Scythopolis or nearby. DSCY 7 is probably of local Scythopolitan manufacture. Additionally, the locally collected clay samples from Scythopolis (BS1 and BS2) have chemical compositions similar to those of the Palestinian discus lamps in Group B of the Y/Cr plot. However, considering the complexities of the geological setting of the Scythopolis area and the difficulties of comparing the chemical compositions of raw clays to those of pottery and lamps, further exhaustive analyses of clays there and of other reference materials will be necessary to substantiate these findings.

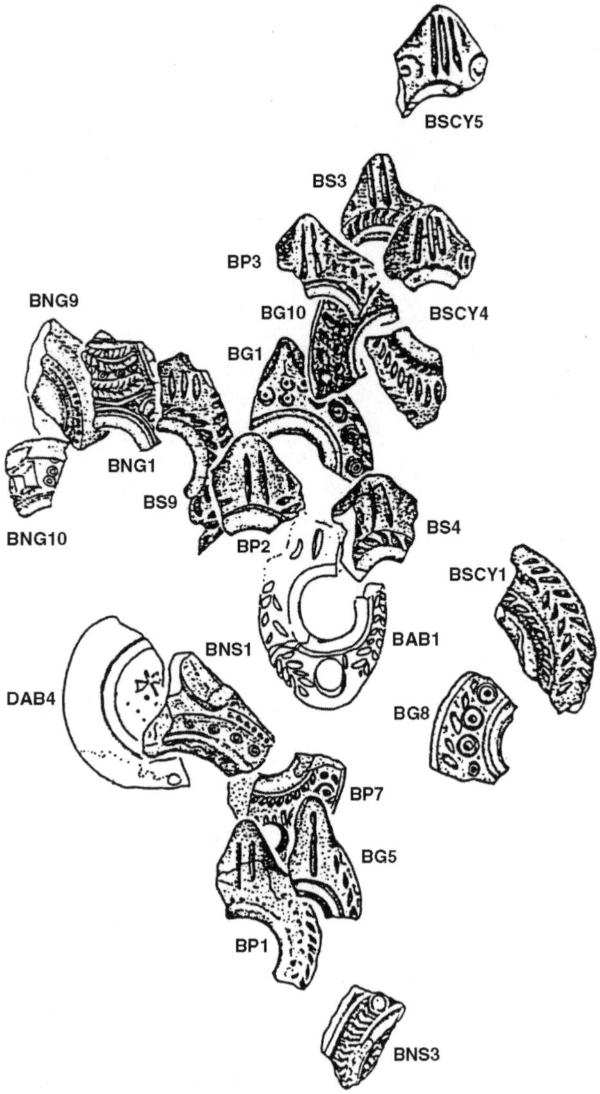

Fig. 18. One-to-one plot of each lamp sample in Group C in relation to its determined chemical position on a trace-element plot of Sc/Sr (in ppm). Drawing by E. C. Lapp.

Chemical Composition Group C. Group C consists of bilanceolate samples collected from the Galilean site of Sepphoris (BS3, BS4, and BS9) and the Decapolis cities of Abila (BAB1), Gadara (BG1, BG5, BG8, and BG10), Pella (BP1, BP2, BP3, and BP7), and Scythopolis (BSCY1, BSCY4, and BSCY5). Beit Nattif lamps found at Sepphoris (BNS1 and BNS3) and Gadara (BNG1, BNG9, and BNG10) and a local version of the Palestinian discus type recovered at Abila (DAB4) also cluster in this group (fig. 18). The most characteristic feature of this chemical composition group is that it contains all 15 lamp samples belonging to the bilanceolate type. That the bilanceolate samples share similar chemical compositions is further corroborated by their similar petrographic characteristics. Because the typological, petrographic, and chemical characteristics of the bilanceolate lamp samples are similar, the lamps likely have the same provenance: a single clay source of foraminifera-rich marly clay exploited by one or more workshops. Taking into account the occurrence of foraminiferous marls in northern Israel and the distribution of this lamp type and heaviest concentration of finds at Scythopolis, Hammat Gader, and Pella, these lamps were probably made at or near Scythopolis.

All of the Beit Nattif lamps from Gadara (with the exception of sample BNG8) and Sepphoris cluster into this composition group. This had been anticipated for samples BNG1, BNG9, and BNG10, which belong to the same petrographic fabric group as do the bilanceolate samples. However, I had not anticipated that samples BNS1 and BNS3, which belong to a different petrographic fabric group, would share similar compositions to the bilanceolate samples. Unless one workshop was using clays from two different sources, this variation in compositions between the Gerasa (and sample BNG8 from Gadara) and Beit Nattif lamps and those from Gadara and Sepphoris suggests two different workshops for the Beit Nattif lamp fragments analyzed in this study. Notably, Gilead samples BNJ2 and BNJ4 do not cluster in Group C. This makes sense because their petrography is distinct from the Beit Nattif samples. Furthermore, no Palestinian discus samples cluster in this group (e.g., DS30). Although sample DAB4 collected from Abila has a relatively similar chemical composition to the other Group C lamps, its generally different petrographic makeup suggests the use of a different clay source for its manufacture.

Chemical Composition Group D. Group D consists of only Jerash bowl samples JB1, JB2, JB3, and JB5. It was anticipated that these samples would share a similar chemical composition and petrographic characteristics because they were recovered from the same kiln. No lamps, however, occur in Group D. I had hoped that at least a few lamp samples would exhibit similar chemical compositions to the Jerash bowl reference material, especially Gilead lamp fragments BNJ2 and BNJ4 (misidentified in my dissertation as bow-shaped nozzle lamps; Lapp 1997a: 159) recovered from excavations at Jerash. Gilead lamps, such as samples BNJ2 and BNJ4, were unearthed in enormous quantities at Jerash, particularly in the hippodrome workshop (da Costa 2003: 1.100). That Gilead lamp molds were also found there (da Costa 2003: 1.100) ties the Gilead lamp group to Jerash as the origin of manufacture. The petrographic characteristics and chemical compositions of the Gilead lamps, however, differ distinctively from those of the Jerash bowl samples. This indicates that two different types of clay were used in the same pottery workshop for the manufacture of these pottery types.

Comparative Petrographic and Chemical Compositions

Palestinian Discus

The Palestinian discus lamp samples clustered into two distinct chemical groups (figs. 114, 117–18). Group A comprises samples DS9, DS10, DS29, DP11, and DA9. The majority of discus lamp fragments of this survey constitute Group B. These include samples DM10, DS24, DS4, DS22, DS21, DS30, DHH2, DSCY6, DSCY7, DSCY8, DP3, DC6, and DA2. The diagnostic characteristics associated with these fragments include the lamp maker's mark *eta* (DS24, DS30, DM10, DC6, and DA3), bichrome slipping (DS4, DHH2, and DA2), the associative motif of ibis-and-crab (DP3), lamp maker's mark of tri-bud floral motif (DS21), lamp maker's mark *alpha* (DS22), lamp maker's mark *nu* (DSCY8), lamp maker's mark *nu alpha* (DSCY6), lamp maker's mark *šin* (?) (DSCY7). Similar to Group A, the samples in Group B were recovered from sites of different regions. These include Meiron (Upper Galilee), Sepphoris and Ḥorvat Ḥazon (Lower Galilee), Scythopolis and Pella (Transjordan), and Caesarea Maritima and Ashkelon (coastal plain).

Among the Palestinian discus lamp fragments, only sample DAB4 from Abila clustered outside both Groups A and B. Based on its rough style and unique maker's mark, sample DAB4 appears to be of local manufacture, at or near Abila. The clay fabric of this sample is also unique and, as far as I am aware, is paralleled by other lamp fragments found at Abila only. It does, however, share a similar chemical composition to the bilanceolate and Beit Nattif lamp fragments forming Group C (fig. 18).

Bichromes. The three bichrome lamp fabrics (samples DA2, DHH2, and DS4) have similar chemical compositions and cluster into the same chemical group. This suggests that the same clay source may have been used by one or more workshops to manufacture these lamps. It is noteworthy that bichrome samples DS4 and DHH2 are chemically similar because they were both recovered from sites located in Lower Galilee. Notably, bichrome sample DA2 also clusters within the same Group B, and it was found in the distant southern coastal site of Ashkelon. That the bichrome lamps cluster with the unfired sample DSCY6 recovered from Scythopolis suggests possible manufacture at or near this Decapolis city. If so, their occurrence at Sepphoris and at Ḥorvat Ḥazon in Lower Galilee and at Ashkelon in the southern coastal plain suggests interregional interactions. It is also very interesting that the fragments bearing the maker's mark *eta* (DS24, DS30, DM10, DC6, and DA3) recovered from sites as far south as Ashkelon and as far north as Sepphoris cluster in the same Group B. If in fact the same clay source was exploited by lamp makers to manufacture these fragments, as their chemical clustering suggests, the significant geographical distribution suggests cultural contact among the urban centers of the coastal plain, Lower Galilee, and Transjordan.

Lamp Maker's Mark, Greek Eta *(*H*).* The five lamp fragments bearing the lamp maker's mark *eta* itself or in combination with another mark or initial clustered together in DCP Group B. These samples include: DA3, DC6, DM10, DS24, and DS30. Taking into account that these lamps also group with the unfired sample DSCY6 recovered from Scythopolis, it is possible that they too were manufactured at or near this city. If so, their occurrence at Sepphoris and Meiron in Galilee and at Caesarea and Ashkelon on the coastal plain suggests interaction among these places. Similarly, samples DS21 and DS22 cluster in Group B, suggesting a similar provenance. Multiple lamp workshops located at or near Scythopolis could have collected clay from a local single source.

Other Lamp Makers' Marks. The three samples from Scythopolis (DSCY6, DSCY7, and DSCY8) cluster within Group B (fig. 17; see also Lapp 1997a: figs. 13–14, 16). Because these samples appear to be of local manufacture in terms of both style and unique makers' marks, the clay source for the lamp fragments belonging to Group B could be at or near Scythopolis. This hypothesis is strengthened by the fact that DSCY6 is an unfired lamp fragment. It is highly unlikely that an unfired lamp traveled far from its source of manufacture. Although far less accurate, the petrographic characteristics of the Group B samples' fabrics are compatible with the local sedimentary geological setting of the Scythopolis area. However, until the origin of DSCY6 is established and more reference materials are analyzed, I cannot confirm that the lamp fragments composing Group B were manufactured at or near Scythopolis. That they cluster into a distinct chemical group suggests that their fabrics may originate from the same clay source.

Tarsus

Tarsus sample DA9 from Ashkelon (formerly identified as a Palestinian discus lamp) clustered with the Palestinian discus samples (DS9, DS10, DS29, and DP11) that constitute DCP chemical Group A (fig. 16; see also Lapp 1997a: figs. 114, 117–

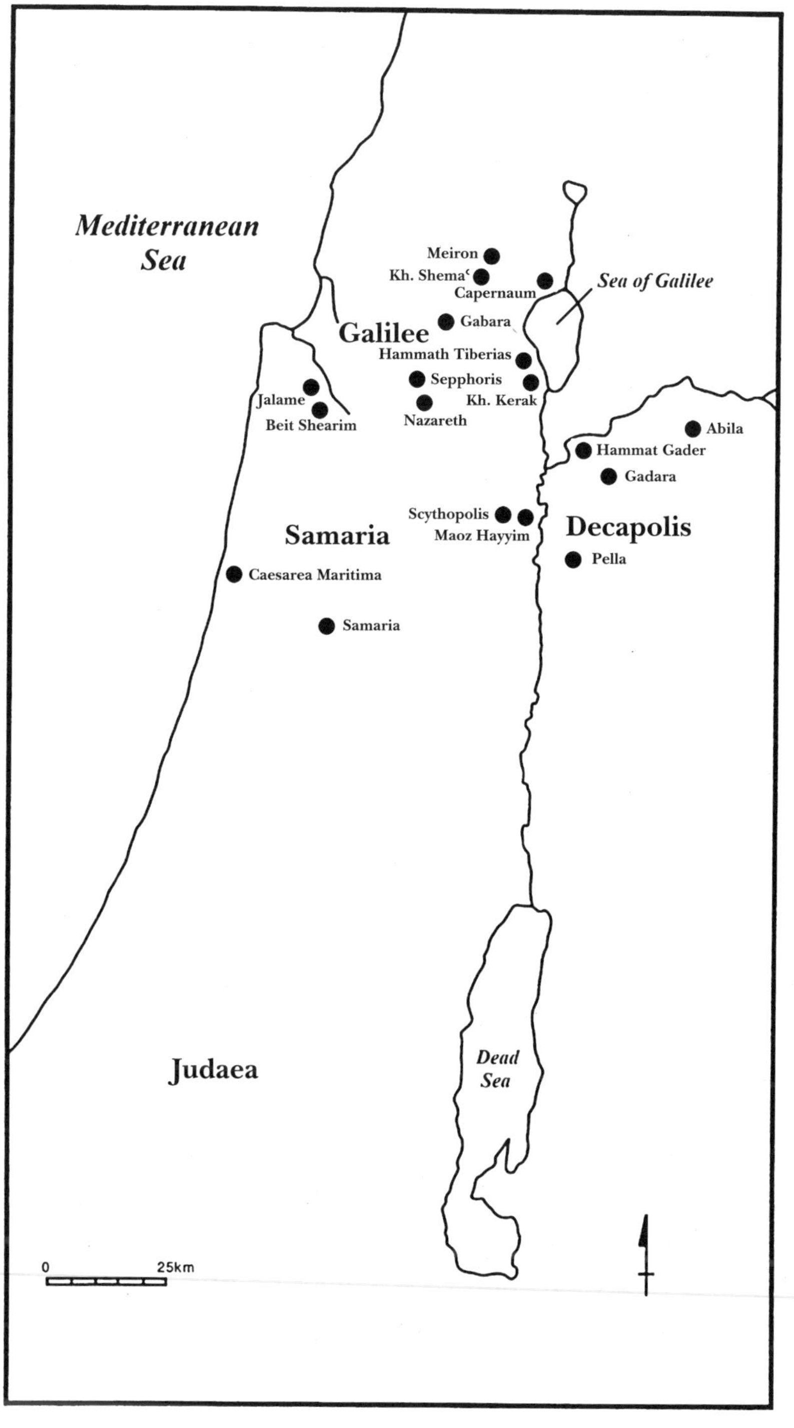

Fig. 19. Map showing distribution of bilanceolate lamps. Map drawn by E. C. Lapp.

18). It has a relatively high abundance of SiO_2 (45.55%) and CaO (30.84%). It is also petrographically similar to Palestinian discus sample DS10. This suggests that the lamp was locally manufactured and did not originate from Cilicia. It is conceivable that Tarsus lamp molds were acquired at Tarsus and brought to Ashkelon or were imported from Cilicia for use in the manufacture of Tarsus-like lamps at Ashkelon.

Bilanceolate

The 15 bilanceolate lamp samples exhibit similar chemical compositions and cluster into composition Group C. These include samples collected from Sepphoris (BS3, BS4, and BS9) and the Decapolis cities of Scythopolis (BSCY1, BSCY4, and BSCY5), Pella (BP1, BP2, BP3, and BP7), and Gadara (BG1, BG5, BG8, and BG10). No bilanceolate lamp fragments cluster among the Palestinian discus lamps in Groups A and B. Multivariate (Principal Components Analysis or PCA) statistical analysis was conducted on the dcp-oes data for the bilanceolate and Beit Nattif lamps presented in my dissertation and included in this volume (da Costa 2003: 1.63–65). The results of the statistical analysis distinguished chemical variation among the bilanceolate and Beit Nattif samples within Group C: 2 main groups and 6 outliers were identified (da Costa 2003: 1.63). Da Costa (2003: 1.64) observes: "This patterning does clearly show a difference between chemical groups." PCA Group 1 includes bilanceolate samples uncovered in Sepphoris (BS9) and Gadara (BG1, BG5, BG8, and BG10; da Costa 2003: 1.63–64). PCA Group 2 contains bilanceolate samples excavated at Sepphoris (BS3 and BS4) and Pella (BP2, BP3, and BP7; da Costa 2003: 1.63–64). My reexamination of the petrographic characteristics of these bilanceolate lamp samples reveals, however, that the PCA Group 1 samples do not consistently fall under the same petrographic group as I would have expected, nor do the PCA Group 2 samples. In some instances, the bilanceolate samples from da Costa's 2 groups even fall into the same or similar petrographic subgroups: BG8 (PCA Group 1) = BS3 (PCA Group 2); and BS9 and BG1 (PCA Group 1) = BP2 (PCA Group 2). Only samples BP3 and BP7 (PCA Group 2) belong to the same petrographic group and thereby corroborate her statistical findings. In light of the dcp findings and the PCA analysis, da Costa (2003: vol. 1.i, 64) best sums up the bilanceolate lamp fabrics: "All are different in a similar way, suggesting a common but heterogeneous source." That bilanceolate lamps found in the western summit of Sepphoris have petrographic and chemical characteristics similar to those unearthed in Gadara and Pella further underscores the cultural interactions among these *poleis* of Galilee and the Decapolis.

The bilanceolate samples cluster roughly into a single group suggesting, though inconclusively, that they come from a similar clay source. This is an insightful finding, taking into consideration that these samples come from five different sites and belong to several different subtypes of this lamp group. Petrographic examination of the bilanceolate sherds revealed micromorphological variation among their respective clay fabrics, thereby reflecting variation in the clay source and the parent raw source materials. The relatively widespread distribution of the bilanceolate samples (fig. 19) made from a similar clay source suggests interregional contacts between these urban centers and possibly even trade of this lamp type throughout Galilee and Transjordan.

The petrographic results show that lamp makers consistently used foraminiferous-rich marls, which are commonly found in central and northern Israel, to make bilanceolate lamps. DCP characterization indicates that the samples have a similar chemical "fingerprint," suggesting a similar clay source: all 15 fabrics recovered from five sites in Galilee (Sepphoris) and in Transjordan (Scythopolis, Pella, Gadara, and Abila) were found to cluster in the same chemical Group C. With the exception of sample DAB4, which also matches the composition of this group, the samples in Group C are chemically distinct from the Palestinian discus samples occurring in Groups A and B. Like that of Group B, the chemical compositions of several Beit Nattif lamp fabrics also clustered in Group C. These samples include 2 from the Galilean city of Sepphoris (BNS1 and BNS3) and 4 from a Decapolis city, Gadara (BNG1, BNG9, and BNG10).

The quantitative distribution of bilanceolate lamps indicates that their origin of manufacture falls somewhere within the geographical area of the Galilee and the northern Jordan Valley where they have been recovered (fig. 19; also see map 9 in da Costa 2003: 2.454). Bilanceolate lamps are found as far north as Meiron, as far south as Samaria, as far west as Caesarea Maritima, and as far east as Abila. The largest quantities come from Gadara and the baths at Hammat Gader, and they are also common at Sepphoris, Capernaum, Scythopolis, and Pella. That they do not occur south of Samaria and Pella rules out a southern provenance. One or more foraminiferous marl clay sources were exploited by one or more lamp workshops located somewhere within the geographical area of their distribution. Foraminiferal marl sources are located at or in proximity to each site where the lamp samples were collected. One would expect lamp fabrics composed of foraminiferous marls to have similar chemical compositions because of similar mineralogical components. The clustering of the 15 samples into one chemical group reflects this chemical similarity rather than indicating a single source

Taking into account the heavy concentration of bilanceolate lamps found in the Decapolis cities of Scythopolis, Pella, and Gadara in addition to those from the nearby baths of Hammat Gader, the DCP results suggesting a similar clay source and the petrographic finding that foraminiferous marl common in central and northern Israel consistently comprises the bilanceolate fabrics, it is likely that one or more lamp workshops were located at or near Scythopolis (which was also known for its pottery manufacture).

The clustering of the bilanceolate samples into a similar chemical group can be explained by the type of clay common to the samples, as determined by petrographic thin-section analysis. The bilanceolate lamp fabrics are consistently made with foraminiferous marls. Access to marly clay sources exists in the vicinity of each archaeological site where the bilanceolate lamp fragments of this survey were recovered. Sepphoris itself is located in a chalk-marl-shale complex. Although Scythopolis is situated within a travertine zone, the Lisan marls are found in the vicinity of the site to the east. Pella, too, is located near the Lisan sediment complex and near marly limestone and chalk formations. Abila is located within the marly limestone and chalk complex, and Gadara too is in the vicinity of this formation. It is also possible that bilanceolate lamps were produced in each of these urban centers and obtained clay from nearby marly sources. A marly clay will have basically the same chemical composition no matter where it is collected, since similar geological processes are responsible for its creation. Hence, a bilanceolate lamp made of marly clay and manufactured at Sepphoris will probably cluster chemically with one produced with the same type of clay at another site, such as Abila. Until reference material from kilns in which bilanceolate lamps were fired is found and analyzed, it will not be possible to determine with certainty whether a single site or others had manufactured this lamp in the region. Because certain settlements were known for manufacturing specific types of common wares (e.g., Kefar Hananiah), it is entirely possible that a specific estate, village, or town was known to have produced a specific lamp type, such as the bilanceolate variety. The chemical clustering of the bilanceolate lamp fragments into a single chemical group may reflect this practice.

Beit Nattif

The quantitative distribution of Beit Nattif lamps suggests two areas of origin: the Beit Nattif area in the south, where molds have been found, and in the north where a different version of this group occurs (see map 12 in da Costa 2003: 2.457). The macroscopic (friable fabric, red slip, jewel-like ornamentation) and petrographic characteristics of sample BNS1 (Sepphoris) are similar to those exhibited by samples of the southern version of Beit Nattif lamps from Roman Aila. This indicates that the Sepphoris sample (BNS1) is not a northern version of the Beit Nattif group but, rather, an actual southern Beit Nattif lamp originating in the Shephelah. The petrographic and micromorphological features of the six Beit Nattif samples of this study are inconsistent. Petrographic differences within the sampling of the Beit Nattif lamps indicate the use of different clay sources for the manufacture of the samples collected at Sepphoris, Gadara, and Scythopolis. The results of comparative chemical analyses were also inconclusive regarding the provenance of the Beit Nattif samples of this study (see also the PIXE results for this lamp group under bow-shaped nozzle type in da Costa 2003: 1.230–32, 235–36). This is further reflected in the various petrographic characteristics occurring within this lamp group. That BNG8 clusters among the Group B lamps suggests that the same clay source had been exploited. Beit Nattif lamp samples BNG1, BNG9, BNG10, BNS1, and BNS3 loosely cluster in DCP Group C with the bilanceolate lamps. This would suggest the possible exploitation of a similar clay source for lamp manufacture.

Gilead

The quantitative distribution of the lamp and mold findspots of the Gilead type point to Gadara/Jerash as their origin of manufacture (see map 14 in da Costa 2003: 2.459). The well-sorted and highly spheric quartz grains composing the fabrics of Gilead samples BNJ2 and BNJ4 (initially identified as bow-shaped Beit Nattif lamps) clearly distinguish them from the Beit Nattif samples. However, the clustering of Gilead samples BNJ2 and BNJ4 (Jerash) with Beit Nattif sample BNG8 (Gadara) and Palestinian discus sample DSCY6 (Scythopolis) in DCP Group B suggests the use of a similar clay source in the northern Jordan Valley, probably at or near Scythopolis. Gilead lamps were manufactured at Jerash (da Costa 2003: 1.100). It is unclear, however, whether samples BNJ2 and BNJ4 were manufactured in Jerash, for their fabrics do not chemically match the reference material JB1, JB2, JB3, JB4, and JB5 recovered from the hippodrome kilns in this Decapolis city. This does not exclude the possibility that another workshop in Jerash or nearby used a different clay source and had produced these lamps.

Reference Materials

The possibility of locating one or more ancient clay sources exploited for the manufacture of any of the lamp samples was certainly remote. The alluvial clay sample L1 collected in the Wadi Jirm at Pella exhibited a chemical composition similar to the Group C lamps. This is not surprising, considering the high suitability for pottery-making attributed to this clay by Edwards and the compatible sedimentary characteristics of this clay to the petrography of the lamp fabrics. Despite the fact that this would suggest a clay source for the manufacture of the Group C lamps, that conclusion must await further, intensive sample collection and analysis of clays from this formation. This clay formation is extensive and therefore does not necessarily indicate that Pella had been the actual manufacturing center of lamps; one or more workshops located near this formation along the northern Jordan Valley, perhaps between Scythopolis and Pella could have been exploiting this clay source. Nonetheless, any future provenance project must include further sampling and analysis of this Quaternary formation to determine whether it consistently matches the chemistry of the clay lamp fabrics. If so, the area between Scythopolis and Pella could have been a major lamp manufacturing center supplying the needs of many cities, towns, and villages in Galilee, the Jezreel Valley, the northern Jordan Valley, and the northern Decapolis.

The chalky clay samples from Pella (L3) and Sepphoris (S1) did not chemically match any of the discus lamp fabrics. Given the fact that the fabrics of several Palestinian discus lamps (e.g., DS10, DS29, and DS30) are similar to the fabrics of the Umayyad biscuit ware (CN206) mentioned by Edwards in that these samples are composed of fine-grained calcareous clays exhibiting pale yellow fired fabrics (Edwards 1992: 293, table 4), I had hoped that these clays would match the lamps in composition. This would have suggested a possible origin. With respect to Scythopolis clays BS1 and BS2, I had anticipated and hoped for similar results but to no avail. Taking into account the pottery-making suitability of clay sample G2, I had hoped for a match with samples collected from Gadara.

Use of Calcium Carbonate-Rich Clays for Lamp Manufacture

Petrographic analysis has identified abundant, nonplastic calcium carbonate ($CaCO_3$) inclusions in the fabrics of the sampled clay lamps. These include calcite and limestone in addition to mollusk shells, oolites, ostracod valves, coral, and other microfossils. The DCP analysis further corroborates the petrographic findings: the relatively high calcium content (in the form of calcium oxide, CaO) on average found in the respective lamp fabrics and clays: Palestinian discus (28.50%); bilanceolate (41.61%); Beit Nattif (44.83%); and clays (62.55%; see table 11). By comparison, the ferruginous Jerash bowls exhibit a significantly lower abundance of calcium

(2.44%). Pottery vessel fabrics that contain abundant calcium carbonate suffer the destructive effects of lime-spalling ("popping") and crackage when fired over 650°C. For this reason, the use of calcareous clays for lamp manufacture by potters and lamp makers of Roman and Byzantine Palestine is surprising. When calcium carbonate ($CaCO_3$) is heated above 650°C, it converts to calcium oxide or lime (CaO) and releases carbon dioxide (CO_2):

$$CaCO_3 + 650\text{–}900^\circ C \rightarrow CaO + CO_2 \uparrow$$

calcium carbonate + heat → calcium oxide ("lime") + carbon dioxide

The release of carbon dioxide causes mottling of the surface of the vessel. More serious, however, is the formation of calcium hydroxide ($Ca[OH]_2$) when calcium oxide is exposed to atmospheric water and reacts to the water. Because the calcium hydroxide crystal has a larger volume than the calcite crystal, its subsequent expansion causes cracking, spalling, and the ultimate destruction of the ceramic vessel (Hoard et al. 1995; Edwards 1992: 286, 294). Given the severe problems of calcium carbonate–rich clays for pottery, why would lamp makers use calcareous-rich clays or tempers that could potentially destroy the products of their craft?

The most plausible explanation is the widespread availability of this type of clay in the region: local and regional lamp makers simply had no other choice than to use calcareous clays. The use of carbonaceous clays for lamp manufacture, however, required the development of clay preparation techniques that would help avoid spalling and cracking, or at least improve upon the firing capabilities of the clay. Thus lamp makers in Galilee, the Decapolis, and other regions of Roman and Byzantine Palestine fired their lamps below 650°C, calcined ("burnt") the limestone temper, added thoroughly levigated lamp clay, or ground the calcium carbonate components into smaller particles. By thoroughly levigating the clays, the lamp maker removed some of the larger lime particles and/or ground the calcium carbonate inclusions to reduce the grain size (Edwards 1992: 286, 294). Calcining calcium carbonate tempering agents above 500°C before adding them to the clay fabrics facilitated the crushing to a finer grain size and prevented the expansion of the calcium carbonate from damaging the fabric of the lamp during firing (Edwards 1992: 286, 294; Feathers 1989).

Despite the destructive effects of calcium carbonate in lamp fabrics, lamp makers may even have preferred using calcareous clays and/or deliberately added calcareous tempering agents—such as ground microfossil-rich limestone—to the fabrics. Simulation tests, for example, have shown the benefits of using tempering agents rich in calcium carbonate (Hoard et al. 1995). Calcium carbonate in the form of calcite increases flocculation with the result of improving clay workability. The fired clay is significantly more resistant to thermal shock during firing, because the calcium carbonate allows for thinner vessels to be made. The thermal expansion rate of calcite is very similar to that of raw clay and thus creates a strong ceramic composite that resists cracking (Hoard et al. 1995). So, despite the inherent difficulties associated with the high concentration of calcium carbonate in the lamp clays from Roman Palestine, the calcareous-rich clays of the region could actually have proven advantageous for lamp manufacture: by producing a more workable clay body, the workers were able to produce a more durable, thinner lamp. This is most evident in our examination of the thin discus lamps, whose fabrics are characteristically rich in carbonaceous microfossils. In order to make a thinner yet durable wall, the lamp maker may have added calcined and well-ground microfossil-rich limestone to the fabrics as a tempering agent. Clays already rich in naturally occurring temper, such as the foraminiferous-rich marls of the region, would not need additional temper. The fact that lamps were successfully manufactured using calcium carbonate–rich clays indicates that such clays were suitable for lamp manufacture.

Firing Temperature

The presence of calcium carbonate ($CaCO_3$) in the fabrics of the sampled lamps is a useful indicator of firing temperature. During firing between 650 and 900°C, calcium carbonate decomposes to form lime (CaO; Rice 1987: 98). Thus the presence of lime in a lamp fabric indicates a firing temperature somewhere between 650 and 900°C. For example, mosaic-like grains observed in a North African Carthaginian lamp fabric using a petrographic microscope were identified as lime through electron microprobe analysis. The presence of lime indicates that the lamp makers of the North African type fired their lamp clays between 650 and 900°C. The decomposed carbonate and vitrified clay matrix observed in the Palestinian discus lamp sample DS10 attests a firing temperature higher than 850°C (Greenberg and Porat 1996: 16). The mottled, disarticulated, poorly preserved microfossils in DS10 may be lime, which is suggestive of a firing temperature of at least 650°C. Microscopy will be necessary to confirm whether the grains are indeed lime. Apart from this sample, no lime was observed in the other Palestinian discus, bilanceolate, Beit Nattif, and Gilead lamp fabrics of this study.

The well-preserved microfossils made of $CaCO_3$ found in the bilanceolate lamp fabrics of this petrographic study do not exhibit any effects of thermal decomposition, thereby indicating that the lamp makers of this group preferred a lower firing temperature below 650°C. Lime is hygroscopic and, as a result, leads to particle expansion. By firing at lower temperatures, lamp makers attempted to avoid the destructive properties of lime-popping and the release of carbon dioxide that explain the spalling on the surfaces of lamps and, with respect to the bilanceolate group, explain the cracks and highly friable condition of the lighting vessels' walls. A firing temperature above 1000°C ensures that calcium melts into other components that fail to absorb free moisture (da Costa 2003: 1.257). Although the iron content found in the sampled lamp fabrics is low, on average–for example, Palestinian discus (5.72%), bilanceolate (4.91%), Beit Nattif (4.56%; table 11)–its presence may have helped the lamp maker achieve a high degree of sintering at lower firing temperatures as a result of the fluxing properties of the iron (Cohen-Weinberger and Goren 2011: 223). No anorthite or gehlenite was observed in the Palestinian discus, bilanceolate, Beit Nattif, Gilead, and Tarsus lamp fabrics, which would indicate a firing temperature higher than 950°C (Maggetti 1981; Maggetti, Westley, and Olin 1984). The occurrence of gehlenite would indicate a firing temperature at approximately 1100°C (Cohen-Weinberger and Goren 2011: 223). Byzantine glazed

Deir ʿAin ʿAbata wares made with relatively refactory, noncalcareous clay could be fired at an exceptionally high temperatures of up to 1150°C (Freestone, Politis, and Stapleton 2001: 200). As such, the lamp makers of these types apparently did not choose to fire their lamps in a reducing atmosphere as a technique for minimizing crackage (see da Costa 2003: 1.257).

Temper

The petrographic study has shown that *terra rossa* nodules were added as temper to several lamp fabrics belonging to the Palestinian discus, bilanceolate, and Beit Nattif types. Petrographic examination of the samples of this study, however, indicates that lamp makers generally did not add temper to the lamp clays because it was not necessary. The high microfossil ($CaCO_3$) content of the bilanceolate lamp fabrics already served as a naturally occurring temper. Except for a single residual basalt fragment identified in sample DSCY8, no basalt grains were observed in the lamp fabrics. The occurrence of basalt temper in the lamp matrices would point to an origin for the temper in the alkali-olivine basalt flows of the Neogene-Pleistocene of the Galilee or the Golan Heights (Cohen-Weinberger and Goren 2011: 218).

Techniques

The petrographic and DCP findings presented in this chapter indicate the use of calcareous-rich clays for the manufacture of the Palestinian discus, Tarsus, bilanceolate, Beit Nattif, and Gilead lamps. The preference for carbonaceous clays among local and regional lamp makers is unremarkable. First, lamp makers were limited for the most part to calcareous clays because of the predominantly sedimentary geomorphology of the region where they operated. They simply had no choice but to use carbonaceous clays. Such clays were not ideal for lamp-making because of the risk of spalling (lime-popping) caused by the formation of calcium hydroxide during firing. However, clays exhibiting a high concentration of calcium carbonate in the mineral form of calcite could actually prove advantageous for lamp manufacture by not only increasing the workability of sticky clay but also being more resistant to thermal shock and possessing an expansion rate similar to that of raw clay.

The clay preparation for lamp manufacture indicates kneading practices similar to those used in other pottery production. The petrographic evidence, however, suggests that lamp makers did not necessarily add temper to the lamp clays of the Palestinian discus, bilanceolate, and Beit Nattif groups: the nonplastic inclusions found in the lamp fabrics of these types represent naturally occurring secondary components of the clay. Next, the clay was pressed into the lamp-shaped cavities of the upper and lower halves of carved molds. Fingerprints often found on the interior surfaces of the oil reservoirs of Palestinian discus lamps indicate that thumbs and fingers—not instruments—were used for this procedure.

Once the lamp was removed from the mold, Palestinian discus and Beit Nattif lamps in particular were slipped by dipping them into a thin wash of clay. In order to keep the clay particles in suspension and to avoid coagulating, a peptizing agent such as potash was added to the solution (Bailey 1972: 15). Rich deposits of potash found along the shoreline of the Dead Sea were exploited in antiquity and were a likely source for the potash used by potters in Roman and Byzantine Palestine. Oil lamps such as the Palestinian discus and Beit Nattif types were slipped not only for decorative and apotropaic purposes but also to reduce lamp fabric permeability and thereby prevent oil seepage of the fuel reservoir.

Lamp-making instruments were used to shave off excess clay or for trimming the contours of molded lamps (Kehrberg 1995: 525, fig. 18). Such instruments were also used for incising decorations and for touching up impressed patterns of molds (such as the impressed herringbone patterns of the bilanceolate lamp fragments in this study). The only known lamp- and pottery-making instruments found in Roman Palestine consist of a group of blades, scrapers, rubbing, and boring implements recovered from the hippodrome kilns at Gerasa (Kehrberg 1992: 451–64; for other lamp-fashioning instruments, see Fremersdorf 1922: 39, fig. 44.1–3). Notably, the instruments were made mostly of local Late Roman ceramic wares and of imported terra sigillata selected for their flaking suitability.

The unfired lamps were then stacked in a dome-shaped furnace, probably in a pyramidal fashion. The most common type of kiln used in Roman Palestine was the vertical, semi-subterranean updraft variety. These consist of a circular fuel chamber (or fire box) sunken into the ground above which the firing floor of the superstructure was constructed. Similar kilns were used in Weisenauer, Germany, where one of the best preserved was excavated (Fremersdorf 1922: 39; Lapp 1997a: 173–74, figs. 124–25). The well-preserved foraminifera that characterize the bilanceolate lamp fabrics indicate that relatively low firing temperatures were used for their manufacture. By contrast, the poorly preserved bioclasts of several Palestinian discus lamp fragments (especially, DCP Group A) indicate high firing temperatures. The lamps were fired in an oxidizing environment. Upon completion of the firing process, the clay dome of the kiln was destroyed and the finished products removed.

9. Summary

This study has demonstrated the importance and usefulness of the petrographic microscope in the characterization of Roman and Early Byzantine lamps. Six distinct petrological fabric groups have been identified among the Palestinian discus, bilanceolate, and Beit Nattif lamp forms examined. The nonplastic inclusions found in the lamp fabrics are characteristic of clays derived from primary, sedimentary source rocks: quartz sand, limestone fragments, calcite, dolomite, and carbonaceous microfossils. The mineralogical constituents of the lamp fabrics match the local sedimentary geology of Israel and Jordan. No lamps of the study were made of raw material from a different geological setting. For example, none of the Palestinian discus, bilanceolate, and Beit Nattif lamp samples were made with micaceous clays, which are unusual in Israel and point to production outside the country (Cohen-Weinberger and Goren 2011: 222).

Petrography has also shown that the nonplastic inclusions in the lamp fabrics represent naturally occurring components of the clay. DCP spectrometry substantiates the petrographic findings that indicate the high concentration of Ca and the highly calcareous nature of the clay fabrics of the Roman Palestinian lamp samples of this study. It has distinguished three distinct chemical groups among the lamps (Groups A, B, and C) and a fourth for the Jerash bowl fragments (Group D). And, although the DCP results in many cases corroborate the petrological groupings of the lamp sherds (e.g., the bilanceolate samples), this study also shows that the fabrics of samples belonging to the same chemical groups do not necessarily exhibit the same petrological-textural characteristics. Nearly 40% of the samples analyzed (especially, BNS1, BNS3, BNG8, BNJ2, BNJ4, DP11, DS9, DS10, DS29, DA9, DS24, DA3, DC6, DSCY6, DSCY7, DSCY8, and DAB4) indicate that the chemical groupings do not always agree with the petrographic findings.

Although the presence of microfossils in many of the samples could be explained as fossil-rich limestone ground and added to the clays by the lamp maker as temper, the predominant very fine to fine sand size and presence of fossil-rich primary source clays in the area suggest that it is naturally occurring. Microfossils include foraminifera (especially globeriginids), disarticulated ostracod valves, and mollusk shells. Apart from a single basalt rock fragment found in the Beit Nattif sample DSCY8, no diagnostic basalt temper, which would indicate a provenance in the volcanic lava flow areas of the Golan, Hauran, and Lower Galilee, was observed in any of the lamp fabrics.

Quantitative distribution maps of the bilanceolate and Beit Nattif lamp forms indicate the local Palestine provenance for these types. The fact that the composition of several Beit Nattif samples (BNG8, BNJ2, and BNJ4) chemically matched that of the majority of discus lamp fragments that compose DCP Group B (DA2, DA3, DC6, DHH2, DM10, DP3, DSCY6, DSCY7, DSCY8, DS4, DS21, DS22, DS24, and DS30) indicates the Palestinian provenance for the Palestinian discus type. In addition, given the fact that the respective fabrics of three different styles of lamps belonging to more or less different chronological periods (Palestinian discus, late first–third centuries C.E.; bilanceolate, late fourth–sixth centuries C.E.; and Beit Nattif, late fourth century C.E.) exhibit uniform compositions in general, such findings further strengthen the conclusion that the discus lamp class was locally made. This attribution is strengthened significantly by the fact that the Beit Nattif sample BNG8 belongs to the same petrographic fabric group as discus lamp fragments DSCY7 and DSCY8 recovered at Scythopolis. Most importantly, the fact that the chemical composition of the unfired discus lamp sample DSCY6 of likely Scythopolis origin matches that of samples DSCY7, DSCY8, BNG8, and the other Palestinian discus lamp fragments found in DCP Group B represents convincing corroborative evidence for a local Palestinian provenance—possibly even at or near Scythopolis—for the Palestinian discus and Beit Nattif lamp samples composing the latter chemical group.

The Palestinian discus lamp samples clustered into two distinct chemical composition groups: Group A (DP11, DS9, DS10, and DS29) and Group B (DA2, DA3, DC6, DHH2, DM10, DP3, DSCY6, DSCY7, DSCY8, DS4, DS21, DS22, DS24, and

DS30). This suggests that at least two distinct clay sources were exploited for the making of Palestinian discus lamps. Tarsus sample DA9 clustered with the Palestinian discus samples of Group A. Notably, however, the Palestinian discus and Tarsus samples found in both chemical groups belong to the same petrographic fabric group. This discrepancy is problematic, given the fact that one would expect samples exhibiting the same petrographic and micromorphological characteristics to share similar chemical compositions. One possible explanation for this discrepancy is that the lamps of Group A do in fact originate from the same deposit as those composing Group B but had been collected from a different location within that source, either at a different height or from another location on the surface.

Petrological and chemical analyses have shown diachronic overlap among the discus and several samples of the Beit Nattif classes. The fact that late fourth-century Beit Nattif lamps (especially, BNJ2, BNJ4, and BNG8) have compositions similar to the earlier first-third-century discus lamps suggests that the same clay source had been exploited for a long period. Since the bilanceolate and Beit Nattif samples date to the Late Roman to Early Byzantine periods, the fact that they share similar compositions suggests the exploitation of the same clay source during these periods for the manufacture of these lamps. The fact that the variation discus lamp fragment DAB4 from Abila belongs to the same chemical Group C as do the bilanceolate and Beit Nattif samples may indicate that it is a Middle–Late Roman version of the earlier Palestinian round with decorated discus lamp class.

Petrographic thin-section analysis has revealed that foraminiferous-rich marls were used for the manufacture of the bilanceolate lamps. Foraminiferous marls occur in central and northern Israel as well as in Jordan. Apart from the southernmost occurrence of the bilanceolate lamp at Samaria, the distribution of this lamp class is confined to sites located in Galilee, the northern Jordan Valley, and northern Transjordan. The quantity of bilanceolate lamps found at Scythopolis and Pella is particularly high and hence points to this area as the possible epicenter for the production of this lamp class. Petrography has shown that the bilanceolate lamp fabrics recovered from five geographically diverse sites belong to the same textural Group 2. Significant is the fact that DCP spectrometry further substantiates this petrographic fabric group: all 15 bilanceolate lamp samples comprising Petrographic Fabric Group 2 have the same chemical fingerprint and cluster in the same DCP Group C. Thus clustering in the same chemical group indicates a single clay source exploited for their manufacture.

The application of direct current plasma-optical emission spectrometry (dcp-oes) has shown the viability, cost-effectiveness, and nonnuclear means of this alternative method for the characterization of clay lamps and pottery. Chemical characterization using DCP spectrometry has demonstrated the ability of this method to distinguish different and distinct clay sources in a confined geographical region.

On the basis of the following evidence and observations, Scythopolis or a place nearby is the likely provenance for the manufacture of the Palestinian discus, bilanceolate, and Beit Nattif lamp fragments of this multiple-site regional survey. The chemical compositions of the unfired sample DSCY6 and samples DSCY7 and DSCY8, which suggest local Scythopolitan manufacture, match other discus and Beit

Nattif lamp forms. As indicated by the distribution maps, the Beit Nattif forms are indigenous to Roman Palestine. However, it still remains to be determined whether these samples chemically match those from the Shephelah (e.g., Beit Nattif and Beth Guvrin). The fabrics of the Beit Nattif lamp samples of this study differ from those found at Beit Nattif and Beth Guvrin. Extensive deposits of foraminiferous marls and variegated calcareous-rich clay deposits were available and easily accessible for exploitation by lamp makers working in the area of Scythopolis; petrography has shown that foraminiferous marls were a common clay used to manufacture the three lamp classes examined (especially, the bilanceolate form). A copious water supply necessary for lamp and pottery manufacture was available in the area (from wadis and the Jordan River). Archaeological evidence (e.g., lamp molds and kilns) indicates active lamp manufacture at Scythopolis, which enjoyed outstanding access to the extensive Roman road and highway system, linking the city to others located along the coast, the Galilee, the Jezreel Valley, and Transjordan. The three chemical composition groups suggest that one or more workshops exploited three clay sources for lamp manufacture. Because the fabric composition of the Pella clay sample L1 matches the Group C lamp fragments, extensive clay sampling and analysis of the alluvium deposits of Wadi Jirm and in the vicinity should be analyzed to determine whether this area was in fact the location of the primary clay beds. The future analysis of additional lamps may also establish that composition Groups B and C represent a single group rather than two separate ones, as suggested in this study. Reference materials collected from actual lamp workshops (lamp wasters, unfired lamp fragments, and kiln clay) are essential for the determination of the exact provenance for the Palestinian discus, Beit Nattif, and bilanceolate types.

CHAPTER 5

Regionalism, Trade, and the Sepphoris Marketplace

1. Introduction

Most of the clay lamps recovered from the western summit played a role in the commercial dynamic of the Sepphoris marketplace. Sepphoreans likely purchased their lamps in the marketplace of the Lower City. The lamps in this market came partly from the recently discovered lamp-production center in the satellite village of Shikhin. Sepphoreans could also have acquired lamps directly from lamp makers operating in Sepphoris as well as from the lamp workshops in Shikhin. Travelers and merchants would have sold their lighting vessels, and these found their way into the domestic quarters of the Upper City, in the urban marketplace.

2. Lamp Regionalism and Continuity

A "lamp regionalism" existed in Galilee in that certain types–specifically the northern stamped, Sepphorean spatulate, "Sepphorean" discus with linear geometrical patterns, and Galilean–tend to be limited to Sepphoris and other sites in Galilee. The relatively restricted geographical distribution of these lamp types suggests their manufacture in Galilee. The types of lamps recovered from Jewish villages of Upper Galilee as compared with those found at the sites of Lower Galilee, including the urban centers of Sepphoris and Tiberias, indicate no regional discontinuity among the lamp forms of Upper and Lower Galilee (cf. E. M. Meyers 1978: 100). This supports the current state of archaeological evidence, which indicates considerable continuity between the pottery of Upper and Lower Galilee (E. M. Meyers 1978; Chancey 2002: 18).

This continuity in lamp types is best exemplified by the Palestinian discus type, which is found in substantial quantities at Sepphoris in Lower Galilee and at the Upper Galilean villages of Gush Ḥalav, Khirbet Shemaʿ, Meiron, and Nabratein.

Richly decorated with a variety of mythological themes including Europos and the Bull and risqué erotic scenes, Palestinian discus lamps had a "pollinating" effect on village life, introducing and spreading Greco-Roman attitudes throughout the rural communities of Galilee. Such lamps offered villagers an alternative to the otherwise conservative decoration of geometrical, floral, or plain patterns of locally produced lamps. Despite the higher levels of literacy in the East (D. Kennedy 2007: 138), the images on Palestinian discus lamps communicated visually, as did images in early Christian art (Brown 1999: 15–16 n. 1; 18 n. 10). Northern stamped lamps also occur at these Upper and Lower Galilean sites. However, bilanceolate lamps are only found at Meiron and Khirbet Shemaʿ. Examples of this type, though, were found at Capernaum and Hammath Tiberias in Lower Galilee. Villagers depended on the city for necessities that only a cosmopolitan city such as Sepphoris could provide. Village potters who made lamps, for example, probably acquired their lamp molds from the lamp-mold makers at Sepphoris or Shikhin. Galilean villagers and others who desired to purchase any imported fine wares—such as the African Red-Slip Ware, Cypriot Red-Slip Ware, and Eastern Terra Sigillata found in abundance at the site (Balouka 2013: 50–51)—would turn to Sepphoris or another important market place or production center.

3. *Identity of Population and Cultural Interactions*

The population of the cities, towns, and villages in Galilee was primarily Jewish in the first century C.E. (Chancey 2002: 22–27, 61–62). Jewish Galileans were acquainted with lamps of the Herodian, Darom, northern collar neck, Judean floral, Palestinian discus, and Sepphorean spatulate types, which were manufactured and purchased in this period and have been found at Sepphoris (cat. nos. 18–142) and at other sites in the region. Ancient sources and archaeological evidence suggest an influx of Gentiles in Galilee—Roman soldiers and their entourage—by the second century C.E. (Chancey 2002: 22–27, 61–62).

Sepphoris was largely Jewish, as were the satellite villages of Shikhin, Yodfat, and Cana and also the more distant villages in Upper and Lower Galilee. The coastal cities of Ashkelon and Caesarea Maritima were mixed—Jews, Christians, and primarily Gentiles. The Byzantine-period synagogues unearthed in Scythopolis and Gerasa indicate the existence of Jewish communities in these largely Gentile urban centers (L. I. Levine 2000: Scythopolis, pp. 198–204, 332–37; Gerasa, pp. 239–40). Abila, Gadara, and Pella were Gentile. A Beit Nattif lamp decorated with a menorah was found at Pella (da Costa 2003: 2.516, no. 92).

4. *Lamp Workshops*

Nearly all the known lamp workshops in Roman and Byzantine Palestine were located in cities (Adan-Bayewitz 1995: 180), Shikhin being the notable exception. This pattern of urban lamp production is consistent with evidence from other sites in the Roman world (Adan-Bayewitz 1995: 180). Although no evidence for an actual lamp or pottery workshop has been found in Sepphoris, the discovery of at least three lamp molds suggests lamp production: the mold of an upper half used to manufacture lamps of the Sepphorean round-bodied type (Adan-Bayewitz 1995: 177–80, fig. 1, showing an alternating oval-triangle pattern on the shoulder, a wide circular body, and a spatulate nozzle); an upper half for making either a Sepphorean round-bodied or Sepphorean spatulate type (reg. no. 86.1114.2X.1, with only the vestige of a cavity used to make a spatulate nozzle with lines following the contours of the nozzle); and an upper half for making a Galilean type lamp (reg. no. 86.1105.2X.1., showing a wide round body with a large discus, a spatulate nozzle, and a pinched pointed handle). In light of these finds, lamp manufacture at Sepphoris, or at least the existence of a shop that produced carved lamp molds for sale to local rural potters and lamp makers is certainly plausible.[1]

Petrographic and NAA analyses indicate that only three sampled Herodian lamps from Sepphoris were probably manufactured with raw soil materials that were available in the Galilee–including brown rendzinic soil–and were not from the Jerusalem area (Adan-Bayewitz et al. 2008: Sepphoris Group 2, nos. 3070, 3072, 3078, table 3; pp. 60–61, 68). A fragment of what was possibly a stone mold was found at nearby Nazareth. The deep incisions found in the cavity of the mold may indicate that it was used for the production of bilanceolate lamps with a chevron pattern (Bagatti 1969: 309). Such incisions, however, would create relief decoration, and relief chevrons do not exist among bilanceolate lamps (da Costa 2003: 1.382). The fragment may have been a matrix for the making of clay molds or may simply have been part of a stone bowl (da Costa 2003: 1.382–83). Furthermore, the chevron pattern was likely made by a rouletting tool after the ceramic vessel was removed from the mold (da Costa 2003: 1.382–83).

Rabbinic sources do not mention Sepphoris and Tiberias as pottery-manufacturing centers, but they do name Kefar Hananiah and Kefar Shikhin as pottery production sites.[2] The sources do not specifically refer to lamp production. If the findspots of lamp molds indicate their place of manufacture and use for the production of lamps, then lamp workshops in Roman Palestine and Arabia were located in cities (Caesarea Maritima, Beth Shean/Scythopolis, Sepphoris, Gerasa, and the Petra "suburb" of Zurabbeh), towns (Beit Nattif, Nazareth, Khirbet Kerak, and Niʿaneh), and rural settings (Mt. Nebo; see da Costa 2003: 1.380–84).

Because of the expertise required to carve limestone molds, as indicated by the sophistication of many examples from Roman Palestinian cities, including Sepphoris, the manufacture of fine-ware lamps required the talents of master artisans, who would most likely be found in a cosmopolitan urban environment or a nearby

1. The lamp maker is synonymous with the potter. Ancient sources do not identify a specialized craftsperson called "lamp maker." Any potter could make a lamp because all it required was the preparation and pressing of clay into a two-part mold and then firing–all skills that any potter possessed. The carving of the mold, however, required a different skill-set.

2. According to Adan-Bayewitz (1993: 32–38), Kefar Hananiah is mentioned with various ceramic features in *y. Pe'ah* 7.4, 20a (*lifsa* = wide-mouthed cooking pot), *b. Ketub.* 112a (version of the latter), *Lam. Zuta* 1.4 (*kedera* = cooking pot), *Gen. Rab.* 86.5 (*krozin* = water pail or pitcher). Shikhin appears in *t. Ter.* 7.14 and *y. Ter.* 8.6, 45d (*ḥavit* = storage jar). Transliterations are those of Adan-Bayewitz.

village. A lamp-mold specialist probably carved molds used to make lamps or statuettes and then sold them to pottery workshops. Molded lamps thus are not only a type of ceramic vessel; they are also a type of sculpture in the round. Although it is improbable that lamp makers made their own molds, lamp workshops may have had an additional area where stone cutters fashioned molds on location. For example, evidence of lime kilns for the production of a gypsum-like substance used to make molds was found in close association with pottery kilns in the hippodrome workshop at Gerasa, indicating that lamp molds were produced on the site (Ostrasz 1993: 500). This evidence, coupled with the difficulty, practicality, and expense of transporting relatively heavy stone molds seems to suggest that lamp molds were generally made on location.

The evidence for actual workshops in Roman Palestine is fragmentary and previously reflected the tendency for archaeologists to excavate urban sites (Adan-Bayewitz 1995: 180–81). In Galilee, lamps were produced mainly in cities, whereas common wares were produced in rural settlements (Adan-Bayewitz 1995: 181, fig. 2). Recent archaeological evidence, however, indicates lamp production in the satellite village of Kefar Shikhin.[3] This indicates that clay lamps were manufactured in both urban centers (e.g., Caesarea Maritima and Gerasa) and small settlements (Kefar Shikhin) as well as in key locations along major roadways. The Mishnah also reports (*m. Beṣah* 4:4) that crude lamps were crafted in the home. As far as I am aware, homemade lamps have not been found at Sepphoris or at any other site in Galilee. Such lamps, however, have been found in significant quantities at sites in the Black Sea area, particularly in Olbia, Chersonesos, Tanais, Belinskoe, and at other Bosporan cities (Žuravlev 2007: 226–27, fig. 19).

Any village potter was capable of making both wheel-thrown (such as the Herodian type; see Lapp 2001c: 296; see Adan-Bayewitz et al. 2008: 74 on the likelihood of Herodian lamp manufacture at Shikhin) and mold-made lamps (with molds acquired from an urban-based mold maker). Excavations of the kiln works of the Tenth Legion Fretensis on the outskirts of Jerusalem show this pattern of production (Arubas and Goldfus 1995). Several types of earthenware vessels, tiles, and bricks were manufactured there and, according to the excavators, this place functioned as a sort of village that supplied Early Roman–period Jerusalem with some of its ceramic needs (Arubas and Goldfus 1995: 107). It is certainly plausible that similar pottery- and lamp-producing satellite villages were located near Sepphoris and supplied lamps to the city's merchants and residents.

This pattern of lamp production in rural settlements in Galilee (as substantiated by the Kefar Shikhin finds) and in the Decapolis and elsewhere in Roman and Byzantine Palestine has parallels outside the region. For example, archaeological evidence has shown that lamp production occurred at rural estates in Roman Tunisia (Peacock, Bejaoui, and Lazreg 1990). The sprawling complex of Sidi Marzouk

3. I thank Professor James Riley Strange of Samford University and Dr. Mordechai Aviam of Kinneret College for informing me about the discovery of lamp molds at Kefar Shikhin and for sharing a photograph of the lower half of a mold discussed above. They will publish a preliminary report on the archaeological evidence suggesting lamp production at the site.

Tounsi, for instance, is situated 5 km west of el-Ala and has the appearance of a villa surrounded by an extensive zone of pottery manufacture. This zone was covered with lamp molds, lamp fragments, red pottery, and kiln furnaces (Peacock, Bejaoui, and Lazreg 1990: 68). The abundant lamp fragments and collection of lamp molds made of white refractory clay indicates the importance of the site for lamp production. Lamp fragments and molds were also found at the farm site of Henchir es-Srira (Peacock, Bejaoui, and Lazreg 1990: 74–76, 78, fig. 12) and at Henchir el Guellal, Djilma (Peacock, Bejaoui, and Lazreg 1990: 76–81, fig. 14). Taking into consideration the evidence from North Africa, rural villas and farms in Roman Palestine should be considered possible production centers for lamps and other ceramic wares.

The location of workshops depended on accessibility to clay sources and a copious water supply. The main sources of water in Syria–Palestine are springs and active wadis. The potters working in the major urban centers of Roman Palestine relied to some extent on the Roman aqueduct system (as at Caesarea Maritima, Sepphoris, and Dor) for their industrial water. The location of the potters' workshop also would have depended on easy access to a plentiful wood supply to provide charcoal fuel for the kilns. Easy access to transportation routes was also necessary for the marketing of lamps.

5. Lamp Supply from Village to Polis

The recent discovery of several limestone lamp molds—mostly lower halves, to make the lamp bases—at Kefar Shikhin indicates lamp production in this village near Sepphoris. Upper halves are preferable for the identification of the type of lamp made with a mold because they are carved with ornamentation and other diagnostic features. The cavity of the lower half of a mold found at Shikhin is distinctive enough to posit its use for the manufacture of the base of lamps belonging to the Sepphorean spatulate type. The carving is indicative of the round and slightly convex base with the three thin concentric rings characteristic of this group. The village supplied the residents of Sepphoris with these lamps, several of which were found on the western summit (cat. nos. 44–52). Taking into account that rabbinic sources identify Kefar Shikhin as a pottery-production center for storage jars (*ḥavit*; Adan-Bayewitz 1993: 37–38), we should not be surprised that clay lamps were also made in the village. The proximity of Shikhin to Sepphoris and the well-developed road network in Lower Galilee (Safrai 1994: 286, figs. 63–66; see also Strange 2014) ensured a steady supply of lamps to the city and other sites in the region.

6. Road Networks, Lamps, and the Sepphoris Marketplace

The variety and abundance of local and regional lamp types discovered on the western summit of Sepphoris reflect the city's advantageous location at the crossroads of the major Lower Galilee trade routes (fig. 20). A Roman road running

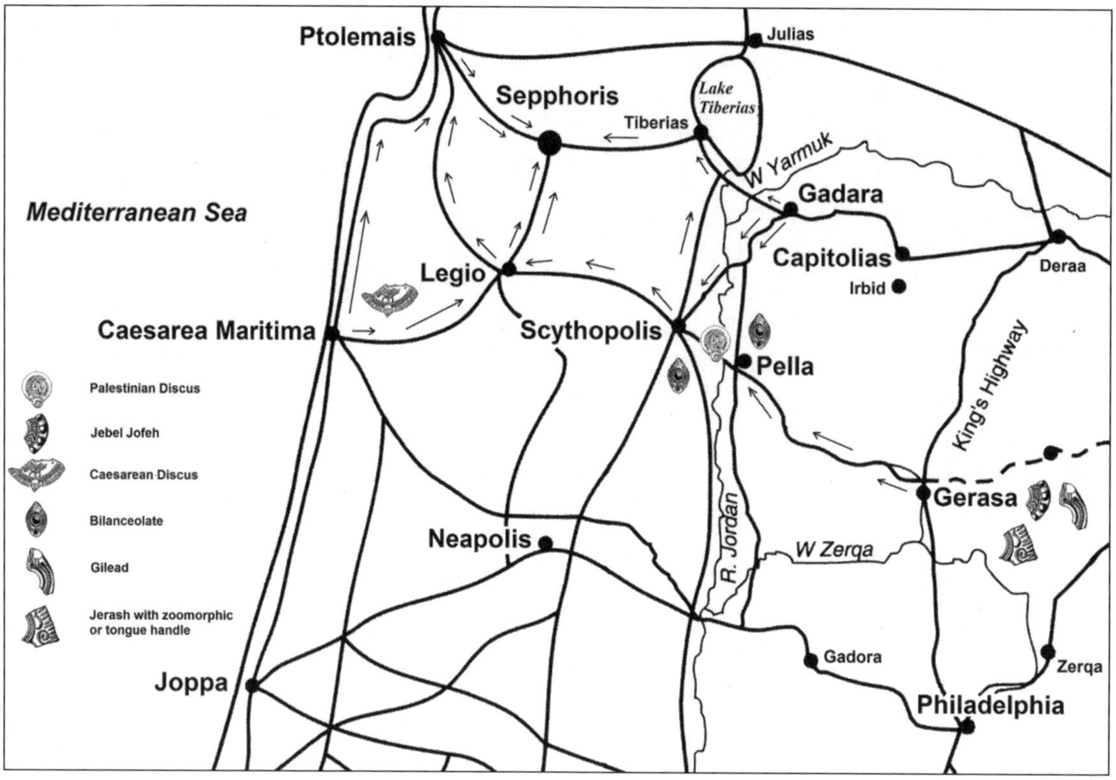

Fig. 20. Map showing Roman road network in Roman Galilee and cities of the Decapolis; also indicating routes used for lamp transport to Sepphoris. Map drawn by E. C. Lapp, based on D. Kennedy 2007: 90.

eastward linked the port of Akko-Ptolemais to Tiberias and continued deep into the Hauran via Julius/Bethsaida north of the Sea of Galilee to Bostra. Another road connected Ptolemais to the Jezreel and Beth Shean valleys, running in the vicinity of Legio and then to Beth Shean, the Galilee's gateway to the Decapolis and all its commercial and cultural resources. A north–south road linked Sepphoris to both highways. The Roman road system connecting the major pottery-manufacturing centers (e.g., Gerasa, Scythopolis, and Caesarea) would have served well the itinerant peddler and donkey and/or camel caravans. Using the Roman highway system and village road network, traders of lamps could travel freely and at low expense to a number of towns and villages. Rural roads and paths leading into the countryside also provided an extensive network connecting villages, towns, and cities. Although transport by land was more expensive than by sea, there was no other way to move goods in the region except at sites along the Sea of Galilee and the Dead Sea. The architectural remains of a Roman-period harbor at Tiberias and seven piers around the shores of the Sea of Galilee (e.g., ʿEin Gev and Capernaum) could attest this means of transporting lamps to local markets. Perhaps the greatest commodity transported on the roads of the region were the ideas and traditions of the various peoples who traveled them.

Table 12. Estimated Standard Mean Dry Weight (Wf) and Quantity of Classic Nabatean Lamps according to Means of Transport

Means of Transport	*Wf (seah)*[a]	*Wf (kg)*	*Wf (lb)*	*Quantity of Lamps*[b]
Porter	5	32	71	642
Donkey	15	96	213	1,926
Camel	30	192	427	3,852
Wagon	90	576	1,280	11,557
Regular ship	180	1,728	3,840	34,671
Large ship	540	5,184	11,520	104,013

a. Based on Safrai 1994: 289.

b. The standard mean dry-weight value of a classic Nabatean lamp equals 49.84 g (Lapp forthcoming a; Barrett 2008: Appendix, figs. 7.22.ii; 7.23; Lapp 2012: 64, 67, sample no. 26, fig. 4, table 4).

7. Lamp Transport to the Sepphoris Marketplace

The transport of goods by donkey was less expensive than wagon transport (Safrai 1994: 289). One must presume, therefore, that the donkey was the ordinary means of transport for clay lamps in Roman Palestine. Calculations based on information found in the Mishnah indicate that a donkey could carry up to 15 *seah*s or 96 kg (Safrai 1994: 289). If a donkey driver were to transport only clay lamps to Sepphoris and only those belonging to the classic Nabatean group, which has a mean standard weight of 48.89 g (Lapp forthcoming a), this would amount to 213 lb of lamps or the equivalent of 1,926 lamps (by weight; see table 12). Whether the driver could arrange and pack this quantity of lamps onto a donkey is another issue. At any rate, a donkey driver could transport a substantial number of clay lamps from a workshop (such as the hippodrome pottery at Gerasa) to the marketplace of Sepphoris. Galilean donkey caravans took part in regional trade and frequented Akko-Ptolemais, Tyre, and Sidon (Safrai 1994: 272). They were probably responsible in part for the transport of Palestinian discus lamps to the marketplace of Tyre, where numerous examples were recovered in the Shrine of Apollo (Marchand 1996: 64–65, nos. 60–68). Palestinian discus lamps most likely arrived via donkey caravan to the port of Ptolemais, where they were loaded on ships bound for the coastal cities of Cyprus (e.g., Salamis, Kourion, and Amathus). This lamp type was found at sites throughout the island, indicating that it was sold in the markets of Cyprus' cities.

Camels were used mainly for overland transport (Safrai 1994: 272, 289). Camels, therefore, should be considered another means for transporting lamps as far westward as Sepphoris from the desert regions of the Hauran. If the camel shown on a discus lamp fragment found at the site is any indication, Sepphoreans were not unaware of this animal. A camel packed with only classic Nabatean lamps could carry 30 *seahs* or 192 kg, which equal 427 lb or 3,852 lamps. A full load of classic Nabatean lamps transported by a wagon is even more impressive: 90 *seahs* (576 kg) or 1,280 lb of lamps, the equivalent of 11,557 lamps (table 12). These figures refer to the weight that each type of transport can carry and, in my opinion, seem high. They do, however, indicate that a sizable number of lamps could be transported to city markets in Galilee. Future experimental archaeology should test these figures by loading lamps of a single type and consistent mean weight onto an accurately reproduced Roman-period wagon.

The lamps manufactured in the workshops of the Decapolis and unearthed by the Duke and Hebrew University excavations at Sepphoris probably made their way to the city's bustling marketplace via the major Roman road running from Gerasa to Pella to Beth Shean/Scythopolis, continuing to Legio and then northward to the city. The lamps originating from the Gerasan workshops and found at the site—the Jebel Jofeh (cat. nos. 145–54; and Gärtner 1999: Roman Type 11, 25–39, pls. 13–14.59–62, 28.138), the Jerash zoomorphic (cat. no. 209; and Gärtner 1999: Islamic Type 1, 87–88, pl. 21.99–100, 29.144), and the Gilead (cat. no. 167) forms—were also probably transported along this route, as were the Palestinian discus (cat. nos. 78–142) and bilanceolate (cat. nos. 168–72) lamps, probably made in the vicinity of Scythopolis and Pella.

Small loads of lamps were carried on a person, donkey, or camel. Larger shipments were conceivably transported in a two-wheeled, donkey-drawn cart or in a four-wheeled wooden wagon of the sort carved in high relief on a Roman tombstone on exhibit in the Archaeological Museum of Arlon in Belgium (museum inv. no. IAL GR/S 051). Wagon ruts have been identified in the pavement of the southernmost road, located approximately 90 m south of the *decumanus* at Sepphoris (Weiss and Netzer 1996b: 33). At least some of the ruts were presumably made by wagons transporting common wares and lamps for sale in the markets of the Lower City. Although the roads of Roman Palestine were fully capable of supporting wagon transport, particularly between *poleis*, wagon transport on village and rural roads was extremely difficult and nearly impossible (Safrai 1994: 289). That said, rural roads were numerous in Galilee and were linked to village neighbors within a 3- to 5-km radius (Safrai 1994: 276–77). The lamp-manufacturing village of Kefar Shikhin thus would have had excellent access to the Sepphoris markets via rural roads. Consequently, it is easy to envision potters arriving in the *cardo* of Sepphoris on their way to the marketplace carrying their lamps and other ceramic wares on their backs or packed on a donkey or in a donkey-drawn cart.

Water transport played a role in delivering imported lamps to Sepphoris. Lamps belonging to the Athenian, Corinthian, and Ephesian types unearthed by the Duke expedition (nos. 215–19) probably arrived by ship at the ports of Caesarea Maritima, Dora, and Akko-Ptolemais. That so few were found at the site suggests they arrived with other major goods as an auxiliary product, or more likely, were brought to Sepphoris by travelers or merchants as personal items. Perhaps Sepphoreans journeying to the Greek mainland or the western coast of Asia Minor brought them home as souvenirs. That so few imported lamps have been found on the western summit of Sepphoris, and for that matter at Beth Shean, Ashkelon, or Caesarea Maritima indicates that regional pottery workshops, such as the shop at Shikhin, satisfied the lamp and lighting needs of local populations throughout Roman Palestine. Although transport by water was less expensive than by land, the cost of transport in general was high and was certainly a factor in impeding the import of lamps from distant lands. Whether they arrived in Sepphoris as a trade good or as the personal item of a traveler, imported lamps would have been transported by land on the Caesarea Maritima-to-Beth Shean/Scythopolis (then traveling northward near Legio to Sepphoris) or Ptolemais-to-Tiberias roads (journeying southward to Sepphoris midway between Ptolemais and Tiberias; see fig. 20). The North African lamp sherd found by the Hebrew University expedition (Gärtner 1999: 106–7, pl. 26.120) most likely arrived from the Carthage area via the Roman road running from Carthage to Lepcis Magna-Alexandria-Gaza-Caesarea Maritima-Legio-Sepphoris. The Galilean lamps found at Capernaum (Loffreda 1974: 131; da Costa 2003: 1.337–38), once arriving in the small port of Tiberias via the Ptolemais–Tiberias road, were then loaded onto small boats that transported them northward on the Sea of Galilee (for the location of the Tiberias piers, see Safrai 1994: 255, fig. 53).

8. *Selling Lamps in City Marketplaces*

Archaeological evidence and rabbinic sources both offer information about how lamps were probably sold in the marketplace of Sepphoris. Clay lamps and other pottery vessels were sold in Shops 1–3 in the marketplace of Beth Shean/ Scythopolis (Agady et al. 2002: 464–65; also see fig. 21b here). The three shops are situated near the nymphaeum and yielded a total of 34 complete lamps belonging to 8 types, including, for example, the Galilean, bilanceolate, candlestick, Beit Nattif version, and Jerash types (Agady et al. 2002: 464–76, figs. 19.2–4, 20.5–8): two complete lamps were found in Shop 1; 27 complete lamps in Shop 2; and 5 complete lamps in Shop 3 (Agady et al. 2002: 464–76). At Beth Shean, 21 complete lamps belonging to 8 types were found inside Byzantine shops on the Street of the Monuments (Agady et al. 2002: 464–76). The lamp types include: Herodian; Palestinian discus; bilanceolate; candlestick; Beit Nattif, northern version; Galilean; Caesarean discus; and Jerash with zoomorphic handle (Agady et al. 2002: 464–76, figs. 19–20, pls. 5–7). That the lamps were found in association with other objects suggests that they were sold with a variety of different commercial goods. The lamp types found in the Beth Shean shops originated from Beth Shean workshops, and types manufactured in regional workshops as far east as Caesarea (the Caesarean discus type), a sister Decapolis city of Gerasa/Jerash (Jerash zoomorphic type), and Galilee (the Galilean type) were also sold there. Perhaps the shop owners obtained their lamps by bargaining: a juglet for a customer offering, in turn, a few lamps. Because no shops contained lamps exclusively, none of the shops can be interpreted as "lamp shops."

A lamp shop has been identified in the residential area of Capernaum (Loffreda 1974: 88, 131, fig. 46.18–27; da Costa 2003: 1.337–38, fig. 59; da Costa 2010: 75;

a

b

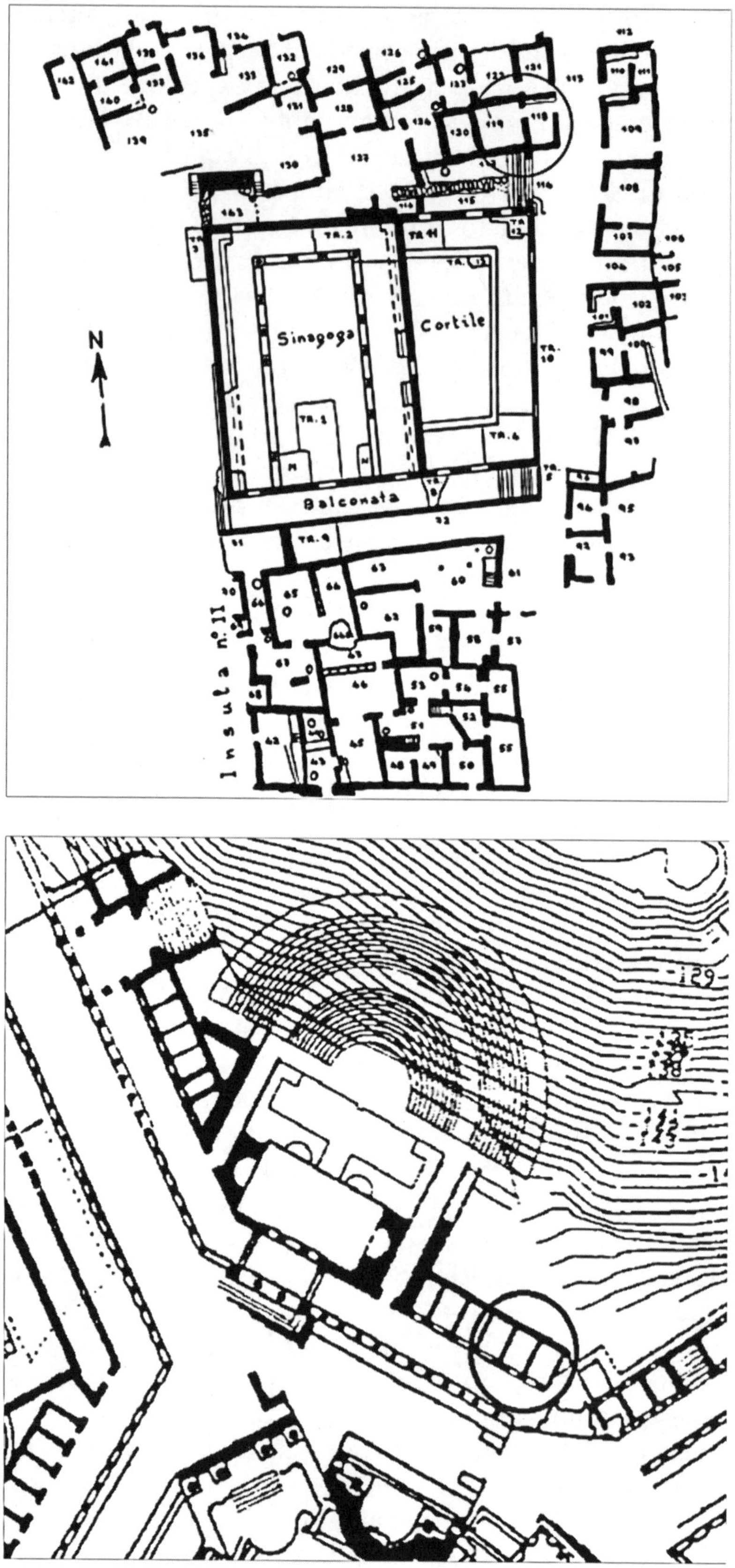

Fig. 21. Location of lamp shops (circled) at Capernaum (a) and Beth Shean/Scythopolis (b). Capernaum plan after Loffreda 1974; Scythopolis plan after Agady et al. 2002: 426.

see also fig. 21a here). A large concentration of unused lamps was recovered from ambiente 119 in Insula V situated only a few meters north of the synagogue's *beth midrash*. Much other pottery, including a Cypriot Red-Slip Form 10 bowl (Loffreda 1974: fig. 46.11; da Costa 2003: 1.338) was found in the structure, indicating that the lamps were sold with other ceramic wares. This offers another scenario regarding how lamps were sold outside the marketplace. The shop was synonymous with the workshop, and the place of work was also called a "store" (Safrai 1994: 225). Rabbinic sources suggest that the workshop-store did not also usually serve as a residence (Safrai 1994: 225). An alternative explanation for the room filled with lamps at Capernaum is that it served as a lamp storeroom for the synagogue. At Ashkelon as well, hundreds of unused discus-type lamps decorated with erotic imagery were found in a room, suggesting a brothel, lamp shop, or lamp storehouse (Stager 1991: 47; Lapp 2001b: 62–65).

The shops in Beth Shean/Scythopolis and Capernaum were not attached to or associated with a pottery workshop, unlike other instances (Peacock 1982: 30, fig. 11.1–4), suggesting that the shopkeepers had acquired their lamps from a supplier (middleman) or directly from a workshop located either inside or outside their settlements. Shopkeepers operating in the Umayyad period at Beth Shean, for example, probably acquired their ceramic wares and possibly lamps from the pottery workshop located just a short distance from the six shops near the nymphaeum (Agady et al. 2002: 426, fig. 2.21).

Shopkeepers who sold lamps in the colonnaded marketplace of the Oval Piazza in Gerasa probably acquired their lamps directly from the hippodrome pottery workshop located a short distance outside the city's South Gate. Because substantial space was required for a pottery- and lamp-making complex, the *scaenae* oriented along the *cardo* of the city would have been insufficient for both the manufacture and sale of wares. Such workshops were probably located away from the *cardo,* where more space was available. An example would be the hippodrome lamp workshop located outside the marketing center at Gerasa. It is possible that each urban workshop had an "outlet" shop on the *cardo*, but this may not have been necessary, depending on the proximity of the workshop to the central market.

Sepphoreans perhaps bought their lamps in the Lower City, where excavations have revealed possible market areas (Weiss and Netzer 1996b: 32–34; also see fig. 22 here). Around the first half of the second century C.E., Sepphoris expanded to accommodate the civic and commercial needs of its rapidly growing population (Weiss and Netzer 1996b: 31). The *cardo* and *decumanus* of the Lower City were a result of this urban development, and one should expect that the lamps found in our corpus that date to this period and later (e.g., Darom, Palestinian discus, Caesarean discus, northern stamped, bilanceolate, etc.) would have been sold by the merchants there. Barrel-vaulted shops were unearthed along a road approximately 50 m north of the *decumanus* (Weiss and Netzer 1996b: 32–34). The city's central market (*macellum*) is believed to be located south of these shops. A possible *agora* (forum) was uncovered across the *cardo* and west of the *macellum*, and numerous shops run along the *decumanus* and the colonnaded *cardo*. Stalls were presumably set up between the columns of the Sepphoris *stoa* (Sperber 1998: 11 n. 16; see *y. B. Bat.* 2.3), as understood by

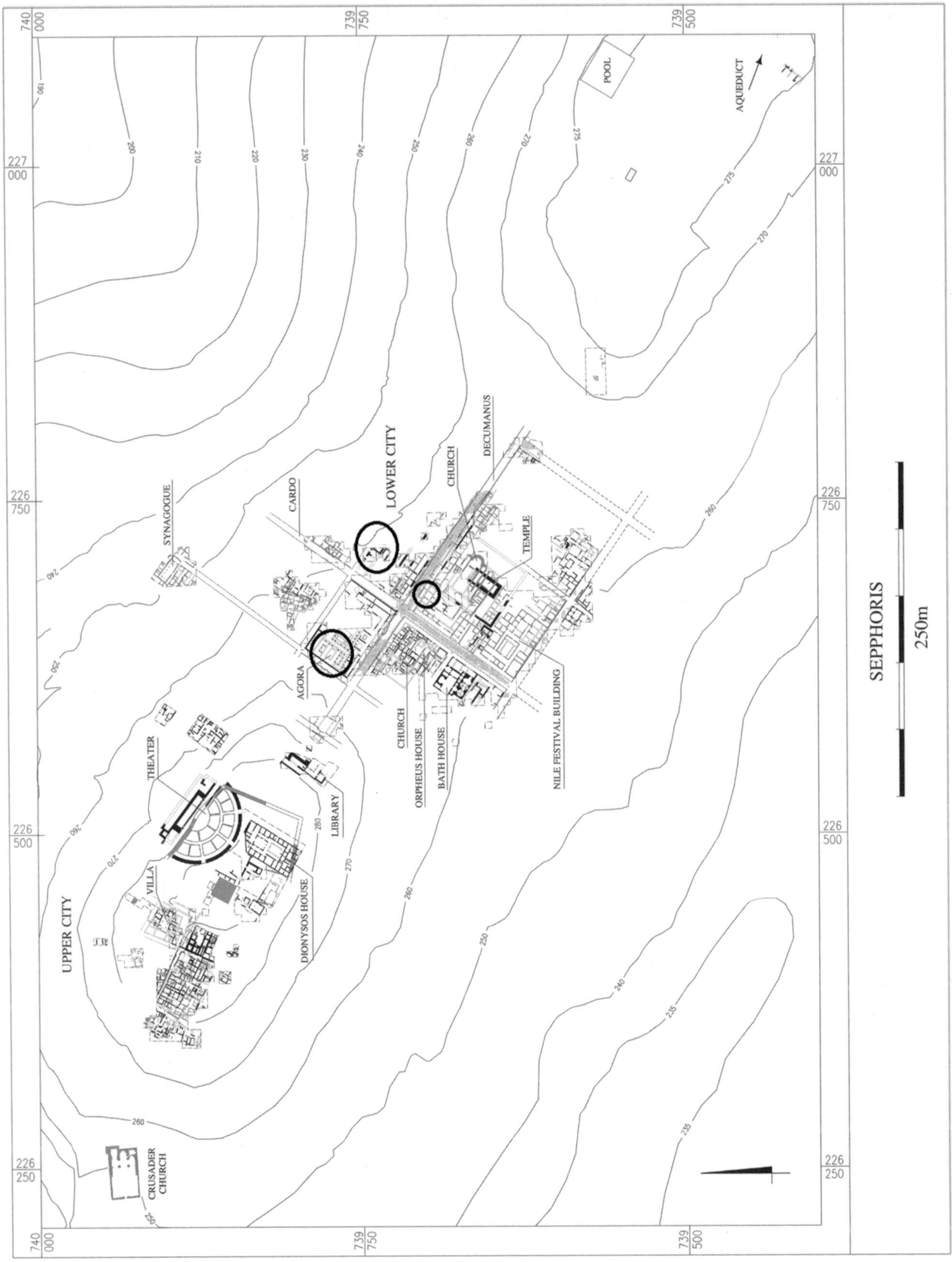

Fig. 22. Plan of Sepphoris showing both Upper and Lower Cities and indicating the possible locations (circled) of shops and markets where

Jewish Sepphorean merchants, and were probably rented by the city to its shopkeepers as suggested by ancient sources (Sperber 1998: 12 n. 29; see Libanus, *Oration* 11.254). Sepphoreans probably purchased their locally and regionally made lamps from shops in these market areas or along the main streets.

Rabbinic sources report the displaying of merchandise outside shops and in some instances in boxes placed in the street (Sperber 1998: 12 n. 37). The shopkeepers who sold lamps in Sepphoris probably displayed their various clay lamps outside the confines of their shops as well. They unquestionably hawked their lamps and wares by hollering out to passersby, as other shop owners are purported to have done (Sperber 1998: 12 n. 32; *Gen. Rab.* 40:3). Some lamps even bear advertisements; for example, an inscription on a lamp reads: "Buy me!" (Walters 1905: 422; for additional lamp advertisements, see Lapp 2007: 372). Lamp makers at Sepphoris may also have advertised their lamps with inscriptions. For example, an inscription on a Hadad Type 36 lamp, a type found at Sepphoris, reads: "In the name of Allah. Blessings to him who [buys it]" (Rosenthal and Sivan 1978: 133, no. 542). At times, merchants were known to clutter public thoroughfares with their merchandise, which probably included clay lamps as well. If lamps and other items displayed in the street were damaged or even destroyed by a passing cart or wagon, the cart driver was not held liable for damages because the street was considered public space (Sperber 1998: 9).

Rabbinic sources indicate that potters and presumably lamp makers also brought their merchandise to market for sale. Shops provided convenience for the ancient consumer, saving a strenuous walk all over town looking for goods. The marketplace centralized and consolidated shopping in the urban environment and thereby facilitated the acquisition of goods, including clay lamps. In addition to merchants and itinerant peddlers, the potters and lamp makers could have been directly responsible for the regional and interregional trade of lamps and wares. For example, ancient sources mention a traveling potter as "selling his wares" from the coastal plain to the Jerusalem region (*m. Ḥag.* 3:5). It is plausible that a potter who made enough lamps to create a surplus packed a donkey and set out for a week or more traveling to various towns and villages in the region. Conceivably, a potter might have attended seasonal fairs held in Caesarea Maritima (Safrai 1994: 220).

9. Lamps and the Sepphoris Agoranomoi

The *agoranomos* (market inspector) was responsible for controlling the marketplace of Sepphoris. Among his various official duties, he was in charge of weights and measures (Sperber 1998: 32–34). The *agoranomos* also collected rents from the shops and stalls in the marketplace (Sperber 1998: 39 n. 67), likely including those that sold lamps in the markets of Sepphoris. Because the *agoranomos* was at times responsible for regulating prices (Sperber 1998: 34–35), the determination of lamp prices perhaps fell under his authority too. Sometime in the Roman period, a Jewish *agoranomos* oversaw the marketplace at Sepphoris. His name, "Simon son of Aianos," is cast in Greek on a lead weight recovered from the western summit (Meshorer

1996; Weiss and Netzer 1996b: 34). A second market inspector, "Justus son of . . . ," presumably also a Jew, is mentioned on the weight (Meshorer 1996). Given the relatively large size of the city, the two perhaps divided their duties, one as the inspector of the upper market and the other of the lower market.

When the Jewish *agoranomoi* of Sepphoris inspected the quality of other market goods (Sperber 1998: 35–36 n. 3; *t. Kelim B. Qam.* 6.19), they conceivably inspected clay lamps as well. An earthenware vessel becomes susceptible to impurity once its manufacture is finished or it is fired (Magness 2011: 61, citing *t. Kelim B. Qam.* 3.12; 69–70, citing *m. Kelim* 2:6). Lamps with broken nozzles spotted by the Jewish *agoranomoi* in a shop or market context would not have been considered ritually impure and unfit for sale (*m. Kelim* 3:2). Samaritan lamps recovered at other sites exhibit possible evidence of the deliberate removal of their central discs, which may represent evidence for a Jewish custom: closed oil lamps were considered ritually pure and were "opened" only prior to use (Sussman 1982: 14).

10. Travelers, Merchants, and Pilgrims

Some lamps found at Sepphoris probably arrived as personal items and not trade goods. Because archaeological materials and ancient sources indicate the primarily Jewish population of Sepphoris, it is not surprising that most of the Herodian lamps found at the site were made in Jerusalem and were brought to the city by Jewish pilgrims returning from the temple, probably as souvenirs (Adan-Bayewitz et al. 2008: 75). Nor is it surprising that Darom lamps with a strong Jewish association were also found at Sepphoris. The use of Jerusalem clay quarried specially for Herodian lamp manufacture must have been particularly cherished by worshipers.

Travelers similar to those portrayed on a tombstone from a Roman cemetery at Arlon, Belgium (Henrotay and Warzée 2009: 20, fig. 18, inv. no. IAL GR/S 045) were responsible in part for the arrival lamps to Sepphoris. The lamps may have accompanied the travelers on their journeys for nighttime and early-morning use or were souvenirs. That lamps were an integral part of Roman-period travel is evidenced in papyrological sources, including packing lists. For example, the *Archive of Theophanes* suggests that it was a practice for travelers to bring their personal lamps with them on their journeys abroad. Theophanes, a legal adviser who journeyed from Upper Egypt to Antioch in the early fourth century C.E., includes a hanging lamp and a lampstand (Roberts and Turner 1952: 104–5, 118, no. 627: "Memoranda and Accounts," col. 3, line 54: λ[ύχνος] ἐν τῷ λυχν(είῳ); col. 3, line 55: λ[ύχν(ος) κρε] μαστός) in the list of items for travel. Officials, merchants, and others who traveled for shorter periods probably took lamps with them as well and at times may have left them behind.

The administrative borders would not have affected the distribution of Herodian lamps entering Galilee–and Sepphoris–from the Jerusalem area (Adan-Bayewitz et al. 2008: 73). Administrative taxes would not have affected the lamps entering Galilee from any region, because the limited quantities of lamps from other regions indicate that they were probably not a traded commodity (like wine- or olive oil–filled amphorae) and likely arrived in Sepphoris as an occasional souvenir or random item for sale. The antiwar stance of the Sepphoreans hinted at by Josephus (*J.W.* 3.30–32) would certainly have facilitated their travel throughout the Roman-controlled areas of Palestine and abroad in the Roman period. This ease of travel is reflected by the types of Roman-period lamps found in the western summit and originating from workshops in different regions, such as Judea (Herodian and Darom), Beth Shean/Scythopolis (Palestinian discus), and Gerasa/Jerash (Jebel Jofeh). These regional contacts continued into the Byzantine period, as evidenced by the discovery at Sepphoris of Caesarean discus lamps (from Caesarea Maritima) and Jerash-manufactured lamps (Jerash lamp prototype and Jerash with zoomorphic handle).

It is difficult to know from the archaeological record how a natural disaster, primarily the earthquake of 363 C.E., affected lamp manufacture and interregional lamp trade and distribution. The effect of the Gallus War, if there were any repercussions at Sepphoris at all, on lamp manufacture and trade is likewise undeterminable. The occurrence of Herodian (cat. nos. 19–43) and Darom (cat. nos. 53–70) lamps, which were probably manufactured in the Jerusalem area, may be evidence for the migration of Jews to Galilee in the aftermath of the destruction of the Second Temple by the Romans in 70 C.E. and of the Bar Kokhba uprising in 132–135 C.E.

In the fifth and sixth centuries C.E., Sepphoris was the seat of a bishopric and, according to historical sources, a bishop of Sepphoris attended the Council of Chalcedon in 451 C.E. (Meyers, Netzer, and Meyers 1992: 17). A fragment of a redware vessel with a stamped cross does indicate a Christian presence at the site after the earthquake of 363 C.E. (Rutgers 1996: 208, no. 85, fig. 28). Taking this into account, I would have expected the discovery of more than the six candlestick lamp fragments reported by Gärtner (Gärtner 1999), as this type is typically decorated with crosses and/or Christian liturgical inscriptions (Magness 1993: 176). No "candlestick" lamps were found by the Duke team, and none of the Caesarean discus lamps uncovered in the western summit were decorated with Christian symbols. The dearth of Christian lamps at Sepphoris, at least with respect to the excavated material culture of the western summit there, may be explained in part by the words of Epiphanes, who observes the absence of Christians in Tiberias and Diocaesarea (Sepphoris) (*Panarion* 30.11.9–10; see discussion in Gafni 1996: 51–52). An abundance of earlier Palestinian discus lamps with mythological or erotic images found in the western summit either suggests a Gentile presence at Sepphoris or the use by Jews of lamps with those images.

11. Indicators of Interactions: Local and Regional Palestinian Lamps Abroad

Few locally and regionally manufactured lamps have been recovered from archaeological contexts outside Roman Palestine and Arabia. This indicates that, in general, Palestinian-manufactured lamps were not a trade good sold on a mass scale to an international market, as were the widespread and mass-marketed Carthaginian lamps. A knife-pared Herodian lamp nozzle at Berenice and a few provincial Palestinian lamps (e.g., the Judean molded, the bilanceolate, Beit Nattif, Caesarean, and Samaritan) have been found at urban centers and rural settlements outside Roman Palestine. However, among lamps manufactured in Roman and Byzantine Palestine, Palestinian discus lamps were traded interregionally and internationally because they are found widely and, in some instances, in substantial quantities at sites throughout Palestine, Arabia, and Cyprus. They also occur in Tyre and Lebanon (Marchand 1996: 64–65, nos. 60–68), and two have been recovered from deposits as far west as the Athenian Agora, though these were likely souvenirs or represent an occasional, coincidental trade good or lamps used during travel.

12. Imported Lamps at Sepphoris and Other Sites of Roman Palestine

In comparison with the substantial quantities of lamps of local and regional manufacture recovered at sites in Roman and Byzantine Palestine and Arabia, imported lamps represent a small percentage of lamps found at Sepphoris and at sites on the coast, in Galilee, and in the Decapolis (e.g., for Ashkelon, see Johnson 2008: 127–31, nos. 380–89; for Scythopolis [Beth Shean], see Hadad 2002: 131–35, nos. 507–21; for Caesarea Maritima, see Sussman 2008: 252–55, nos. 217–29). This can be explained by the well-developed and strong lamp-manufacturing tradition in Palestine and Arabia and also by the expense of transporting lamps from far to near. The general lack of imported lamps in Galilee, Samaria, Judea, and Arabia testifies to the strength of the highly competitive local and regional lamp-production centers, which manufactured high-quality lamps that appealed to the religious sensibilities of their clientele by sufficiently fulfilling the population's lamp-lighting needs.

Three imported lamp types from Western workshops in Greece and Asia Minor were found on the western summit of Sepphoris (cat. nos. 215–19). Their occurrence indicates possible direct contact with Athens, Corinth, and Ephesus, though they could have just as easily changed many hands before arriving at the site. The imported lamps found at Sepphoris most likely arrived via the Roman roads running from the coastal commercial port cities of Caesarea Maritima and Ptolemais, specifically via the Caesarea Maritima–Scythopolis and Ptolemais–Tiberias routes.

Major pottery factories operating throughout Roman North Africa, Egypt, Anatolia, Italy, and Greece marketed their lamps and wares on an international scale, according to the archaeological record. That so few were found at Sepphoris and at other sites in the region– notably, the port of Ashkelon, where only nine examples have been reported (Johnson 2008: 127–31, nos. 380–89)–indicates that the major international workshops were not marketing their lamps to the cities, towns, and villages of Roman and Byzantine Palestine. The occurrence of imported lamps in Roman and Byzantine Palestine indicates interactions between regions and not lamp trade itself. An occasional foreign lamp may have shown up in the urban marketplace, but the lamp shops of Scythopolis and Capernaum contained local Galilean and Decapolean types and not imports. The imported lamps may have been brought by foreign travelers visiting or conducting commerce in Syria–Palestine. This may help explain the occurrence of Italian lamps and the popular western *Firmalampen* (though the pale yellow fabrics of a number of these suggest local manufacture) at Masada and Jerusalem, which perhaps arrived in the region with Roman soldiers and their auxiliaries. Similarly, the occurrence of imported lamps at Syro-Palestinian sites may reflect travel by residents of Palaestina Secunda returning from foreign lands: they had carried lamps with them either for their use on the homeward journey or as a possible gift or souvenir. In contrast to the paucity of imported lamps in Galilee and Golan, imported fine wares (e.g., African Red-Slip Ware, Cypriot Red-Slip Ware, and Eastern Terra Sigillata, etc.) are common.

13. Summary

Current archaeological evidence indicates that lamp molds were generally manufactured in the cities of Roman Palestine and that lamp production took place both in urban (Caesarea Maritima and Gerasa) and in village settings (Kefar Shikhin). It is plausible that the potters of Shikhin acquired their lamp molds in nearby Sepphoris and then returned with them to fashion lamps in their workshops. But it is equally plausible, given the large number of molds found at the site, to hypothesize that there were artisans in Shikhin capable of making molds. Subsidiary industries were often practiced at estates and farmsteads in the Roman world. Although few farmsteads have been excavated in Palestine, literary references hint at their existence. Military installations may also have produced their own lamps and pottery. The excavation of the kiln works of *Legio X Fretensis* and the small-scale smithing installations found in the barrack of the *Legio IV Martia* legionary fortress at el-Lejjun (Lapp 2006) suggest self-sufficiency in these industrial activities.

Galilean lamps, such as the Sepphorean spatulate, "Sepphorean" discus, and Galilean discus, were probably made in workshops located in Sepphoris and Shikhin. Their limited geographical distribution (to Sepphoris and a few other sites) suggests that they were traded on a local and interregional level. The distribution of bilanceolate lamps indicates their trade or interactions on a local and regional level. The distribution of Palestinian discus lamps indicates that they were probably traded on a local, regional, and international level. The paucity of Palestinian lamps elsewhere in the Roman East and Mediterranean suggests that provincial Palestinian lamps were not traded on an international scale. The occurrence of several Syro-Palestinian lamps in Greece and North Africa does not necessarily indicate trade of these forms but, rather, is evidence of travelers from the Near East. The presence of imported lamps found at sites in Galilee, the Decapolis, and elsewhere in Roman Palestine and Arabia should not be explained as merchandise marketed from abroad but, rather, as an indicator of foreign merchants or travelers from afar or Palestinian travelers returning from abroad.

Pottery workshops in Kefar Shikhin and probably Sepphoris and other places supplied the lighting needs of the city's inhabitants and surrounding populations. Occasional regional lamps arrived in the Sepphoris marketplace from Judea (Darom and some Herodian lamps), the coastal port of Caesarea Maritima (Caesarean discus), and the Decapolis cities of Scythopolis (probably Palestinian discus and bilanceolate) and Gerasa areas (Jerash zoomorphic, Jebel Jofeh, and Gilead). They would have been transported to Sepphoris by individuals, donkeys, or camels, or hauled by a donkey-drawn, two-wheeled cart or a four-wheeled wagon for large shipments. The recovery of clay lamps in shops excavated in Scythopolis and Capernaum indicates that this type of lighting vessel was sold in shops. The shops did not sell only lamps, as other ceramic wares were found in association with these commercial buildings. Clay lamps were probably sold in shops in the upper and lower markets of Sepphoris. The quality of lamp merchandise and the prices of lamps were regulated by the *agoranomoi* of Sepphoris.

CHAPTER 6

Conclusion

1. "Culture of Light"

A "culture of light" permeated the daily lives of Sepphoreans and other inhabitants of Roman Galilee. Light itself cannot be excavated, but the evidence for its existence can be recovered in word, image, and form. In word, the Mishnah reveals a number of ways that light and its main vehicle, the clay lamp, were used in Jewish religious practices. It mentions the customs of lighting household lamps in the evening to mark the beginning of the Sabbath (*m. Šabb.* 2:6,7). Prior to the onset of Passover, a ceremonial search using lamp light was conducted throughout one's home to locate and destroy any leaven (*hametz*) found inside (*m. Pesaḥ.* 1:1). In image, menorah branches support glass lamps with radiant flames of light in the Sepphoris synagogue mosaic. The eternal light (*ner tamid*) is also pictured hanging from the aedicula of the Torah Shrine. The sun (Helios?) is shown leading the chariot in the center of the zodiac mosaic of the Sepphoris synagogue (Weiss and Netzer 1996a: 26–27, 35–36; Weiss 2005: 104–10). In form, light was created and sustained by the many clay lamps unearthed by the Duke and Hebrew University excavations on the western summit of Sepphoris, where they were the chief means of artificial lighting. The discus lamps were a popular medium for depicting religious symbols and keeping religious beliefs alive in the public consciousness.

2. Connections

The clay lamps unearthed in the western summit of Sepphoris indicate that local and regional pottery workshops sufficiently fulfilled the lighting needs of their Sepphorean clientele. Twenty-four local and regional types are represented in this corpus, covering a continuous chronological range from the Late Hellenistic through Early Islamic periods. Recent archaeological evidence indicates lamp production at Kefar Shikhin, a small village near Sepphoris, which was a chief supplier of locally manufactured lamps intended for the marketplace of Sepphoris. The large quantities of lamps, lamp molds, lamp fragments, and wasters found in Shikhin substantiate this claim. The discovery of several lamp molds at Sepphoris suggests that the city may also have had commercial lamp workshops. One or more residential lamp workshops in the city might also have made lamps for the local population.

Most lamps probably arrived in the Sepphorean marketplace via donkey, the most common means of land transport. Two-wheeled carts and four-wheeled wagons drawn by donkeys would have been necessary for larger shipments of lamps that accompanied other goods from afar (such as the pottery, spices, oil, and wine produced in the Decapolis). Wagons required the wider and more solidly constructed Roman roads for long-distance passage. The itinerant peddler was also probably responsible for the occasional locally, regionally, or even internationally produced lamps entering the city's markets. Sepphoreans purchased their clay oil lamps in the marketplace (*macellum*) of the Lower City, including in the shops and stalls of the colonnaded *cardo*, the *decumanus maximus*, and the narrow streets. Upon entering the cosmopolitan setting of the Sepphoris marketplace, travelers also could purchase lamps and after leaving the city would ensure the broader geographical distribution of the lamps in the region.

The widespread network of Roman roads, rural and village roads, and footpaths that crisscrossed the region meant that the villages of Galilee were not isolated from the cosmopolitan influences of Sepphoris and Tiberias (Strange 2014). Galilean villages specializing in the manufacture of specific goods, such as Kefar Shikhin (storage jars and lamps) and Kefar Hananiah (cooking pots and possibly lamps), sold their products in the cities. Contrary to previous beliefs, Galilean villagers were not poor peasants; they prospered from the various industries carried out in their settlements (Aviam 2013: 43). Trade interactions made possible by the road network meant that Palestinian discus lamps decorated with mythological images found their way into nearby Jewish villages, spreading a Greco-Roman "pollinating" effect across village culture and daily life. Thus the occurrence of Palestinian discus lamps bearing images of Helios at the Jewish village of Nabratein and at the primarily Jewish city of Sepphoris is not extraordinary.

The occurrence of Darom and gray-fabric Herodian lamps represents evidence for visitors or newcomers from Judea, where these forms were manufactured. The large quantities of Darom lamps recovered at ʿIblin as well as the examples found at synagogue sites in Upper Galilee suggest manufacture in the region, using molds brought by Jewish emigrants from Judea who settled in Galilee after the destruction of the Second Temple. Only by means of trace-element analysis will we be able to determine whether the clays originated in Galilee or in the south. The lamps of the Sepphoris corpus suggest contact with certain cities of the Decapolis. The relatively high quantities of Palestinian discus lamps indicate connections with the Beth Shean and Pella area, where the results of the study's preliminary DCP spectrometric and petrographic analyses suggest their possible manufacture (Lapp 1997a: 178–79). The significant quantities of Palestinian discus lamps found at Sepphoris indicate that they were marketed to Sepphoreans and may be evidence that Sepphoreans conducted trade in the central Jordan Valley and Decapolis. One Jerash lamp fragment belonging to the form with zoomorphic handle and several Jebel Jofeh lamps with round bodies and pronounced nodules manufactured at Jerash suggest contact with this Decapolis city.

The Gerasa-to-Pella-Scythopolis-Caesarea Maritima road was the main cultural and commercial artery for Sepphoreans into the Decapolis, providing them easy access to its abundant agricultural and industrial resources. Scythopolis was Sepphoris's gateway into the Decapolis and its Greco-Roman culture. The commonality and continuity of local and regional lamp types found at Sepphoris and in the cities of the Decapolis—the Herodian, Darom, Palestinian discus, Galilean, bilanceolate, Jerash zoomorphic, Hauranian types—indicate that Sepphoris was part of the cosmopolitan Greco-Roman world. These lamp forms are found in Sepphoris and in the Decapolis cities of Scythopolis, Pella, Gerasa, Gadara, and Abila—cities with a strong Greco-Roman culture.

Four imported types originate from workshops in western Asia Minor and Greece: from Asia Minor, an Ephesian form with its characteristically sharply pointed spatulate nozzle (cat. nos. 215–16); from the Peloponnese, a first-century C.E. Corinthian round-bodied lamp (cat. no. 217); and from Athens, the later fifth-century C.E. Attic form that corresponds to Broneer Type XXVIII (cat. nos. 218–19). Similar to Beth Shean, Caesarea Maritima, and Gaza in Palestine, and Aila in Nabataea, a North African lamp from the Carthaginian production center was found in Sepphoris (Gärtner 1999: pl. 26.120; Hadad 2002: type 8, nos. 133–34, 518–19; Lapp 2012: 59–60, table 1, nos. 78–103) but not on the western summit. Because a bishop of Sepphoris attended the Council of Chalcedon in 451 C.E., I had expected to find more Byzantine candlestick lamps than the one example found by the Hebrew University team (Gärtner 1999: no. 94, pl. 20). These lamp types were closely associated with Christian pilgrimages to Jerusalem churches and were commonly decorated with Christian liturgical passages and symbols, indicating target-marketing to a Christian clientele (Lapp 2007: 376).

The few imported lamps at Sepphoris reflect social, economic, and religious interactions rather than actual lamp trade. The lamps from Athens, Corinth, and Ephesus were probably brought to Sepphoris by travelers, such as the merchants of fine wares who arrived in the ports of Caesarea Maritima, Dora, and Akko-Ptolemais and continued on to sell their goods in the city markets of Galilee, Decapolis, and elsewhere. As the *Archive of Theophanes* suggests, travelers journeyed with personal lamps and, presumably, the visitors to Sepphoris did as well. Sepphoreans returning to Galilee from distant lands, for example, probably purchased lamps in the markets of Athens, Corinth, and Ephesus for use on their journey home or as souvenirs. The Syrian discus lamp decorated with a menorah and found in the western summit perhaps indicates limited contact between the Jewish communities of Sepphoris and Syria.

3. Material Sourcing

The petrographic, micromorphological, and chemical characteristics of the lamp fabrics analyzed in this volume are similar to those exhibited by other locally and regionally produced lamps. The results of the petrographic and dcp-oes analyses indicate that the ancient lamp makers of the Palestinian discus, bilanceolate, and Beit Nattif types used highly calcareous clays. Carbonaceous clays are poor for making ceramic vessels because of the property of spalling, or lime-popping, which cracks the fabric of pottery. But lamp makers and potters made the best use of their clay resources and developed ways to produce durable lamps by calcining their clays to avoid breakage. This petrographic analysis shows variation among the clay fabrics of the lamps. Fabric variation occurs among the five groups, and characteristics are generally distinctive to each type. The fabrics of the Palestinian discus lamps generally exhibit few inclusions—primarily a low quantity of quartz sand grains—and some examples have microfossils. By contrast, bilanceolate lamp fabrics contain a high abundance of foraminifera and bioclastic material. Beit Nattif lamp fabrics contain a relatively high abundance of *terra rossa* clay balls, quartz sand, and some microfossils. Lamp makers and potters added *terra rossa* clay balls as temper to several lamp fabrics, including the locally manufactured version of the Palestinian discus lamp sample (DAB4) from Abila. The use of *terra rossa* clay nodules as a tempering agent was also observed in the fabrics of select Herodian lamps examined by the Lawrence Berkeley National Laboratory. No basalt temper was identified among the fabrics, which would point to an origin (at least for the temper) in the basalt areas southwest and northeast of the Sea of Galilee.

The results of the dcp-oes analysis indicate that clays were quarried from at least two distinct sources for use in Palestinian discus lamp production. Group A consists of chemically similar lamps belonging to the Palestinian discus (collected at Sepphoris and Pella) and Tarsus types (found at Ashkelon). Group B includes chemically similar lamps of the Palestinian discus (found at Ashkelon, Caesarea Maritima, Sepphoris, Meiron, Ḥorvat Ḥazon, Scythopolis, and Pella) and Beit Nattif groups (recovered at Gadara). That the Tarsus sample (DA9) is chemically similar to Palestinian discus lamps suggests the use of a lamp mold that was conceivably carved in a shop in Tarsus and acquired by a potter in the Ashkelon area, who then manufactured Tarsus-type lamps. The petrographic characteristics of the sample (carbonaceous and foraminifera) suggest clay from a sedimentary source, but this of course, could be located in the region of Tarsus or Ashkelon. That some lamps of the Palestinian discus type were manufactured with calcareous and foraminiferous clays (DS10) and others with ferrous clays with quartz sand and *terra rossa* nodules (DM10, DC6, DAB4) indicates the use of different clay sources and suggests the trade of lamp molds (rather than lamps) to various workshops located on the coast, in Galilee, and in the Decapolis. The Jerash bowl samples, both fired and unfired were collected from the hippodrome kiln site in Gerasa; they are chemically similar as anticipated. The clay fabrics of the Palestinian discus, bilanceolate, and Beit Nattif samples were chemically dissimilar to the Jerash bowl reference materials, suggesting that they were not manufactured by the hippodrome pottery in Gerasa.

4. Cosmopolitanism

The Sepphoris lamps accurately reflect what one would expect to find in an open-market, cosmopolitan city of Lower Galilee. The city's "lamp culture" was defined by a visual language communicated by the very images and symbols portrayed on the lighting vessels themselves. The mythological themes found on the Palestinian discus lamps—for example, Europa riding the bull (Zeus) and Medusa—surely resonated with the Gentile population of the city. Such imagery also contributed to the worldly atmosphere experienced daily by the Jewish residents of Sepphoris, including the community's most revered rabbis and scholars. Gentile themes also appear in the various panels of the contemporaneous Dionysos mosaic floor of the villa, presumed to have been the residence of the Jewish patriarch of the city (Bowersock 2006: 39) and later in the zodiac of the city's synagogue.[1] That a substantial quantity of Palestinian discus lamps was recovered from the western summit indicates that Jewish residents had purchased them for use as well. The Jewish symbols, such as the menorah and Torah shrine, depicted on the locally produced (possibly at Kefar Shikhin) "Sepphorean" discus lamps would have appealed to the city's predominantly Jewish population and were manufactured for the community. The lamps decorated with simple geometrical and floral patterns would have been attractive to a mass-market audience, regardless of religious affiliation (Lapp 2007: 372–74). The lamp finds indicate that Sepphoris was not an isolated and culturally conservative city but was fully engaged in commercial, administrative, and cultural interactions with other urban centers, towns, and villages both near and far.

The pragmatic, antiwar stance shown by Sepphoreans toward Roman rule during the reign of Nero certainly benefited the city's population economically. Josephus hints at these attitudes (*J.W.* 3.30–32), and two bronze coins minted in the city in the year 67/68 C.E. indicate that the city did not participate in the Great Revolt. The coins rename the city *Eirenopolis*, the "City of Peace" (Meshorer 1996: 195–96, nos. 43–44). A later medallion of Caracalla (198–217 C.E.) commemorates a treaty of friendship between Rome and the *boulē* (council) of Sepphoris (see Meyers, Hendin, and Meyers 2015). With peace often comes commercial stability, which in turn promotes prosperity. In the case of Sepphoris, peace probably facilitated the city's trade links with other regions, such as the cities of the Decapolis, and the travel of Sepphoreans to and from the urban centers of Roman Palestine and beyond. This is reflected in the diversity of regional lamp types found at the site and presented in this volume. Although it is difficult to gauge to what extent the multicultural atmosphere of Sepphoris is reflected in the range of artifacts discovered there, the variety of lamp forms from the site seems to reflect a diverse population over several centuries. The myriad geometric and floral patterns and symbols decorating the lamps indicate that the lighting vessels were marketed to a diverse clientele—Gentiles, Jews, Christians, and Muslims—which in turn points to a degree of cosmopolitanism in this Galilean city.

1. There may have been a number of synagogues at Sepphoris in addition to the one with the zodiac mosaic.

APPENDIX A

Reports of Select Loci

Prepared with the Assistance of the Series Editors

This appendix provides descriptions of all loci that yielded 3 or more lamps published in the catalogs of chaps. 2–3. The first two loci (L84.1068 and L85.1195) are exceptional in this regard, having contributed a total of 78 (59 and 19, respectively) of the 221 published specimens. Like those two loci, many of the others consist of refuse deposits heaped up in cisterns and other subterranean spaces. Included in the data given below are field readings for ceramic vessels, cross-references to published vessels, and readings of legible coins. This appendix anticipates the publication of similar locus reports in the forthcoming volume on the stratigraphy of the western summit (Meyers, Meyers, and Gordon forthcoming).

L84.1068

Description of Locus	The extraordinarily rich fill of a refuse pit accumulated in a cistern (L84.1066) under the floors of a Roman-period house. The refuse pit contains large numbers of Early Roman and Middle Roman ceramics, with a few Late Roman fragments and coins.
Location	Area 84.1, Squares K10–K11, J10–K11
Number of Cataloged Lamps	59
Cataloged Lamps	20, 22–23, 25, 29, 45, 47, 51–52, 54, 77–81, 83–84, 86–87, 89, 91–93, 95–97, 99–104, 106–9, 111–26, 129–32, 135–37
Field Readings for Ceramics	Persian (infrequent), Early Hellenistic (few), Late Hellenistic, Early Roman (abundant), Middle Roman (abundant), Late Roman (few)
Published Ceramics	Balouka 2013: pls. 5.3, 6, 11; 8.2–9, 20; 9.3–4, 15; 10.9; 11.1, 10, 13–14, 23–24; 12.7; 13.1–7, 12; 14.2, 4–9, 11; 15.1–2, 4–6, 8–9, 12–13, 15–16; 16.1, 3–5, 7–9, 13–15; 17.6–8; 18.1–2, 6; 19.2, 4–8; 20.1, 5, 9, 11; 21.1–2, 4–5, 8–12; 22.1–6, 8–9, 11–14; 23.1–4, 8–11; 24.1–10, 12–13; 25.1–3, 7–10, 13, 16; 26.1–6, 9, 12–14, 16; 27.1–6; 34.1, 4–5, 7–9; 35.1–3, 6, 8–9; 36.1–6; E. Meyers 2013: pl. 39.7; Vincenz 2013: pl. 52.4

Coins	Ariel and Bijovsky forthcoming: IAA 113999, 104–76 B.C.E. IAA 116164, 104–76 B.C.E. IAA 113993, second century C.E. IAA 116165, 341–346 C.E. IAA 113998, 337–341 C.E. IAA 113714, 383–395 C.E.

L85.1195

Description of Locus	The fill of a cistern (L85.1029) under the northern rooms of a Roman-period house whose use continued into the Byzantine period. Dozens of Byzantine jugs were retrieved from the cistern.
Location	Area 85.1, Square L16
Number of Cataloged Lamps	19
Cataloged Lamps	189–91, 193–208
Field Readings for Ceramics	Middle Roman (few), Late Roman (abundant), Early Byzantine
Published Ceramics	Vincenz 2013: pl. 53.1–27
Coins	Ariel and Bijovsky forthcoming: IAA 99613, 351–361 C.E. IAA 99555, 393–395 C.E. IAA 99558, second half of fifth century C.E. IAA 99553, fourth century C.E. IAA 99551, fourth century C.E. IAA 99552, fourth century C.E. IAA 99557, fourth–fifth centuries C.E. IAA 99554, second half of fifth century C.E. IAA 99556, fifth–sixth centuries C.E.

L84.1094

Description of Locus	One of several fill loci comprising a rich refuse deposit, with many Middle Roman ceramics, in the central and southern rooms of a Roman-period house. The deposit appears to have been stratified during an alterations phase of the house. See L84.1120 below.
Location	Area 84.1, Square L11
Number of Cataloged Lamps	5
Cataloged Lamps	49, 74–75, 128, 134
Field Readings for Ceramics	Iron IIb, Persian, Early Hellenistic, Late Hellenistic, Early Roman, Middle Roman
Published Ceramics	Balouka 2013: pls. 13.9, 17.4, 19.1, 20.6, 22.7, 26.8
Coins	Ariel and Bijovsky forthcoming: IAA 113716, 98–117 C.E. IAA 113717, 219–221 C.E.

L84.4142b

Description of Locus	The fill of a *miqveh* (L84.4190) in a Roman-period house that continued in use into the Byzantine period.
Location	Area 84.4, Square N14
Number of Cataloged Lamps	5
Cataloged Lamps	180, 184, 187–88, 192
Field Readings for Ceramics	Late Hellenistic, Late Roman, Early Byzantine (abundant), Late Byzantine
Published Ceramics	Vincenz 2013: pls. 45.10; 48.11; 51.2–3; 55.5, 7
Coin	Ariel and Bijovsky forthcoming: cat. no. 516 (IAA 48596, 383–395 C.E.)

L84.1056

Description of Locus	The fill of a cistern or underground space (L84.1053) under a Roman-period house that continued to be occupied into the Byzantine period. This fill and adjacent loci contained large numbers of coins from the 330s and 340s C.E., suggesting a scattered hoard.
Location	Area 84.1, Square K11
Number of Cataloged Lamps	4
Cataloged Lamps	71a, 142, 158, 171
Field Readings for Ceramics	Early Roman, Middle Roman, Late Roman (abundant), Early Byzantine
Published Ceramics	Balouka 2013: pls. 20.2; 21.3; 28.6, 8–9; 29.8, 10; 30.5, 8; 31.7
Coins	Ariel and Bijovsky forthcoming: c IAA 113679, 218–222 C.E. IAA 113664, 315–316 C.E. IAA 113669, 324–330 C.E. IAA 113662, 335–341 C.E. IAA 113676, 341–346 C.E. IAA 113670, 341–346 C.E. IAA 113667, 341–346 C.E. IAA 113671, 341–346 C.E. IAA 113678, 335–337 C.E. IAA 113674, 341–346 C.E. IAA 113661, 341–346 C.E. IAA 113663, 341–346 C.E. IAA 113668, 341–346 C.E. IAA 113673, 341–346 C.E. IAA 113675, 341–346 C.E. IAA 113672, 341–346 C.E. IAA 113677, 402–408 C.E. IAA 113666, fourth century C.E. IAA 113665, fourth century C.E.

L84.1120

Description of Locus	One of several fill loci comprising a rich refuse deposit, with many Middle Roman ceramics, in the central and southern rooms of a Roman-period house. The deposit appears to have been stratified during an alterations phase of the house. This locus was excavated in a doorway of an underground chamber (L84.1127). See L84.1094 above.
Location	Area 84.1, Squares L11–L12
Number of Cataloged Lamps	4
Cataloged Lamps	19, 53, 59, 62
Field Readings for Ceramics	Late Hellenistic (abundant), Early Roman (abundant), Middle Roman
Published Ceramics	Balouka 2013: pls. 5.8, 14; 8.1, 8, 14–17; 9.1, 5, 7, 11–12, 16; 10.5–6; 11.8, 17; 12.1–3, 5
Coins	None

L84.2077

Description of Locus	The fill of an underground space (L84.2078) that was hewn out for use as a cistern under a house but was later joined to a small complex of three subterranean chambers.
Location	Area 84.1, Square L5
Number of Cataloged Lamps	4
Cataloged Lamps	156, 162–63, 172
Field Readings for Ceramics	Iron Age II, Early Hellenistic, Early Roman, Middle Roman (abundant), Late Roman (abundant), Early Byzantine, Late Byzantine, Early Islamic (?)
Published Ceramics	Vincenz 2013: pls. 56.10, 11; 57.5
Coins	Ariel and Bijovsky forthcoming: IAA 113756, 99/100 C.E. IAA 113754, 215–216 C.E. IAA 113748, 270–275 C.E. IAA 113749, 270–275 C.E. IAA 113752, 272–275 C.E. IAA 113753, 337–341 C.E. IAA 113757, 351–361 C.E. IAA 113750, 383–395 C.E. IAA 113761, 402–408 C.E. IAA 113759, 383–395 C.E. IAA 113763, 383–395 C.E. IAA 113755, fourth century C.E. IAA 113762, fourth century C.E. IAA 113758, fourth century C.E. IAA 113764, fourth century C.E. IAA 113760, fourth–fifth century C.E.

L84.2172

Description of Locus	A tamped-earth layer, possibly a surface, stratified within the ruins of a room in a poorly preserved area on the southwestern edge of the site.
Location	Area 84.2, Squares M4, N3–N4
Number of Cataloged Lamps	4
Cataloged Lamps	82, 85, 94, 98
Field Readings for Ceramics	Early Roman, Middle Roman, Late Roman
Published Ceramics	Balouka 2013: pl. 20.10 de Vincenz 2013: pls. 42.5, 52.7
Coins	Ariel and Bijovsky forthcoming: IAA 116258, first–second centuries C.E. IAA 116259, 239–240 C.E.

L84.5085

Description of Locus	A thick pottery layer, likely a refuse deposit, excavated in the space of a collapsed cistern (L85.4084).
Location	Area 84.5, Square P9
Number of Cataloged Lamps	3
Cataloged Lamps	64–66
Field Readings for Ceramics	Persian (?), Early Hellenistic, Late Hellenistic, Early Roman (abundant), Middle Roman, Late Roman, Early Byzantine
Published Ceramics	None
Coins	Ariel and Bijovsky forthcoming: IAA 99548, 222–235 C.E.

L85.3037b

Description of Locus	An unsealed fill in one of the northwestern squares of Area 85.3.
Location	Area 85.3, Square C
Number of Cataloged Lamps	3
Cataloged Lamps	4, 9, 217
Field Readings for Ceramics	Early Hellenistic, Late Hellenistic, Early Roman, Middle Roman
Published Ceramics	Balouka 2013: pls. 2.8–10; 4.4, 6
Coins	Ariel and Bijovsky forthcoming: IAA 48585, second century C.E.

APPENDIX B

Macroscopic Descriptions of Lamp Samples

The term *macroscopic* refers to all features of the samples that are visible to the naked eye. The samples included in this appendix were analyzed by both petrographic thin-section and DCP analysis. With respect to the Palestinian discus lamps, I include the category "Diagnostic Characteristic," which refers to special features such as a certain type of discus motif, lamp feature, or lamp maker's mark. Such features help distinguish certain lamp specimens from others. Using the standard Munsell Soil Color Charts (Munsell 1971), we determine the clay fabric color of each sample from a broken edge. If slip is present, its color is also noted. The hardness of each sample is based on the Mohs hardness scale. Following standard archaeological convention, the types of inclusions appearing in the fabric are referred to as "grit" by color. The size and frequency of the inclusions are noted. The sorting of inclusions follows the scale suggested by Barraclough (as described by Orten, Tyers, and Vince 1993). Traces of carbon are noted because they indicate burning and use of the lamp. Since a fragment is only a small representative portion of the complete lamp, the presence or absence of carbon does not always provide us with an accurate picture as to whether the complete lamp from which it originated had ever been used. The carbon, for example, may have resulted from secondary burning not associated with the function of the lamp. The "fabric condition" refers to the degree of friability of the samples' clay, high friability being considered "poor" and low "good." The degree of oxides and other contaminates on the surfaces of the lamp fragments are taken into further consideration with respect to the condition of a sample.

Samples by Region and Site

Coastal Plain

Ashkelon[1]

1. DA2 Type: Palestinian discus
Date: Late first to third centuries C.E.
Period: Early Roman to Middle Roman
Stratigraphic Context: Street-front room of house (fill)
Greatest Dimension: 4.3 cm
Part: shoulder and discus fragment
Diagnostic Feature: bichrome slip
Fabric Color (Munsell): 10YR 8/4 very pale brown
Slip Color (Munsell): 7.5YR 4/3 brown (outer surface)
Hardness (Mohs): 3.0
Inclusion Type(s): Red grits (very fine sand; low frequency)
Sorting: Poor
Carbon Traces: None
Fabric Condition: Good (-)

2. DA3 Type: Palestinian discus
Date: Late first to third centuries C.E.
Period: Early Roman to Middle Roman
Stratigraphic Context: Street-front room of house (fill)
Greatest Dimension: 3.2 cm
Part: Base fragment
Diagnostic Feature: Lamp maker's mark (*eta*)
Fabric Color (Munsell): 7.5YR 7/4 pink
Slip Color (Munsell): 2.5YR 5/6 red
Hardness (Mohs): 2.0
Inclusion Type(s): Red grits (very fine sand; low frequency); white grits (very fine sand; low frequency)
Sorting: Fair
Carbon Traces: None
Fabric Condition: Good

3. DA9 Type: Cilician (Tarsus Group XVI)
Date: Second–third centuries C.E. (Hayes 1980: 72–73)
Period: Middle Roman
Stratigraphic Context: Street-front room of house (fill)
Greatest Dimension: 3.1 cm
Part: Shoulder fragment
Diagnostic Feature: Distinctive "white" fabric; radial lines; import
Fabric Color (Munsell): 2.5Y 8/3 pale yellow
Slip Color (Munsell): 10YR 5/3 brown

---

1. The Ashkelon samples were recovered from a small room on the north side of a house where hundreds of mold-made discus lamp sherds were found. None of them showed any evidence of use. Although the room in which they were found does front a street, Stager doubts that the oil lamps were being sold from the ground floor of the house. See Stager 1991: 45, 47; Rose 1997.

Hardness (Mohs): < 2.0
Inclusion Type(s): None visible
Sorting: Not visible
Carbon Traces: None
Fabric Condition: Good

Caesarea Maritima

4. DC6 Type: Palestinian discus
Date: Late first to third centuries C.E.
Period: Early Roman to Middle Roman
Stratigraphic Context: Possible synagogue[2]
Greatest Dimension: 5.6 cm
Part: Base fragment
Diagnostic Feature: Lamp maker's mark (*eta* and tri-bud floral)
Fabric Color (Munsell): 2.5YR 6/6 light red (core); 2.5Y 7/3 pale yellow (outer boundaries of core)
Slip Color (Munsell): None
Hardness (Mohs): 2.0
Inclusion Type(s): White grits (very fine sand; medium frequency); red grits (fine sand to very fine sand; low frequency)
Sorting: Fair
Carbon Traces: None
Fabric Condition: Good (-); traces of oxides on the surface

Galilee

Meiron

5. DM10 Type: Palestinian discus
Date: Late first to third centuries C.E.
Period: Early Roman to Middle Roman
Stratigraphic Context: Cistern 4 (C-7), Lower City, Area MVII[3] R5236/Spec. 80–385
Greatest Dimension: 3.9 cm
Part: Base fragment
Diagnostic Feature: Lamp maker's mark: *eta*
Fabric Color (Munsell): 5YR 6/6 reddish yellow (core); 10YR 7/6 yellow (outer)
Slip Color (Munsell): None
Hardness (Mohs): 2.5
Inclusion Type(s): White grits (fine sand, medium frequency)

---

2. Although the lamp sherd was uncovered in Caesarea Maritima, its context is unknown. Rivka Berger of the Israel Antiquities Authority informed me in June 1997 that it may have been recovered by Avi-Yonah's excavations of the Caesarea synagogue in the late 1950s and early 1960s.

3. This cistern is located in a courtyard at the end of a street that leads south from MI (first insula excavated in the Lower City) into MVII (insula south of MI). The cistern fill contained pottery from the Late Roman through Early Byzantine periods, and eight coins date the fill securely to the fourth century. See Meyers, Strange, and Meyers 1981: 103–4, fig. 6.6.

Sorting: Good
Carbon Traces: None
Fabric Condition: Good

Ḥorvat Ḥazon

6. DHH2 Type: Palestinian discus
Date: Late first to third centuries C.E.
Period: Early Roman to Middle Roman
Stratigraphic Context: Pit[4]
Greatest Dimension: 4.0 cm
Part: Shoulder and discus fragment
Diagnostic Feature: Bichrome slip
Fabric Color (Munsell): 5YR 7/6 reddish yellow
Slip Color (Munsell): 5YR 6/6 reddish yellow
Hardness (Mohs): 2.0
Inclusion Type(s): Gray grits (fine sand; very low frequency); red grits (coarse sand to fine sand; medium frequency); white grits (fine sand; very low frequency)
Sorting: Poor
Carbon Traces: None
Fabric Condition: Good (-)

Sepphoris[5]

7. DS4 Type: Palestinian discus
Date: Late first to third centuries C.E.
Period: Early Roman to Middle Roman
Stratigraphic Context: Fill (L84.1068) in cistern (L84.1066)
Greatest Dimension: 4.2 cm
Part: Shoulder, discus, and nozzle-bridge fragment
Diagnostic Feature: Bichrome slip
Fabric Color (Munsell): 5YR 7/4 pink
Slip Color (Munsell): 10R 4/6 red (discus)/7.5YR 3/1 very dark gray (shoulder)
Hardness (Mohs): 2.0
Inclusion Type(s): Red grits (fine sand to very fine sand; medium frequency); white grits (very fine sand; low frequency)
Sorting: Fair
Carbon Traces: None
Fabric Condition: Good

8. DS9 Type: Palestinian discus
Date: Late first to third centuries C.E.
Period: Early Roman to Middle Roman
Stratigraphic Context: Fill (L84.1068) in cistern (L84.1066)
Greatest Dimension: 3.8 cm
Part: Shoulder and discus fragment
Diagnostic Feature: Erotic Type 1 motif

4. Bahat 1974: no. 4, fig. 5 (from Pit 1, second to early third centuries C.E.).
5. See appendix A for the context of select Sepphoris samples.

Fabric Color (Munsell): 2.5Y 8/4 pale yellow
Slip Color (Munsell): 10YR 3/2 very dark grayish brown
Hardness (Mohs): 2.0
Inclusion Type(s): Red grits (fine sand; very low frequency)
Sorting: Poor
Carbon Traces: None
Fabric Condition: Good (–)

9. DS10
Type: Palestinian discus
Date: Late first to third centuries C.E.
Period: Early Roman to Middle Roman
Stratigraphic Context: Fill (L84.1068) in cistern (L84.1066)
Greatest Dimension: 4.7 cm
Part: Shoulder and discus fragment
Diagnostic Feature: Bi-associative motif (Bacchus and chalice)
Fabric Color (Munsell): 2.5Y 8/4 pale yellow
Slip Color (Munsell): 10YR 3/2 very dark grayish brown
Hardness (Mohs): 2.5
Inclusion Type(s): Red grits (very fine sand; high frequency)
Sorting: Fair
Carbon Traces: None
Fabric Condition: Fair (+)

10. DS21
Type: Palestinian discus
Date: Late first to third centuries C.E.
Period: Early Roman to Middle Roman
Stratigraphic Context: Fill (L84.1068) in cistern (L84.1066)
Diagnostic Feature: Lamp maker's mark (tri-bud floral)
Fabric Color (Munsell): 10YR 7/3 very pale yellow (core); 5YR 7/3 (outer surfaces of poorly defined core)
Slip Color (Munsell): 10YR 3/1 very dark gray
Hardness (Mohs): 2.0
Inclusion Type(s): Red grits (very fine sand; very high frequency)
Sorting: Good
Carbon Traces: None
Fabric Condition: Good

11. DS22
Type: Palestinian discus
Date: Late first to third centuries C.E.
Period: Early Roman to Middle Roman
Stratigraphic Context: Fill (L84.1068) in cistern (L84.1066)
Greatest Dimension: 7.5 cm
Part: Base fragment
Diagnostic Feature: Lamp maker's mark (*alpha*)
Fabric Color (Munsell): 2.5Y 8/4 pale yellow
Slip Color (Munsell): 10YR 3/2 very dark grayish brown
Hardness (Mohs): 2.0
Inclusion Type(s): Red grits (fine sand to very fine sand; medium frequency); black grits (fine sand to very fine sand; low frequency)
Sorting: Fair
Carbon Traces: None

Fabric Condition: Fair

12. DS24 Type: Palestinian discus
Date: Late first to third centuries C.E.
Period: Early Roman to Middle Roman
Stratigraphic Context: Fill (L84.1068) in cistern (L84.1066)
Greatest Dimension: 7.8 cm
Part: Base fragment
Diagnostic Feature: Lamp maker's mark (*eta*)
Fabric Color (Munsell): 7.5YR 7/4 pink
Slip Color (Munsell): 5YR 5/4 reddish brown
Hardness (Mohs): 2.0
Inclusion Type(s): Red grits (coarse sand to very fine sand; high frequency); gray grits (very find sand; very high frequency)
Sorting: Poor
Carbon Traces None
Fabric Condition: Very good

13. DS29 Type: Palestinian discus
Date: Late first to third centuries C.E.
Period: Early Roman to Middle Roman
Stratigraphic Context: Fill (L84.1068) in cistern (L84.1066)
Greatest Dimension: 4.8 cm
Part: Base fragment
Diagnostic Feature: Lamp maker's mark (*gamma alpha*)
Fabric Color (Munsell): 2.5Y 7/3 pale yellow
Slip Color (Munsell): 10YR 3/2 very dark grayish brown
Hardness (Mohs): 2.0–2.5
Inclusion Type(s): White grits (coarse sand; very low frequency; fine sand to very fine sand; high frequency); red grits (very fine sand; high frequency)
Sorting: Good
Carbon Traces: None
Fabric Condition: Good

14. DS30 Type: Palestinian discus
Date: Late first to third centuries C.E.
Period: Early Roman to Middle Roman
Stratigraphic Context: Fill (L84.1068) in cistern (L84.1066)
Greatest Dimension: 4.5 cm
Part: Base fragment
Diagnostic Feature: Lamp maker's mark (*eta*)
Fabric Color (Munsell): 7.5YR 8/3 pink (core); 2.5Y 8/3 pale yellow (poorly defined outer boundaries of core)
Slip Color (Munsell): 5YR 3/4 dark reddish brown
Hardness (Mohs): 2.0
Inclusion Type(s): White grits (very fine sand; medium frequency); red grits (fine sand; very low frequency)
Sorting: Poor
Carbon Traces: None
Fabric Condition: Good

15. BNS1
Type: Beit Nattif with bow-shaped nozzle
Date: End of third to late fifth/early sixth centuries C.E. (da Costa 2003: 1.68)
Period: Late Roman–Byzantine
Stratigraphic Context: Fill (L84.2077) in cistern (L84.2078)
Greatest Dimension: 4.8 cm
Part: Shoulder, discus, and nozzle-bridge fragment
Diagnostic Feature: Circles and nodules decorate shoulder
Fabric Color (Munsell): 7.5YR 7/3 pink
Slip Color (Munsell): 10R 4/4 weak red or 2.5YR 4/4 reddish brown
Hardness (Mohs): < 2.0
Inclusion Type(s): Gray grits (fine sand; very low frequency); grog (?) (coarse sand; medium frequency); red grits (fine sand to very fine sand; low frequency)
Sorting: Poor
Carbon Traces: None.
Fabric Condition: Fair

16. BNS3
Type: Beit Nattif with bow-shaped nozzle
Date: End of third to late fifth/early sixth centuries C.E. (da Costa 2003: 1.68)
Period: Late Roman–Byzantine
Stratigraphic Context: Fill (L84.2077) in cistern (L84.2078)
Greatest Dimension: 3.5 cm
Part: Shoulder fragment
Diagnostic Feature: Herringbone and a nodule decorate shoulder
Fabric Color (Munsell): 10YR 7/3 very pale brown
Slip Color (Munsell): 7.5YR 4/3 brown
Hardness (Mohs): 2.5
Inclusion Type(s): None visible
Sorting: None
Carbon Traces: None
Fabric Condition: Fair; friable

17. BS3
Type: Bilanceolate
Date: Third to early fifth centuries C.E. (da Costa 2003: 1.57, 66–67)
Period: Late Roman to Early Byzantine
Stratigraphic Context: L85.2041
Greatest Dimension: 3.2 cm
Part: Shoulder and handle fragment
Diagnostic Feature: Acanthus leaf pattern decorates shoulder
Fabric Color (Munsell): 5YR 6/4 light reddish brown
Slip Color (Munsell): None
Hardness (Mohs): 2.0
Inclusion Type(s): White grits (FS; low frequency)
Sorting: Poor
Carbon Traces: Handle
Fabric Condition: Good; oxides on surface

18. BS4
Type: Bilanceolate
Date: Third to early fifth centuries C.E. (da Costa 2003: 1.57, 66–67)

Period: Late Roman to Early Byzantine
Stratigraphic Context: L84.2232
Greatest Dimension: 3.5 cm
Part: Shoulder and handle fragment
Diagnostic Feature: Acanthus leaf pattern decorates shoulder
Fabric Color (Munsell): 10YR 7/2 light gray
Slip Color (Munsell): 7.5YR 5/3 brown
Hardness (Mohs): 3.0
Inclusion Type(s): Gray grits (very fine sand; low frequency)
Sorting: Fair
Carbon Traces: Handle
Fabric Condition: Fair; slight traces of oxides on inner surface

19. BS9 — Type: Bilanceolate
Date: Third to early fifth centuries C.E. (da Costa 2003: 1.57, 66–67)
Period: Late Roman to Early Byzantine
Stratigraphic Context: L86.1021
Greatest Dimension: 6.5 cm
Part: Shoulder and nozzle-bridge fragment
Diagnostic Feature: Acanthus leaf pattern decorates shoulder
Fabric Color (Munsell):10YR 5/2 grayish brown
Slip Color (Munsell): 5YR 3/2 dark reddish brown
Hardness (Mohs): 2.0
Inclusion Type(s): None
Sorting: N/A
Carbon Traces: Slight areas of wick-hole edge
Fabric Condition: Good; slight encrustation of inner surface

Decapolis

Scythopolis (Beth Shean)

20. DSCY6 — Type: Palestinian discus
Period: Early Roman to Middle Roman
Stratigraphic Context: Shop no. 11, western stoa, foundation trench[6]
Greatest Dimension: 2.9 cm
Part: Base fragment
Diagnostic Feature: Unfired fabric; lamp maker's mark (*nu eta*)
Fabric Color (Munsell):10YR 5/2–6/2 grayish brown, light brownish gray
Slip Color (Munsell): None
Hardness (Mohs): < 2.0
Inclusion Type(s): White grits (very fine sand; very low frequency)
Sorting: Poor
Carbon Traces: None
Fabric Condition: Poor; friable; unfired

6. The foundation trench is of a wall of Shop No. 11 (E6) on Paladios Street (Area EK; Unit PSSWII). The context of the Scythopolis lamp samples was discussed with Deborah Sandhouse, Israel Department of Antiquities, Rockefeller Museum, 23 June 1997.

21. DSCY7 Type: Palestinian discus, Scythopolis version
Date: Late first to third centuries C.E.
Period: Early Roman to Middle Roman
Stratigraphic Context: Surface of Area GK, location of theater and Roman temples
Greatest Dimension: 4.2 cm
Part: Base fragment
Diagnostic Feature: Unique lamp maker's mark: possible *šin* (?); local variant of Broneer XXV
Fabric Color (Munsell): 5YR 5/6 yellowish red
Slip Color (Munsell): None
Hardness (Mohs): 2.0
Inclusion Type(s): None visible
Sorting: None visible
Carbon Traces: None
Fabric Condition: Good

22. DSCY8 Type: Palestinian discus
Date: Late first to third centuries C.E.
Period: Early Roman to Middle Roman
Stratigraphic Context: West bathhouse, dressing room, fill on bedrock[7]
Greatest Dimension: 4.4 cm
Part: Base fragment
Diagnostic Feature: Lamp maker's mark (*nun*)
Fabric Color (Munsell): 5YR 5/4 reddish brown
Slip Color (Munsell): Present but unable to determine due to oxides covering surface
Hardness (Mohs): 2.0
Inclusion Type(s): Gray grits (very fine sand; medium frequency); white grits (very fine sand; low frequency)
Sorting: Good
Carbon Traces: None
Fabric Condition: Fair (+); traces of oxides on surface; slightly friable

23. BSCY1 Type: Bilanceolate
Date: Third to early fifth centuries C.E. (da Costa 2003: 1.57, 66–67)
Period: Late Roman to Early Byzantine
Stratigraphic Context: Lower market, outside row of shops, fill[8]
Greatest Dimension: 5.6 cm
Part: Shoulder fragment
Diagnostic Feature: Acanthus leaf pattern decorates shoulder
Fabric Color (Munsell): 2.5YR 4/1 dark gray (core); 2.5YR 6/4 light yellowish brown (well-defined core boundaries)
Slip Color (Munsell): 7.5YR 5/3 brown
Hardness (Mohs): 2.5
Inclusion Type(s): Red grits (fine sand; low to medium frequency)

7. Area AK, Unit WTBH8B.

8. This sample and BSCY5 below were recovered from the fill above the remains of the Roman-period buildings located outside the lower market shops (Area EK). The pottery from this deposit dates to the end of the fourth to the beginning of the fifth centuries C.E.

Sorting: Fair to good
Carbon Traces: None
Fabric Condition: Fair (-)

24. BSCY4 Type: Bilanceolate
Date: Third to early fifth centuries C.E. (da Costa 2003: 1.57, 66–67)
Period: Late Roman to Early Byzantine
Stratigraphic Context: Lower Market, Shop No. 9 (cleaning of floor)[9]
Greatest Dimension: 3.6 cm
Part: Shoulder and handle fragments
Diagnostic Feature: Acanthus leaf pattern decorates shoulder
Fabric Color (Munsell): 7.5YR 5/6 strong brown
Slip Color (Munsell): None
Hardness (Mohs): 2.0
Inclusion Type(s): None
Sorting: N/A
Carbon Traces: None
Fabric Condition: Good

25. BSCY5 Type: Bilanceolate
Date: Third to early fifth centuries C.E. (da Costa 2003: 1.57, 66–67)
Period: Late Roman to Early Byzantine
Stratigraphic Context: Lower market, outside row of shops, fill[10]
Greatest Dimension: 3.8 cm
Part: Shoulder fragment
Diagnostic Feature: Circles decorate shoulder
Fabric Color (Munsell): 2.5YR 3/2 very dark grayish brown (core); 5YR 6/6 reddish yellow (outer core boundaries)
Slip Color (Munsell): None
Hardness (Mohs): < 2.0
Inclusion Type(s): White grits (fine sand; low frequency)
Sorting: Fair
Carbon Traces: Filling-hole rim; handle
Fabric Condition: Poor; oxides on surface; possible waster

Pella-Tabaqat Fahl[11]

26. DP3 Type: Palestinian discus
Date: Late first to third centuries C.E.
Period: Early Roman to Middle Roman
Stratigraphic Context: Room of house, east cut, Area III, Plot P. Pit
Greatest Dimension: 4.8 cm
Part: Shoulder and discus fragment
Diagnostic Feature: Bi-associative motif (crane and crab)

9. The pottery from this deposit in Area EK dates to the end of the fourth to the beginning of the fifth centuries C.E.

10. See note on BSCY1, above.

11. Several of the samples from the site (DP3, DP11, BP26–BP29) come from a 6-m-deep pit located in the room of a house (Area III Plot P). Most of the finds recovered from the pit, including the lamps, are fragmentary and belong to the second–third centuries C.E. See Smith and McNicol 1992: 119.

Fabric Color (Munsell): 2.5Y 8/3 pale yellow
Slip Color (Munsell): 7.5YR 4/3 brown (outer surface)
Hardness (Mohs): 2.5–3.0
Inclusion Type(s): Red grits (very fine sand; high frequency)
Sorting: Good
Carbon Traces: None
Fabric Condition: Good

27. DP11 Type: Palestinian discus
Period: Early Roman to Middle Roman
Stratigraphic Context: Room of house, east cut, Area III, Plot P. Pit
Greatest Dimension: 5.8 cm
Part: Shoulder and discus fragment
Diagnostic Feature: Lamp maker's mark (*nu alpha*)
Fabric Color (Munsell): 5YR 7/4 pink
Slip Color (Munsell): 10R 4/6 red (inner and outer surfaces)
Hardness (Mohs): 2.0
Inclusion Type(s): Gray grits (very fine sand; very high frequency); red grits (fine sand to very fine sand; low frequency)
Sorting: Good
Carbon Traces: None
Fabric Condition: Good (-); traces of oxides on surface

28. BP1 Type: Bilanceolate
Date: Third to early fifth centuries C.E. (da Costa 2003: 1.57, 66–67)
Period: Late Roman to Early Byzantine
Stratigraphic Context: (Locus 372)
Greatest Dimension: 5.8 cm
Part: Shoulder and handle fragment
Diagnostic Feature: Acanthus leaf pattern decorates shoulder
Fabric Color (Munsell): Inner core 5YR 6/6 reddish yellow, outer surface 10YR 7/2 light gray
Slip Color (Munsell): 5YR 5/4 reddish brown
Hardness (Mohs): 2.0
Inclusion Type(s): None
Sorting: N/A
Carbon Traces: None
Fabric Condition: Fair; slightly friable

29. BP2 Type: Bilanceolate
Date: Third to early fifth centuries C.E. (da Costa 2003: 1.57, 66–67)
Period: Late Roman to Early Byzantine
Stratigraphic Context: (Locus 313)
Greatest Dimension: 3.4 cm
Part: Handle and shoulder fragment
Diagnostic Feature: Acanthus leaf pattern decorates shoulder
Fabric Color (Munsell): 7.5YR 6/3 light brown
Slip Color (Munsell): Altered by firing
Hardness (Mohs): 2.0
Inclusion Type(s): None
Sorting: N/A

Carbon Traces: Handle
Fabric Condition: Fair (-); friable

30. BP3 Type: Bilanceolate
Date: Third to early fifth centuries C.E. (da Costa 2003: 1.57, 66–67)
Period: Late Roman to Early Byzantine
Stratigraphic Context: (Locus 373)
Greatest Dimension: 3.9 cm
Part: Shoulder and handle fragment
Diagnostic Feature: Acanthus leaf pattern decorates shoulder
Fabric Color (Munsell): 10YR 7/2 light gray (core); 7.5YR 7/6 reddish yellow (outer boundaries of well-defined core)
Slip Color (Munsell): 10R 4/6 red
Hardness (Mohs): 3.0
Inclusion Type(s): Red grits (coarse sand to fine sand; high frequency); white grits (coarse sand to fine sand; low frequency)
Sorting: Good
Carbon Traces: None
Fabric Condition: Good

31. BP7 Type: Bilanceolate
Date: Third to early fifth centuries C.E. (da Costa 2003: 1.57, 66–67)
Period: Late Roman to Early Byzantine
Stratigraphic Context: (Locus 368)
Greatest Dimension: 4.0 cm
Part: Shoulder and handle fragment
Diagnostic Feature: Circles and acanthus leaf pattern decorate shoulder
Fabric Color (Munsell): 5YR 6/4 light reddish brown
Slip Color (Munsell): 2.5YR 4/4 reddish brown
Hardness (Mohs): 2.0
Inclusion Type(s): Red grits (very fine sand; medium frequency)
Sorting: Good
Carbon Traces: Wick hole
Fabric Condition: Poor (+); very friable

Gadara (Umm Qeis)[12]

32. BNG1 Type: Beit Nattif with bow-shaped nozzle
Date: End of 3rd to late 5th/early 6th centuries C.E. (da Costa 2003: 1.68)
Period: Late Roman to Byzantine
Stratigraphic Context: Mausoleum (Locus XI.E.433)
Greatest Dimension: 4.4 cm
Part: Shoulder and nozzle-bridge fragment
Diagnostic Feature: Circles and acanthus leaf pattern

12. All of these samples (except for BNG10 and BG1) were recovered from a rubbish dump that filled in the entrance hall of the underground mausoleum. The fill yielded over 1,600 lamp fragments, most of which belong to the bilanceolate class. See da Costa 1990: 332–36, fig. 2. On BNG10 and BG1, see Weber and Hoffman 1990: 323–25, fig. 2. The locus numbers for these samples were supplied by Kate da Costa.

Fabric Color (Munsell): 2.5YR 6/6 light red
Slip Color (Munsell): None
Hardness (Mohs): 2.0
Inclusion Type(s): Grog (?) (coarse sand; very low frequency); red grits (coarse to very fine sand; medium frequency)
Sorting: Poor
Carbon Traces: None
Fabric Condition: Good

33. BNG8 Type: Beit Nattif with bow-shaped nozzle
Date: End of third to late fifth/early sixth centuries C.E. (da Costa 2003: 1.68)
Period: Late Roman–Byzantine
Stratigraphic Context: Mausoleum (Locus XI.289)
Greatest Dimension: 4.4 cm
Part: Shoulder fragment
Diagnostic Feature: Round nodules and circles decorate shoulder
Fabric Color (Munsell): 5YR 5/4 reddish brown
Slip Color (Munsell): 5YR 6/4 light reddish brown
Hardness (Mohs): 3.0
Inclusion Type(s): Gray grits (coarse sand to fine sand; medium to high frequency); grits (fine sand to very fine sand; very low frequency) (grog?)
Sorting: Good
Carbon Traces: None
Fabric Condition: Good

34. BNG9 Type: Beit Nattif with bow-shaped nozzle
Period: Late Roman–Byzantine
Stratigraphic Context: Mausoleum (Locus XI.E.112)
Greatest Dimension: 4.2 cm
Part: Shoulder fragment
Diagnostic Feature: Round nodules decorate shoulder
Fabric Color (Munsell): 2.5YR 6/6 light red
Slip Color (Munsell): None
Hardness (Mohs): 2.0
Inclusion Type(s): Grog (?) (coarse sand; very low frequency); red grits (coarse sand to very fine sand; medium frequency)
Sorting: Poor
Carbon Traces: None
Fabric Condition: Good (-)

35. BNG10 Type: Beit Nattif with bow-shaped nozzle
Date: End of third to late fifth/early sixth centuries C.E. (da Costa 2003: 1.68)
Period: Late Roman to Byzantine
Stratigraphic Context: Tiberias Gate (Locus XII.77)
Greatest Dimension: 2.7 cm
Part: Shoulder and handle fragment
Diagnostic Feature: Circles decorate shoulder
Fabric Color (Munsell): 5YR 6/4 light reddish brown
Slip Color (Munsell): None

Hardness (Mohs): 2.0
Inclusion Type(s): Red grits (fine sand to very fine sand; very low frequency); white grits (very fine sand; very low frequency)
Sorting: Poor
Carbon Traces: None
Fabric Condition: Good (-)

36. BG1
Type: Bilanceolate
Date: Third to early fifth centuries C.E. (da Costa 2003: 1.57, 66–67)
Period: Late Roman to Early Byzantine
Stratigraphic Context: Tiberias Gate (Locus XII.140)
Greatest Dimension: 5.7 cm
Part: Shoulder and handle fragment
Diagnostic Feature: Circles decorate shoulder
Fabric Color (Munsell): Y 4/1 dark gray
Slip Color (Munsell): 5YR 4/2 dark reddish gray
Hardness (Mohs): 2.0
Inclusion Type(s): Gray grits (very fine sand; high frequency)
Sorting: Good
Carbon Traces: None
Fabric Condition: Good

37. BG5
Type: Bilanceolate
Date: Third to early fifth centuries C.E. (da Costa 2003: 1.57, 66–67)
Period: Late Roman to Early Byzantine
Stratigraphic Context: Mausoleum (Locus XI.E.16)
Greatest Dimension: 4.4 cm
Part: Shoulder and handle fragment
Diagnostic Feature: Acanthus leaf pattern decorates shoulder
Fabric Color (Munsell): 5YR 6/4 light reddish brown
Slip Color (Munsell): 2.5YR 5/6 red
Hardness (Mohs): 2.0
Inclusion Type(s): Red grits (coarse sand; very low frequency; very fine sand; medium to high frequency)
Sorting: Good
Carbon Traces: None
Fabric Condition: Good

38. BG8
Type: Bilanceolate
Date: Third to early fifth centuries C.E. (da Costa 2003: 1.57, 66–67)
Period: Late Roman to Early Byzantine
Stratigraphic Context: Mausoleum (Locus XI.457)
Greatest Dimension: 3.4 cm
Part: Shoulder fragment
Diagnostic Feature: Circles decorate shoulder
Fabric Color (Munsell):10YR 7/2 light gray (core); 7.5YR 7/4 pink (outer boundaries of poorly defined core) Slip Color (Munsell): 2.5YR 5/6 red
Hardness (Mohs): 2.0
Inclusion Type(s): None
Sorting: N/A
Carbon Traces: None

Fabric Condition: Good (-)

39. BG10 Type: Bilanceolate
Date: Third to early fifth centuries C.E. (da Costa 2003: 1.57, 66–67)
Period: Late Roman to Early Byzantine
Stratigraphic Context: Mausoleum (Locus XI.339)
Greatest Dimension: 4.1 cm
Part: Shoulder fragment
Diagnostic Feature: Acanthus leaf pattern decorates shoulder
Fabric Color (Munsell): 7.5YR 6/4 light brown
Slip Color (Munsell): 2.5YR 4/4 reddish brown
Hardness (Mohs): < 2.0
Inclusion Type(s): None
Sorting: N/A
Carbon Traces: None
Fabric Condition: Fair

Gerasa/Jerash[13]

40. BNJ2 Type: Gilead
Date: fourth to fifth centuries C.E. (da Costa 2003: 1.97)
Period: Late Roman to Early Byzantine
Stratigraphic Context: Locus XII.19
Greatest Dimension: 3.5 cm
Part: Shoulder fragment
Diagnostic Feature: Acanthus leaf pattern decorates shoulder
Fabric Color (Munsell): 5YR 7/4 pink
Slip Color (Munsell): 10R 6/6 light red
Hardness (Mohs): 2.0
Inclusions: Gray grits (very fine sand; very low frequency); red grits (very fine sand; very low frequency)
Sorting: Poor
Carbon Traces: None
Fabric Condition: Fair (+)

41. BNJ4 Type: Gilead
Date: fourth–fifth centuries C.E. (da Costa 2003: 1.97)
Period: Late Roman to Early Byzantine
Stratigraphic Context: Locus XII.19
Greatest Dimension: 3.1 cm
Part: Shoulder fragment
Diagnostic Feature: Acanthus leaf pattern decorates shoulder
Fabric Color (Munsell): 2.5YR 6/4 light reddish brown
Slip Color (Munsell):10R 5/6 red
Hardness (Mohs): 2.0
Inclusions: None
Sorting: None

13. Four of these samples (JB1, JB2, JB3, JB5) were found among hundreds of unfired and misfired sherds belonging to ceramic types of the late sixth to early seventh centuries C.E. See Kehrberg and Ostrasz 1997: 171–72.

Carbon Traces: None
Fabric Condition: Good (+)

42. JB1
Type: Jerash Bowl
Period: Late Byzantine to Early Islamic (Lapp 2001a)
Stratigraphic Context: Chamber E8, hippodrome
Greatest Dimension: 1.9 cm
Fabric Color (Munsell): 5YR 6/6 reddish yellow
Slip Color (Munsell): None
Ornamentation: None
Vessel Part: Rim
Relative Hardness: 3.0
Inclusion Type(s): Red grits (very fine sand; high frequency); white grits (very fine sand; high frequency)
Sorting: Good
Fabric Condition: Good; fired

43. JB2
Type: Jerash Bowl
Period: Late Byzantine to Early Islamic
Stratigraphic Context: Chamber E8, hippodrome
Greatest Dimension: 3.2 cm
Fabric Color (Munsell): 2.5YR 6/6 light red
Slip Color (Munsell): None
Ornamentation: White line (paint)
Vessel Part: Body
Hardness (Mohs): 3.0
Inclusion Type(s): White grits (coarse sand; very low frequency)
Sorting: Poor
Fabric Condition: Good; fired; gritty

44. JB3
Type: Jerash Bowl
Period: Late Byzantine to Early Islamic
Stratigraphic Context: Chamber E8, Hippodrome
Greatest Dimension: 2.0 cm
Fabric Color (Munsell): 2.5YR 6/6 light red
Slip Color (Munsell): None
Ornamentation: None
Vessel Part: Base
Hardness (Mohs): 3.0
Inclusion Type(s): Red grits (very fine sand; high frequency); white grits (fine sand; very low frequency)
Sorting: Good
Fabric Condition: Good

45. JB5
Type: Jerash Bowl
Period: Late Byzantine to Early Islamic (Lapp 2001a)
Stratigraphic Context: Chamber E8, hippodrome
Greatest Dimension: 2.9 cm
Fabric Color (Munsell): 7.5YR 5/4 brown
Slip Color (Munsell): None
Ornamentation: None
Vessel Part: Body

Hardness (Mohs): < 2.0
Inclusion Type(s): None
Sorting: N/A
Fabric Condition: Fair; unfired

Abila

46. DAB4
Type: Palestinian discus
Date: Late first to third centuries C.E.
Period: Early Roman to Middle Roman
Stratigraphic Context: Unavailable
Greatest Dimension: 7.2 cm
Part: Base fragment
Diagnostic Feature: Distinctive red fabric; lamp maker's mark: double-axe
Fabric Color (Munsell): 2.5YR 6/6 light red
Slip Color (Munsell): 2.5YR 5/6 red
Hardness (Mohs): 2.0
Inclusion Type(s): Red grits (coarse sand to very fine sand; very high frequency); white grits (fine sand to very fine sand; very low frequency)
Sorting: Fair
Carbon Traces: None
Fabric Condition: Good; traces of oxides on surface

47. BAB1
Type: Bilanceolate
Date: Third to early fifth centuries C.E. (da Costa 2003: 1.57, 66–67)
Period: Late Roman to Byzantine
Stratigraphic Context: Locus A92.H32.007.1114
Greatest Dimension: 7.5 cm
Part: Shoulder and nozzle fragment
Diagnostic Feature: Circles and acanthus leaf pattern decorate shoulder
Fabric Color (Munsell): 10YR 7/2 light gray (core); 7.5YR 7/4 pink (outer boundaries of poorly defined core)
Slip Color (Munsell): None
Hardness (Mohs): 2.0
Inclusion Type(s): Red grits (coarse sand; very low frequency)
Sorting: Poor
Carbon Traces: None
Fabric Condition: Poor; very friable; traces of oxides on surface

Bibliography

Abu Dalu, R.

1993 Final Report of the Excavations in the Area of the Pottery Kilns: Jarash 1991. *Annual of the Department of Antiquities in Jordan* 37: 23–34.

Adan-Bayewitz, D.

1993 *Common Pottery in Roman Galilee: A Study of Local Trade.* Bar-Ilan Studies in Near Eastern Languages and Culture. Ramat Gan: Bar-Ilan University Press.

1995 A Lamp Mould from Sepphoris and the Location of Workshops for Lamp and Common Pottery Manufacture in Northern Palestine. Pp. 177–82 in *The Roman and Byzantine Near East: Some Recent Archaeological Research*, ed. J. H. Humphrey. Journal of Roman Archaeology Supplement 14. Ann Arbor, MI: Journal of Roman Archaeology.

Adan-Bayewitz, D.; Asaro, F.; Wieder, M.; and Giauque, R. D.

2008 Preferential Distribution of Lamps from the Jerusalem Area in the Late Second Temple Period (Late First Century B.C.E.–70 C.E.). *Bulletin of the American Schools of Oriental Research* 350: 37–85.

Adan-Bayewitz, D., and Perlman, I.

1990 The Local Trade of Sepphoris in the Roman Period. *Israel Exploration Journal* 40: 153–72.

Adan-Bayewitz, D., and Wieder, M.

1992 Ceramics from Roman Galilee: A Comparison of Several Techniques for Fabric Characterization. *Journal of Field Archaeology* 19: 189–205.

Adler, N.

2004 *Oil Lamps of the Holy Land: The Adler Collection.* Jerusalem: Old City Press.

Agady, S.; Arazi, M.; Arubas, B.; Hadad, S.; Khamis, E.; and Tsafrir, Y.

2002 Byzantine Shops in the Street of the Monuments at Beth Shean (Scythopolis). Pp.423–506 in *What Athens Has to Do with Jerusalem: Essays on Classical, Jewish, and Early Christian Art and Archaeology in Honor of Gideon Foerster,* ed. L. V. Rutgers. Leuven: Peeters.

Ali, W.

1999 *The Arab Contribution to Islamic Art: From the Seventh to the Fifteenth Centuries.* Cairo: The American University in Cairo Press.

Al-Khouly, M.

2001 Nouvelles lampes inscrites de la période omeyyade. Pp. 193–96 in *La Céramique Byzantine et Proto-Islamique en Syrie-Jordanie (IVe–VIIe Siècles Apr. J.-C.),* ed. E. Villeneuve and P. P. Watson. Bibliothèque Archéologique et Historique 159. Beirut: Institut Français d'Archéologie du Proche-Orient.

Alliata, E.

1988 La ceramica dello scavo della capella delprete Giovanni a Kh. el-Mukhayyat. *Liber Annuus* 38: 317–60.

Ambar-Armon, E.

2014 Oil Lamps. Pp. 33–37 in *The Excavations of Maresha Subterranean Complex 57: The 'Heliodorus' Cave,* ed. I. Stern. British Archaeological Reports International Series 2652. Oxford: British Archaeological Reports.

Amiran, R., and Eitan, A.

1970 Excavations in the Courtyard of the Citadel, Jerusalem 1968–1969 (Preliminary Report). *Israel Exploration Journal* 20: 9–17.

ʿAmr, K.

1987 *The Pottery from Petra: A Neutron Activation Analysis of Pottery and Clay from the Zurrabah Kiln Complex.* British Archaeological Reports International Series 34. Oxford: British Archaeological Reports.

Ariel, D. T.; Sharon, I.; Gunneweg, J.; and Perlman, I.

1985 A Group of Stamped Hellenistic Storage-Jar Handles from Dor. *Israel Exploration Journal* 35: 135–52.

Arubas, B., and Goldfus, H.

1995 The Kilnworks of the Tenth Legion Fretensis. Pp. 95–107 in *The Roman and Byzantine Near East: Some Recent Archaeological Research,* ed. J. H. Humphrey. Journal of Roman Archaeology Supplement 14. Ann Arbor, MI: Journal of Roman Archaeology.

Asaro, F.

1981 Neutron Activation Analysis. Pp. 139–44 in E. M. Meyers, J. F. Strange, and C. L. Meyers. *Excavations at Ancient Meiron, Upper Galilee, Israel, 1971–72, 1974–77, 1977.* Meiron Excavation Project 3. Cambridge, MA: American Schools of Oriental Research.

Aviam, M.

2013 People, Land, Economy, and Belief in First-Century Galilee and Its Origins: A Comprehensive Archaeological Synthesis. Pp. 5–48 in *The Galilean Economy in the Time of Jesus,* ed. D. A. Fiensy and R. K. Hawkins. Early Christianity and Its Literature 11. Atlanta: Society of Biblical Literature.

Avigad, N.

1956 The Sixth Season of Excavations at Beth Sheʿarim 1954. *Eretz-Israel* 4: 85–103. [Hebrew]

1970 Excavations in the Western Summit of the Old City of Jerusalem, 1970. Second Preliminary Report. *Israel Exploration Journal* 20: 129–40.

1976 *Beth Sheʿarim: Report on the Excavations during 1953–1958,* vol. 3: *Catacombs 12–23.* New Brunswick, NJ: Rutgers University Press.

Bagatti, P. B.

1963–64 Oggetti inediti di Cafarnao. *Liber Annuus* 14: 261–72.

1969 *Excavations in Nazareth,* vol. 1: *From the Beginning till the XII Century.* Trans. E. Hoade. Publications of the Studium Biblicum Franciscanum 17. Jerusalem: Franciscan Printing Press.

Bagatti, P. B., and Milik, J. T.

1958 *Gliscavi del "Dominus Flevit,"* vol. 1: *La necropoli del periodo romano.* Publications of the Studium Biblicum Franciscanum 13. Jerusalem: Franciscan Printing Press.

Bahat, D.

1974 A Roof Tile of the Legio VI Ferrata and Pottery Vessels from Ḥorvat Ḥazon. *Israel Exploration Journal* 24: 167–68.

Bailey, D. M.
1972 *Greek and Roman Pottery Lamps in the British Museum*. Oxford: Oxford University Press.
1975 *A Catalogue of Lamps in the British Museum*, vol. 1: *Greek, Hellenistic and Early Roman Pottery Lamps*. London: British Museum.
1985 *Excavations at Sidi Khrebish Benghazi (Berenice)*, vol. 3/2: *The Lamps*. Tripoli: Secretariat of Education, Department of Antiquities, Socialist People's Libyan Arab Jamahiriya.
1988 *A Catalogue of Lamps in the British Museum*, vol. 3: *Roman Provincial Lamps*. London: Bath.
1994 Imported Lamps and Local Copies. Pp. 79–106 in *Masada 4: The Yigael Yadin Excavations 1963–1965 Final Reports*, ed. J. Aviram, G. Foerster, and E. Netzer. Jerusalem: Israel Exploration Society.

Balouka, M.
2013 Roman Pottery. Pp. 13–141 in *The Pottery from Ancient Sepphoris*, ed. E. M. Meyers and C. L. Meyers. [Duke] Sepphoris Excavation Report I. Winona Lake, IN: Eisenbrauns.

Barag, D.
1973 A Jewish Burial Cave on Mount Scopus. *Eretz-Israel* 11: 101–3. [Hebrew]
1978 *Ḥanita, Tomb XV: A Tomb of the Third and Early Fourth Century* CE. *ʿAtiqot* 13 (English Series). Jerusalem: Israel Antiquities Authority.

Barag, D., and Hershkovitz, M.
1994 Lamps. Pp. 7–78, 99–106, and 125–47 in *Masada 4: The Yigael Yadin Excavations 1963–1965 Final Reports*, ed. J. Aviram, G. Foerster, and E. Netzer. Jerusalem: Israel Exploration Society.

Barag, D.; Porat, Y.; and Netzer, E.
1981 The Synagogue at ʿEn-Gedi. Pp. 116–19 in *Ancient Synagogues Revealed*, ed. L. I. Levine. Jerusalem: Israel Exploration Society.

Baramki, D. C.
1936 Two Roman Cisterns at Beit Nattīf. *Quarterly of the Department of Antiquities in Palestine* 5: 3–10, pls. 2–12.

Barrett, D.
2008 *The Ceramic Oil Lamp as an Indicator of Cultural Change within Nabataean Society in Petra and Its Environs Circa* CE *106*. Gorgias Dissertation Series 32. Near Eastern Studies 8. Piscataway, NJ: Gorgias Press.

Ben-Arieh, R., and Coen-Uzzielli, T.
1996 The Pottery. Pp. 73–94 in *The Akeldama Tombs: Three Burials Caves in the Kidron Valley, Jerusalem*, ed. G. Avni and Z. Greenhut. Israel Antiquities Authority Report 1. Jerusalem: Israel Antiquities Authority.

Bender, F.
1968 *Geologie von Jordanien*. Beiträge zur Regionalen Geologie der Erde 7. Berlin: Bornträger.

Berlin, A. M.
1999 The Archaeology of Ritual: The Sanctuary of Pan at Banias/Caesarea Philippi. *Bulletin of the American Schools of Oriental Research* 315: 27–46.

Berthier, S.
1985 Sondage dans le secteur des thermes sud à Buṣrā (Syrie). *Berytus* 33: 5–45.

Bezzola, S.

2004 *Lucerne fittili dagli scavi di Palaepaphos (Cipro).* Ausgrabungen in Alt-Paphos auf Cypern 5. Mainz: von Zabern.

Bisheh, G.

1972 A Cave Burial Tomb from Jabal Jofeh el-Sherqi in Amman. *Annual of the Department of Antiquities in Jordan* 17: 8–83.

1982 The Second Season of Excavations at Hallabat, 1980. *Annual of the Department of Antiquities in Jordan* 26: 133–43.

Bliss, F. J., and Macalister, R. A. S.

1902 *Excavations in Palestine during the Years 1898–1900.* London: Palestine Exploration Fund.

Bowersock, G. W.

2006 *Mosaics as History: The Near East from Late Antiquity to Islam.* Cambridge: Harvard University Press.

Broneer, O.

1930 *Corinth 4.* Cambridge: Harvard University Press.

Brown, P.

1999 Images as a Substitute for Writing. Pp. 15–34 in *East and West: Modes of Communication: Proceedings of the First Plenary Conference at Merida,* ed. E. Chrysos and I. Wood. Leiden: Brill.

Büyükkolanci, P.

2002 Objekte: Kleinfunde mit Gladiatorendarstellungen im Ephesos Museum. Pp. 93–96 in *Gladiatoren in Ephesos: Tod am Nachmittag,* ed. S. Erdemgil and F. Krinzinger. Istanbul: Dogan Gümüş.

Bull, R. L.; Krentz, E.; and Storvick, O.

1986 The Joint Expedition to Caesarea Maritima: Ninth Season, 1980. *Bulletin of the American Schools of Oriental Research Supplement* 24: 31–55.

Campbell, G.

2007 *The Grove Encyclopedia of Classical Art and Architecture.* Oxford: Oxford University Press.

Cantarella, E.

1992 *Bisexuality in the Ancient World.* New Haven, CT: Yale University Press.

Chancey, M.

2002 *The Myth of a Gentile Galilee.* Cambridge: Cambridge University Press.

Chrzanovski, L.

2011 Lampes tardo-antiques et byzantines. Pp. 264–330 in *Antiquités paléochrétiennes et byzantines, IIIe–XIVe Siècles: Collections du Musée d'art et d'histoire–Genève,* ed. M. Martiniani-Reber. Milan: 5 Continents Editions.

Clamer, C., and Magness, J.

1997 La céramique romaine tardive de la période Byzantine ancienne. Pp. 81–90, 144–55, and 176–77 in C. Clamer, *Fouilles archéologiques de ʿAïn Ez-Zâra/Callirrhoé: Villégiature hérodienne.* Bibliothèque archéologique et historique 147. Beirut: Institut Français d'Archéologie du Proche-Orient.

Clarke, J. R.

2003 *Roman Sex.* New York: Abrams.

Clermont-Ganneau, C.

1900 Nouveau lychnarion à inscription coufique. *Recueil d'archéologie orientale* 3: 283–85.

Coen-Uzzielli, T.
1997 The Oil Lamps. Pp. 319–46 in Y. Hirschfeld, *The Roman Baths of Hammat Gader*. Jerusalem: Israel Antiquities Authority.
Cohen-Weinberger, A., and Goren, Y.
2011 The Clay Sources of the Theater Pottery Workshop: A Petrographic Study. Pp. 215–28 in *Bet She'an II, Baysān: The Theater Pottery Workshop*, ed. R. Bar-Nathan and W. Atrash. Israel Antiquities Authority Report 48. Jerusalem: Israel Antiquities Authority.
Cormack, S. H.
1996 Roman-Period Sculpture. Pp. 171–72 in *Sepphoris in Galilee: Crosscurrents of Culture*, ed. R. M. Nagy, C. L. Meyers, E. M. Meyers, and Z. Weiss. Raleigh: North Carolina Museum of Art.
forthcoming Two Bronze Figurines: Prometheus, and Pan (?) or a Satyr. In *Artifacts and Other Finds from the Western Summit of Sepphoris*, ed. E. M. Meyers, C. L. Meyers, and B. D. Gordon. Duke Sepphoris Excavation Report IV. Winona Lake, IN: Eisenbrauns.
Costa, K. da
1990 Some Lamps from Excavations at Umm Qeis. *Annual of the Department of Antiquities in Jordan* 34: 332-36.
2003 *Byzantine and Early Umayyad Ceramic Lamps from Palaestina/Arabia (ca. 300–700 AD)*. 2 vols. Ph.D. dissertation. University of Sydney.
2010 Economic Cycles in the Byzantine Levant: The Evidence from Lamps at Pella in Jordan. *Levant* 42: 70–87.
Crowfoot, J. W., and Fitzgerald, G. M.
1929 *Excavations in the Tyropoeon Valley, Jerusalem, 1927*. Palestine Exploration Fund Annual 5. London: Palestine Exploration Fund.
Crowfoot, J. W.; Crowfoot, G. M.; and Kenyon, K. M.
1957 *The Objects from Samaria*. Samaria–Sebaste 3: Reports of the Work of the Joint Expedition in 1931–1933 and of the British Expedition in 1935. London: Palestine Exploration Fund.
Day, F. E.
1942 Early Islamic and Christian Lamps. *Berytus* 7: 65–79.
Dilke, O. A. W.
1985 *Greek and Roman Maps*. Baltimore: Johns Hopkins University Press.
Dobbins, J. J.
1977 *Terracotta Lamps of the Roman Province of Syria*. Ph.D. dissertation. University of Michigan.
Dothan, M.
1983 *Hammath Tiberias: Early Synagogues and the Hellenistic and Roman Remains*. Jerusalem: Israel Exploration Society.
Edwards, I. W.
1992 Appendix 10: A Progress Report on the Pella Pottery Studies, Section II: A Programme Designed to Assess the Suitability of the Clays in the Vicinity of Pella Tell for Pottery Making. Pp. 286–98 in A. W. McNicoll et al., *Pella in Jordan*, vol. 2: *The Second Interim Report of the Joint University of Sydney and College of Wooster Excavations at Pella 1982–1985*. Mediterranean Archaeology Supplement 2. Sydney: Meditarch.

Elgavish, J.
1962 *The Art of the Lamps in Palestine in the Roman and Byzantine Periods*. Ph.D. dissertation. Hebrew University of Jerusalem. [Hebrew]
1994 *Shiqmona: On the Seacoast of Mount Carmel.* Tel Aviv: Hakibbutz Hameuchad. [Hebrew].

Feathers, J. K.
1989 Effects of Temper on Strength of Ceramics: Response to Bronitsky and Hamer. *American Antiquity* 54: 580–81.

Feig, N.
1990 Burial Caves at Nazareth. *ʿAtiqot* 10: 67–79. [Hebrew]

Fernandez, F. D.
1983 *Cerámica común romana de la Galilea.* Madrid: Escuela Biblica.

Fischer, M.
1989 Hellenistic Pottery (Strata V–III). Pp. 177–87 in *Excavations at Tel Michal, Israel,* ed. Z. Herzog, G. Rapp, and O. Negbi. Publications of the Institute of Archaeology 8. Minneapolis: University of Minnesota Press.

Fitzgerald, G. M.
1931 *Beth-Shan Excavations 1921–1923: The Arab and Byzantine Levels.* Publications of the Palestine Section of the Museum of the University of Pennsylvania 3. Philadelphia: University Press for the University of Pennsylvania Museum.

Fossey, J. M., and Zoitopoulou, E. P.
1992 Vers une typologie de lampes égyptiennes de l'époque hellénistique in *Syneiphorà McGill,* ed. J. M. Fossey. *Monographies en archéologie et histoire classiques de l'Université McGill* 6: 117–24.

Freestone, I. C.
1995 Ceramic Petrography. *American Journal of Archaeology* 99: 111–17.

Freestone, I.; Politis, K. D.; and Stapleton, C. P.
2001 The Byzantine Glazed Pottery from Deir ʿAin ʿAbata, Jordan. Pp. 197–206 in *La Céramique Byzantine et Proto-Islamique en Syrie–Jordanie (IVe–VIIe Siècles Apr. J.-C.),* ed. E. Villeneuve and P. P. Watson. Bibliothèque Archéologique et Historique 159. Beirut: Institut Français d'Archéologie du Proche-Orient.

Fremersdorf, F.
1922 *Römische Bildlampen.* Bonn: Schröder.

Fuller, M. J.
1987 *Abila of the Decapolis: A Roman–Byzantine City in Transjordan.* Ph.D. dissertation. Washington University.

Gafni, I.
1996 Daily Life in Galilee and Sepphoris. Pp. 51–57 in *Sepphoris in Galilee: Crosscurrents of Culture,* ed. R. M. Nagy et al. Raleigh, NC: North Carolina Museum of Art.

Gärtner, J.
1999 *Oil Lamps from the Hellenistic through Mamluk Periods at Sepphoris.* M.A. thesis. Hebrew University of Jerusalem. [Hebrew].

Gawlikowski, M., and Musa, A.
1986 The Church of Bishop Marianos. Pp. 137–62 in *Jerash Archaeological Project 1981–1983* 1, ed. F. Zayadine. Amman: Department of Antiquities.

Gihon, M.
1972 The Plan of a Roman Camp Depicted upon a Lamp from Samaria. *Palestine Exploration Quarterly* 104: 38–57.

1993 The Ground Plan of a Military Camp on a Jewish Lamp. Pp. 181–94 in *Studies in the Archaeology and History of Ancient Israel in Honour of Moshe Dothan*, ed. A. Segal and D. Kaufman. Haifa: Haifa University Press.

Gitin, S.

1990 *Gezer III: A Ceramic Typology of the Late Iron II, Persian and Hellenistic Periods at Tell Gezer*. Jerusalem: Hebrew Union College.

Goethert-Polaschek, K.

1985 *Katalog der römischen Lampen des Rheinischen Landesmuseums Trier*. Mainz: von Zabern.

Goldberg, P.; Singer-Avitz, L.; and Horowitz, A.

1989 Petrographic Analysis of Persian Period Pottery. Pp. 264–66 in *Excavations at Tel Michal, Israel*, ed. Z. Herzog, G. Rapp Jr., and O. Negbi. Publications of the Institute of Archaeology 8. Minneapolis: University of Minnesota Press.

Goldman, H., and Jones, F .F.

1950 The Lamps. Pp. 84–134 in *Excavations at Gözlü Kule, Tarsus, I: The Hellenistic and Roman Periods*, ed. H. Goldman. Princeton: Princeton University Press.

Goren, Y.

1995 Appendix A: Petrographic Analyses of Horvat Teiman (Kuntillet ʿAjrud) Pottery. *Tel Aviv* 22: 206–7.

Goren, Y.; Finkelstein, I.; and Naʾaman, N.

2004 *Inscribed in Clay: Provenance Study of the Amarna Letters and Other Ancient Near Eastern Texts*. Emery and Claire Yass Publications in Archaeology Monograph 23. Tel Aviv: Insitute of Archaeology, Tel Aviv University.

Grabar, O.

1987 *The Formation of Islamic Art*. New Haven, CT: Yale University Press.

Greenberg, R., and Porat, N.

1996 A Third Millennium Levantine Pottery Production Center: Typology, Petrography and Provenance of the Metallic Ware of Northern Israel and Adjacent Regions. *Bulletin of the American Schools of Oriental Research* 301: 5–24.

Gunneweg, J., and Perlman, I.

1985–86 The Origin of the Herodian Lamp. *Bulletin of the Anglo-Israel Archaeological Society* 1984–85: 79–83.

Hachlili, R.

1988 *Ancient Jewish Art and Archaeology in the Land of Israel*. Leiden: Brill.

1998 *Ancient Jewish Art and Archaeology in the Diaspora*. Leiden: Brill.

Hadad, S.

2002 *Bet Shean 1: The Oil Lamps from the Hebrew University Excavations at Bet Shean*. Qedem Report 4. Jerusalem: Israel Exploration Society.

Hajjar, J.

1965 Un hypogée Romain a Debʿaal dans la region de Tyr. *Bulletin du Musée de Beyrouth* 18: 61–104.

Hamer, F., and Hamer, J.

1993 *The Potter's Dictionary of Materials and Techniques*. 3rd ed. London: Black.

Hamilton, R. W.

1940 Excavations against the North Wall of Jerusalem, 1937–8. *Quarterly of the Department of Antiquities in Palestine* 10: 1–53.

Harding, G. L.

1946 A Nabatean Tomb at Amman. *Quarterly of the Department of Antiquities in Palestine* 12: 58–62.

1950 A Roman Family Vault on Jebel Jofeh, Amman. *Quarterly of the Department of Antiquities in Palestine* 14: 81–94.

Hayes, J. W.

1980 *Ancient Lamps in the Royal Ontario Museum*, vol. 1: *Greek and Roman Clay Lamps: A Catalogue.* Toronto: Royal Ontario Museum.

Heimerl, A.

2001 *Die römischen Lampen aus Pergamon*. Pergamenische Forschungen 13. Berlin: de Gruyter.

Hoard, R. J.; O'Brien, M. J.; Khorasgany, M. G.; and Gopalaratnam, V. S.

1995 A Materials-Science Approach to Understanding Limestone-Tempered Pottery from the Midwestern United States. *Journal of Archaeological Science* 2: 824–25.

Hodder, I., and Hutson, S.

2003 *Reading the Past: Current Approaches to Interpretation in Archaeology*. 3rd ed. Cambridge: Cambridge University Press.

Holum, K. G.; Hohfleder, L.; Bull, R. E.; and Raban, A.

1988 *King Herod's Dream: Caesarea on the Sea*. New York: Norton.

Henrotay, D., and Warzée, G.

2009 *Arlon la gallo-romaine*. Namur: Institut du Patrimonie wallon.

Howland, R. H.

1958 *The Athenian Agora, IV: Greek Lamps and Their Survivals*. Princeton: American School of Classical Studies at Athens.

Hübinger, U.

1993 *Die antiken Lampen des Akademischen Kunstmuseums der Universität Bonn*. Berlin: Deutsches Archäologisches Institut.

Humphries, D. W.

1992 *The Preparation of Thin Sections of Rocks, Minerals, and Ceramics*. Microscopy Handbook 24. Oxford: Oxford University Press, Royal Microscopial Society.

Illiffe, J. H.

1934 A Tomb at El Bassa of c. A.D. 396. *Quarterly of the Department of Antiquities in Palestine* 3: 81–91.

1945 Imperial Art in Trans-Jordan: Figurines and Lamps from a Potter's Store at Jerash. *Quarterly of the Department of Antiquity of Palestine* 11: 1–26.

Israeli, Y., and Avida, U.

1988 *Oil-Lamps from Eretz-Israel: The Louis and Carmen Warschaw Collection at the Israel Museum Jerusalem*. Tel Aviv: Sabinsky.

Johnson, B. L.

2008 *Ashkelon 2: Imported Pottery of the Roman and Late Roman Periods*. Final Reports of The Leon Levy Expedition to Ashkelon 2. Winona Lake, IN: Eisenbrauns and Harvard Semitic Museum.

Kahane, P. P.

1961 Rock-Cut Tombs at Huqoq: Note on the Finds. *ʿAtiqot* 3: 126–47.

Karivieri, A.

1996 *The Athenian Lamp Industry in Late Antiquity*, vol. 1. Papers and Monographs of the Finnish Institute at Athens 5. Helsinki: Foundation of the Finnish Institute at Athens.

Kehrberg, I.

1989 Selected Lamps and Pottery from the Hippodrome at Jerash. *Syria* 66: 85–97.

1992 Flaked Glass and Pottery Sherd Tools of Late Roman and Byzantine Periods from the Hippodrome of Jerash. *Syria* 49: 451–64.

1995 Jerash/Gerasa, Hippodrome. *American Journal of Archaeology* 99: 525.

2001 Ceramic Lamp Production in the Fourth and Early Fifth Centuries AD at the Hippodrome of Gerasa: An Empirical Study. Pp. 231–39 in *La Céramique Byzantine et Proto-Islamique en Syrie–Jordanie (IVe–VIIe Siècles Apr. J.-C.)*, ed. E. Villeneuve and P. P. Watson. Bibliothèque Archéologique et Historique 159. Beirut: Institut Français d'Archéologie du Proche-Orient.

Kehrberg, I., and Ostrasz, A. A.

1997 A History of Occupational Changes at the Site of the Hippodrome of Gerasa. *Studies in the History and Archaeology of Jordan* 6: 167–73.

Kennedy, C. A.

1963 The Development of the Lamp in Palestine. *Berytus* 14: 67–115.

Kennedy, D.

2007 *Gerasa and the Decapolis: A 'Virtual Island' in Northwest Jordan.* Duckworth Debates in Archaeology. Bath: Duckworth.

Kingsley, S.A.

1999 The Sumaqa Pottery Assemblage: Classification and Quantification. Pp. 263–329 in *A Roman and Byzantine Village on Mount Carmel, Israel*, ed. S. Dar. British Archaeological Reports International Series 815. Oxford: Archaeopress.

Klein, E. M.; Langmuir, C. H.; and Staudigel, H.

1991 Geochemistry of Basalts from the Southeast Indian Ridge, 115°E–138°E. *Journal of Geophysical Research* 96: 2089–2107.

Kloner, A., and Hess, O.

1985 A Columbarium in Complex 21 at Maresha. *ʿAtiqot* 17: 124–33.

Kochavi, M.

1989 Excavations at Aphek-Antipatris. *Qadmoniot* 22: 2–20. [Hebrew]

Ladstätter, S.

2002 Objekte: Lampen mit Gladiatorendarstellungen aus den Hanghäusern in Ephesos. Pp. 97–102 in *Gladiatoren in Ephesos: Tod am Nachmittag*, ed. S. Erdemgil and F. Krinzinger. Istanbul: Dogan Gümüş.

Lapp, E. C.

1991 Zwei spätantike jüdische Tonlampen aus Kleinasien. *Jahrbuch für Antike und Christentum* 35: 165–68.

1992 Jüdische Tonlampen aus der Spätantikeim Landesmuseum Mainz. *Zeitschrift des Deutschen Palästina Vereins* 108: 171–73.

1993 Jewish Archaeological Evidence from the Roman Rhineland. *Journal of Jewish Studies* 44: 70–82.

1995 Byzantine and Early Islamic Oil Lamp Fragments from House 119 at Umm al-Jimal (Jordan). *Annual of the Department of Antiquities of Jordan* 39: 437–45.

1996 Clay Oil Lamps. Pp. 217–24 in *Sepphoris in Galilee: Crosscurrents of Culture*, ed. R. M. Nagy et al. Raleigh, NC: North Carolina Museum of Art.

1997a *The Archaeology of Light: The Cultural Significance of the Oil Lamp from Roman Palestine.* Ph.D. dissertation. Duke University.

1997b Regional Petrography Project. *American Journal of Archaeology* 101: 524–25.

2001a A Comparative Clay Fabric Analysis of Fired and Unfired Jerash Bowl Fragments by Means of Petrography and Direct Current Plasma (DCP) Spectrometry. Pp. 129–37 in *La Céramique Byzantine et Proto-Islamique en Syrie–Jordanie (IVe–VIIe Siècles Apr. J.-C.)*, ed. E. Villeneuve and P. P. Watson. Bibliothèque Archéologique et Historique 159. Beirut: Institut Français d'Archéologie du Proche-Orient.

2001b Les lampes à huile érotiques en Terre Sainte. *Archéologia* 377: 62–66.

2001c Lamps: Israel/Palestine. Pp. 295–97 in *Archaeological Encyclopedia of the Holy Land*, ed. A. Negev and S. Gibson. London: Continuum.

2006a Lampes. Pp. 310–14 in *Dictionnaire Archéologique de la Bible*, ed. A. Negev and S. Gibson. Paris: Hazan.

2006b Lampes: Israël/Palestine. Pp. 314–17 in *Dictionnaire Archéologique de la Bible*, ed. A. Negev and S. Gibson. Paris: Hazan.

2007 Marketing Religious Difference in Late Antique Palestine. Pp. 371–80 in *The Archaeology of Difference: Gender, Ethnicity, Class and the "Other" in Antiquity: Studies in Honor of Eric M. Meyers*, ed. D. R. Edwards and C. T. McCollough. Annual of the American Schools of Oriental Research 60–61. Boston: American Schools of Oriental Research.

2009 Material Culture: Lamps. Pp. 248–72 in E. M. Meyers and C. L. Meyers, *Excavations at Ancient Nabratein: Synagogue and Environs*. Meiron Excavation Project 6. Winona Lake, IN: Eisenbrauns.

2012 A Water Absorption Analysis of Nabatean, North African and Other Clay Lamp Fabrics Unearthed at the Red Sea Port of Roman *Aila* (Aqaba, Jordan). *Archaeometry* 54: 56–79.

forthcoming a Roman *Aila*: Clay Oil Lamps of a Nabatean Port City on the Red Sea. In *The Roman Aqaba Project Final Report*, vol. 2, ed. S. T. Parker.

forthcoming b A Jerash Lamp with a Kufic Inscription Recovered from House 119 at Umm al-Jimal.

under review Exploring Ancient Clay Lamp Breakage and Sherd Creation in Light of Literary Sources, Archaeological Evidence, and Mechanical Experimentation. *Journal of Archaeological Science: Reports*. London: Elsevier.

in press a A Jewish Oil Lamp Unearthed at the Red Sea Port of Roman *Aila* (Aqaba, Jordan). In *Viewing Ancient Art and Archaeology: VeHinnei: Rachel – Essays in Honor of Rachel Hachlili*, ed. A. E. Killebrew and G. Fassbeck. Leiden: Brill.

in press b Encountering Photoamulets and the Use of Apotropaic Light in Late Antiquity. In *The Oxford Handbook on Light in Archaeology*, ed. C. Papadopoulos and G. Earl. Oxford: Oxford University Press.

Lapp, E., and Nicoli, J.

2014 Exploring 3D Modeling, Fingerprint Extraction, and Other Scanning Applications for Ancient Clay Oil Lamps. *Digital Applications in Archaeology and Cultural Heritage* 1/2: 34–44.

Lecomte, C.

2001 *Le mythe d'Europe: Voyage en Béotie, en Crète et en Phénicie*. Ph.D. dissertation. Université de Liège.

Levine, L. I.

2000 *The Ancient Synagogue: The First Thousand Years*. New Haven, CT: Yale University Press.

Levine, T.

2003 Pottery and Small Finds from Subterranean Complexes 21 and 70. Pp. 73–130 in *Maresha Excavations Final Report I: Subterranean Complexes 21, 44, 70*, ed. A. Kloner. Israel Antiquities Reports 17. Jerusalem: Israel Antiquities Authority.

Loeschcke, S.

1919 *Lampenaus Vidonissa: Ein Beitragzur Geschichte von Vindonissa und des antiken Beleuchtungswesens*. Zurich: Beer.

Loffreda, S.

1974 *Cafarnao, II: La Ceramica.* Jerusalem: Franciscan Printing Press.

1976 Alcune osservazioni sulla ceramica di Magdala. *Studia Hierosolymitana in onore di P. Bellarmino Bagatti*, vol. 1: *Studiarcheologici* 22: 338–54.

Lyon-Caen, C., and Hoff, V.

1986 Lampes grecques en terre cuite. Pp. 23–67 in *Musée du Louvre: Catalogue des lampes en terre cuite grecques et chrétiennes*, ed. C. Metzger. Paris: Ministère de la Culture et de la Communication.

Macalister, R. A. S.

1912 *The Excavation of Gezer 1902–1905 and 1907–1909.* Vols. 2–3. London: Palestine Exploration Fund.

MacDonnell, A. M.

1988 Terracotta Lamps. Pp. 116–36 in G. D. Weinberg, *Excavations at Jalame: Site of a Glass Factory in Late Roman Palestine*. Columbia: University of Missouri Press.

Maggetti, M.

1981 Composition of Roman Pottery from Lousanna (Switzerland). *British Museum Occasional Papers* 19: 33–49.

Magetti, M.; Westley, H.; and Olin, J. O.

1984 Provenance and Technical Studies of Mexican Majolica Using Elemental and Phase Analysis. Pp. 151–91 in *Archaeological Chemistry III.* Advances in Chemistry Series 205. Washington, DC: American Chemistry Society.

Magness, J.

1992 Late Roman and Byzantine Pottery: Preliminary Report, 1990. Pp. 129–53 in *Caesarea Papers: Stratton's Tower, Herod's Harbour, and Roman and Byzantine Caesarea*, ed. R. L. Vann. Journal of Roman Archaeology Supplement 5. Ann Arbor, MI: Journal of Roman Archaeology.

1993 *Jerusalem Ceramic Chronology Circa 200–800 C.E.* Journal for the Study of the Old Testament / American Schools of Oriental Research Monograph 9. Sheffield: JSOT Press.

2008 The Oil Lamps from the South Cemetery. Pp. 121–78 in *The Necropolis of Bet Guvrin-Eleutheropolis*, ed. G. Avni, U. Dahari, and A. Kloner. Israel Antiquities Authority Report 36 Jerusalem: Israel Antiquities Authority.

2011 *Stone and Dung, Oil and Spit: Jewish Daily Life in the Time of Jesus*. Grand Rapids, MI: Eerdmans.

Marchand, J.

1996 The Lamps. Pp. 57–67 in P. M. Bikai, W. J. Fulco, and J. Marchand, *Tyre: The Shrine of Apollo*. Amman: National Press.

Mazar, A.

1994 Four Thousand Years of History at Tel Beth-Shean. *Qadmoniot* 27: 66–83.

Mazar, B.

1971 *The Excavations in the Old City of Jerusalem near the Temple Mount: Preliminary Report of the Second and Third Seasons 1969–1970.* Jerusalem: Institute of Archaeology, the Hebrew University of Jerusalem.

1973 *Beth She'arim: Report on the Excavations during 1936–1940*, vol. 1: *Catacombs 1–4*. New Brunswick, NJ: Rutgers University Press.

McNicoll, A. W.; Edwards, P. C.; Hanbury-Tenison, C. J.; Hennessy, J. B.; Potts, T. F.; Smith, R. H.; Walmsley, A. G.; and Watson, P.

1992 *Pella in Jordan*, vol. 2: *The Second Interim Report of the Joint University of Sydney and College of Wooster Excavations at Pella 1982–1985*. Mediterranean Archaeology Supplement 2. Sydney: Meditarch.

McNicoll, A. W., and Smith, R. H.

1980 The 1979 Season at Pella of the Decapolis. *Bulletin of the American Schools of Oriental Research* 240: 63–84.

Meshorer, Y.

1996 Market Weight (67). P. 201 in *Sepphoris in Galilee: Crosscurrents of Culture*, ed. R. M. Nagy et al. Raleigh: North Carolina Museum of Art.

Meyers, C.

forthcoming Menorot Incised or Painted on Ceramic Vessels. In *Artifacts and Other Finds from the Western Summit of Sepphoris*, ed. E. M. Meyers, C. L. Meyers, and B. D. Gordon. Duke Sepphoris Excavation Report 4. Winona Lake IN: Eisenbrauns.

Meyers, C., and Meyers, E. M.

2013 Sepphoris. Pp. 336–48 in *The Oxford Encyclopedia of the Bible and Archaeology*, vol. 1, ed. D. M. Master. New York: Oxford University Press.

in press Images and Identity: Menorah Representations at Sepphoris. In *Ve'Zot Rachel: Festschrift for Rachel Hachlili*, ed. A. E. Killebrew and A. Segal. Leiden: Brill.

Meyers, E. M.

1978 Galilean Regionalism: A Reappraisal. Pp. 115–31 in *Approaches to Ancient Judaism* 5, ed. W. S. Green. Missoula, MT: Scholars Press for Brown University.

2013 Additional Roman Pottery. Pp. 130–41 in Meyers and Meyers 2013b.

Meyers, E. M.; Hendin, D.; and Meyers, C. L.

2015 Further Reflections on Sepphoris and Rome: Numismatics and Archaeology. *Eretz-Israel* 31: 132*–140*.

Meyers, E. M.; Kraabel, A. T.; and Strange, J. F.

1976 *Ancient Synagogue Excavations at Khirbet Shemaʿ, Upper Galilee, Israel 1970–1972*. Meiron Excavation Project 1. Durham, NC: Duke University Press on behalf of the American Schools of Oriental Research.

Meyers, E. M., and Meyers, C. L.

1990 *Excavations at the Ancient Synagogue of Gush Ḥalav*. Meiron Excavation Project 5. Winona Lake, IN: Eisenbrauns.

2009 *Excavations at Ancient Nabratein: Synagogue and Environs*. Meiron Excavation Project 6. Winona Lake, IN: Eisenbrauns.

2013a Introduction. Pp. 1–12 in Meyers and Meyers 2013b.

2013b eds., *The Pottery from Ancient Sepphoris*. [Duke] Sepphoris Excavation Reports I. Winona Lake, IN: Eisenbrauns.

Meyers, E. M.; Meyers, C. L.; and Gordon, B. D., eds.

forthcoming *The Architecture and Stratigraphy of the Western Summit of Sepphoris*. Duke Sepphoris Excavation Report IV. Winona Lake, IN: Eisenbrauns.

Meyers, E. M.; Netzer, E.; and Meyers, C.

1992 *Sepphoris*. Winona Lake, IN: Eisenbrauns.

Meyers, E. M.; Strange, J. F.; and Meyers, C. L.

1981 *Excavations at Ancient Meiron, Upper Galilee, Israel 1971–72, 1974–75, 1977*. Meiron Excavation Project III. Cambridge, MA: American Schools of Oriental Research.

Młynarczyk, J.
1997 *Alexandrian and Alexandria-Influenced Mold-Made Lamps of the Hellenistic Period.* British Archaeological Reports International Series 677. Oxford: Archaeopress.
Modrzewska-Marciniak, I.
1977 Lampes d'Anab Safina (Syrie): L'étude typologique et chronologique. *Archeologia* 28: 134–52.
1988 *Studio iconological delle Lucerne Siro-Palestinesi del IV–VII sec d.c.* Revista di Archaeologia, Supplementi 4. Rome: Bretschneider Giorgio.
Munsell Color Company
1971 *Munsell Soil Color Charts.* Baltimore: Munsell Color Company.
Nagy, R. M.; Meyers, C. L.; Meyers, E. M.; and Weiss, Z., eds.
1996 *Sepphoris in Galilee: Crosscurrents of Culture.* Raleigh: North Carolina Museum of Art.
Nahshoni, P.
2002 A Rock-Cut Burial Cave from the Second Temple Period at Horbat Zefiyya, Judean Shephelah. *ʿAtiqot* 43: 48–71.
Naveh, J., and Shaked, S.
1985 *Amulets and Magic Bowls: Aramaic Incantations of Late Antiquity.* Jerusalem: Magness.
Negev, A.
1971 The Nabatean Necropolis of Mampsis (Kurnub). *Israel Exploration Journal* 21: 110–29.
Neidinger, W.
1982 A Typology of Oil Lamps from the Mercantile Quarter of Antipatris. *Tel Aviv* 9: 157–69.
Nevo, D.
1985 *Sde Boqer and the Central Negev, Seventh–Eighth Century* AD. Jerusalem: Israel Publication Services.
Nielsen, I.; Andersen, F. G.; and Holm-Nielsen, S.
1993 *Gadara–Umm Qēs III: Die byzantinischen Thermen.* Wiesbaden: Harrassowitz.
Oren, E. D., and Rappaport, U.
1984 The Necropolis of Maresha–Beth Govrin. *Israel Exploration Journal* 34: 114–53.
Orten, C.; Tyers, P.; and Vince, A.
1993 *Pottery in Archaeology.* Cambridge: Cambridge University Press.
Ostrasz, A.
1993 Gerasa, Hippodrome. Pp. 498–500 in B. de Vries and P. Bikai, eds. Archaeology in Jordan, *American Journal of Archaeology* 97: 457–520.
Oziol, T.
1977 *Salamine de Chypre: Les Lampes du Musée de Chypre VII.* Paris: Boccard.
1993 *Les Lampes au Musée de la Fondation Piéridès.* Nicosia: Zavallis Litho.
Papanicolaou Christensen, A.; Thomsen, R.; and Ploug, G.
1986 *Hama: Fouilles et Recherches, 1931–1938,* vol. 3/3: *The Graeco-Roman Objects of Clay, the Coins and the Necropolis.* Copenhagen: National Museum.
Parker, S. T.
2006 *The Roman Frontier in Central Jordan: Final Report in the Limes Arabicus Project, 1980–1989.* 2 vols. Washington, DC: Dumbarton Oaks.
Patrich, J., and Arubas, B.
1989 A Juglet Containing Balsam Oil (?) from a Cave near Qumran. *Israel Exploration Journal* 39: 42–59.

Peacock, D. P. S.; Bejaoui, F.; and Ben Lazreg, N.
1990 Roman Pottery Production in Central Tunisia. *Journal of Roman Archaeology* 3: 59–84.

Peleg, M.
1989 Domestic Pottery. Pp. 31–113 in V. Tzaferis, *Excavations at Capernaum*, vol. 1: *1978–1982*. Winona Lake, IN: Eisenbrauns.

Perlzweig, J.
1961 *The Athenian Agora 7: Lamps of the Roman Period*. Princeton, NJ: Princeton University Press.

Perlman, I., and Asaro, F.
1969 Pottery Analysis by Neutron Activation. *Archaeometry* 11: 21–52.

Pettijohn, F. J.; Potter, P. E.; and Siever, R.
1973 *Sand and Sandstone*. Berlin: Springer.

Pierobon, R.
1983 Sanctuary of Artemis: Soundings in the Temple-Terrace, 1978–1980. *Mesopotamia* 18–19: 85–111.

Potts, P. J.
1995 *A Handbook of Silicate Rock Analysis*. London: Chapman and Hall.

Ravani, B.
1961 Rock-Cut Tombs at Huqoq: The Excavations. *ʿAtiqot* 3: 121–25.

Regev, E.
2006 Ancient Jewish Style: Why Were Ossuaries and Southern Oil Lamps Decorated? *Levant* 38: 171–86.

Rey-Coquais, J.-P.
1964 Lampes antiques de Syrie et du Liban. *Mélanges de l'Université Saint Joseph* 39: 147–65.

Rice, P. M.
1987 *Pottery Analysis: A Sourcebook*. Chicago: University of Chicago Press.

Roberts, C. H., and Turner, E. G.
1952 *Catalogue of the Greek and Latin Papyri in the John Rylands Library, Manchester*, vol. 4: *Documents of the Ptolemaic, Roman, and Byzantine Periods* (nos. 552–717). Manchester: University of Manchester Press.

Rose, M.
1997 Ashkelon's Dead Babies. *Archaeology* 50: 12–13.

Rosenthal, R., and Sivan, R.
1978 *Ancient Lamps in the Schloessinger Collection*. Qedem 8. Jerusalem: Hebrew University of Jerusalem.

Rosenthal-Heginbottom, R.
1981 *Römische Bildlampen aus östlichen Werkstätten*. Wiesbaden: Harrassowitz.
1995 Imported Hellenistic and Roman Pottery. Pp. 183–288 in Ephraim Stern, *Excavations at Dor, Final Report. Volume IB. Areas A and C: The Finds*. Qedem Report 2. Jerusalem: Israel Exploration Society.

Rutgers, L. V.
1996 Fragment of a Redware Vessel with Stamped Cross. P. 208 in *Sepphoris in Galilee: Crosscurrents of Culture*, ed. R. M. Nagy et al. Raleigh, NC: North Carolina Museum of Art.

Rutkowski, B.
1986 *The Cult Places of the Aegean*. New Haven, CT: Yale University Press.

Safrai, Z.
1994 *The Economy of Roman Palestine*. London: Routledge.
Saller, S. J.
1957 *Excavations at Bethany* (1949–1953). Jerusalem: Franciscan Printing Press.
Schaefer, J., and Falkner, R. K.
1986 An Umayyad Potters' Complex in the North Theatre, Jerash. Pp. 411–59 in *Jerash Archaeological Project 1981–1983*, vol. 1, ed. F. Zayadine. Amman: Department of Antiquities.
Schneider, G.
1993 X-ray Fluorescence Analysis and the Production and Distribution of Terra Sigillata and Firmalampen. Pp. 129–37 in *The Inscribed Economy: Production and Distribution in the Roman Empire in the Light of Instrumentum Domesticum*, ed. W. V. Harris. Journal of Roman Archaeology Supplement 6. Ann Arbor, MI: Journal of Roman Archaeology.
Schneider, G., and Wirz, E.
1992 Chemical Answers to Archaeological Questions: Roman Terracotta Lamps as Documents of Economic History. Pp. 13–48 in *Sciences de la Terre et Céramiques Archéologiques: Expérimentations, Applications*, ed. S. Mery. Documents et Travaux 16. Cergy: Institut Géologique Albert de Lapparent.
Scholl, T.
1991 The Chronology of Jerash Lamps: The Umayyad Period. *Archeologia* 42: 65–84.
Searle, A. B., and Grimshaw, R. W.
1959 *The Chemistry and Physics of Clay and Other Ceramic Materials*. New York: Interscience.
Sellers, O. R.
1933 *The Citadel of Beth-Zur*. Philadelphia: Westminster.
Sellers, O. R., and Baramki, D. C.
1953 A Roman–Byzantine Burial Cave in Northern Palestine. *Bulletin of the American Schools of Oriental Research Supplementary Studies* 15–16: 7–55.
Shepard, A. O.
1957 *Ceramics for the Archaeologist*. Publication 609. Washington, DC: Carnegie Institution of Washington.
Shier, L. A.
1978 *Terracotta Lamps from Karanis, Egypt*. Kelsey Museum of Archaeology Studies 3. Ann Arbor, MI: University of Michigan Press.
Singer, A.
2007 *The Soils of Israel*. Berlin: Springer.
Siegelmann, A.
1992 Roman and Byzantine Remains in the Northern Central Plain. *ʿAtiqot* 21: 65.
Sivan, R.
1982 Lamps with a Broken Discus. Pp. 115–16 in *Studien zur spätantiken und frühchristlichen Kunst und Kultur des Orients*, ed. G. Koch. Göttinger Orientforschungen, 2nd series: Studien zur spätantiken und frühchristlichen Kunst 6. Wiesbaden: Harrassowitz.
Slane, K. W.
1990 *The Sanctuary of Demeter and Kore: The Roman Pottery and Lamps*. Corinth 18/2. Princeton, NJ: American School of Classical Studies.

Smith, A. W., and McNicoll, A. W.
1992 The Roman Period. Pp. 119–44 in A. W. McNicoll et al., *Pella in Jordan*, vol. 2: *The Second Interim Report of the Joint University of Sydney and College of Wooster Excavations at Pella 1982–1985*. Mediterranean Archaeology Supplement 2. Sydney: Meditarch.

Smith, R. H.
1961 The "Herodian" Lamp of Palestine: Types and Dates. *Berytus* 14: 53–65.
1964a The Household Lamps of Palestine in Old Testament Times (part 1). *Biblical Archaeologist* 27: 2–31.
1964b The Household Lamps of Palestine in Intertestamental Times (part 2). *Biblical Archaeologist* 27: 101–24.
1966 The Household Lamps of Palestine in New Testament Times (part 3). *Biblical Archaeologist* 29: 2–27.
1973 *Pella of the Decapolis*, vol. 1: *The 1967 Season of the College of Wooster Expedition to Pella*. Wooster, OH: College of Wooster.

Smith, R. H., and Day, L. P.
1989 *Pella of the Decapolis*, vol. 2: *Final Report on the College of Wooster Excavations in Area IX, the Civic Complex, 1979–1985*. Wooster, OH: College of Wooster.

Sperber, D.
1998 *The City in Roman Palestine*. Oxford: Oxford University Press.

Stager, L. E.
1991 *Ashkelon Discovered: From Canaanites and Philistines to Romans and Moslems*. Washington, DC: Biblical Archaeology Society.

Stern, E. M.
1978 *Excavations at Tel Mevorakh (1973–1976), Part One: From the Iron Age to the Roman Period*. Qedem 9. Jerusalem: Hebrew University of Jerusalem, Institute of Archaeology.
1981 Hellenistic Dor. *Qadmoniot* 14: 103–10. [Hebrew]

Strange, J. F.
2014 The Galilean Road System. Pp. 263–71 in *Galilee in the Late Second Temple and Mishnaic Periods*, ed. D. A. Fiensy and J. R. Strange. Minneapolis: Fortress.

Sussman, V.
1969 Ancient Burial Cave at Rehovot. *ʿAtiqot* 5: 69–71. [Hebrew]
1970 A Shabbat Lamp. *ʿAtiqot* 6: 80–81 [Hebrew]; 10–11* [English].
1978 Samaritan Lamps of the Third–Fourth Centuries A.D. *Israel Exploration Journal* 28: 238–50.
1980 Molds for Lamps from a Caesarea Workshop. *ʿAtiqot* 14: 76–79.
1982 *Ornamented Jewish Oil-Lamps from the Destruction of the Second Temple through the Bar-Kokhba Revolt*. Warminster: Aris and Philips.
1989 Northern Stamped Oil Lamps and Their Typology. *Michmanim* 4: 23–58.
1990 The Lamp. Pp. 97–98 in S. Wachsmann, *The Excavations of an Ancient Boat in the Sea of Galilee (Lake Kinneret)*. ʿAtiqot 19. Jerusalem: Israel Antiquities Authority.
2008 The Oil Lamps. Pp. 207–92 in *Archaeological Excavations at Caesarea Maritima: Areas CC, KK and NN, Final Report*, vol. 1: *The Objects*, ed. J. Patrich. Jerusalem: Israel Exploration Society.
2009 *Greek and Hellenistic Wheel- and Mold-Made Closed Oil Lamps in the Holy Land: Collection of the Israel Antiquities Authority*. British Archaeological Reports International Series 2015. Oxford: Archaeopress.

2012 *Roman Period Oil Lamps in the Holy Land: Collection of the Israel Antiquities Authority*. British Archaeological Reports International Series 2447. Oxford: Archaeopress.

Tal, O., and Bastos, M. T.

2012 Intentionally Broken Discus Lamps from Roman Apollonia: A New Interpretation. *Tel Aviv* 39: 104–14.

Thalmann, J.-P.

1978 Tell ʿArqa (Libannord), campagnes I–III (1972–1974), vol. 1: Rapport préliminaire. *Syria* 55: 1–152.

Tucker, M. E.

1981 *Sedimentary Petrology: An Introduction*. Geoscience Text 3. Oxford: Blackwell.

Turquety-Pariset, F.

1982 Fouille de la Municipalité de Beyrouth (1977): Les objets. *Syria* 59: 27–76.

Tushingham, A. D.

1985 *Excavations in Jerusalem 1961–1967*, vol. 1. Toronto: Royal Ontario Museum.

Tzaferis, V.

1982 The Ancient Synagogue at Mao's Hayyim. *Israel Exploration Journal* 32: 215.

Uscatescu, A.

1992 Un ejemplo de tradición en las producciones cerámicas: Las lucernas bizantinas y omeyas de Gerasa (Ŷaraš, Jordania). *Caesaraugusta* 69: 183–218.

Vincenz, A. de

2013 Fine Wares; Byzantine–Early Islamic Wares. Pp. 142–215 in *The Pottery from Ancient Sepphoris*, ed. E. M. Meyers and C. L. Meyers. [Duke] Sepphoris Excavation Report I. Winona Lake, IN: Eisenbrauns.

Vine, K., and Hartelius, G.

2000 The Corpus of Terracotta Lamps from Caesarea Maritima, Israel, 1971–1980. http://digcaesarea.org/images/CaesareaLamps/.

Waagé, F. O.

1941 Lamps. Pp. 55–82 in *Antioch-on-the-Orontes, III: The Excavations 1937–1939*, ed. R. Stillwell. Princeton, NJ: Princeton University Press.

Waliszewski, T.

1995 Lampes proto-islamiques de Syrie–Palestine dans les collections du Musée National de Varsovie. *Archeologia* 46: 23–29.

Walters, H. B.

1905 *History of Ancient Pottery*, vol. 2. London: Murray.

1914 *Catalogue of the Greek and Roman Lamps in the British Museum*. London: Trustees of the British Museum.

Watson, O.

2004 *Ceramics from Islamic Lands*. London: Thames & Hudson.

Weber, T., and Hoffman, A.

1990 Gadara of the Decapolis Preliminary Report of the 1989 Season at Umm Qeis. *Annual of the Department of Antiqutites in Jordan* 34: 321–42.

Weinberg, S. S.

1971 Tel Anafa: The Hellenistic Town. *Israel Exploration Journal* 21: 86–109.

Weiss, Z.

2005 *The Sepphoris Synagogue: Deciphering an Ancient Message through Its Archaeological and Socio-Historical Contexts*. Jerusalem: Israel Exploration Society.

Weiss, Z., and E. Netzer

1996a *Promise and Redemption: A Synagogue Mosaic from Sepphoris*. Jerusalem: Israel Museum.

1996b Hellenistic and Roman Sepphoris: The Archaeological Evidence. Pp. 28–37 in *Sepphoris in Galilee: Crosscurrents of Culture*, ed. R. M. Nagy et al. Raleigh: North Carolina Museum of Art.

Widemann, F.; Picon, M.; Asaro, F.; Michel, H. V.; and Perlman, I.

1975 A Lyons Branch of the Pottery-Making Firm of Ateius of Arezzo. *Archaeometry* 17: 50–57.

Wilkens, H.; Turner, F. J.; and Gilbert, C. M.

1982 *Petrography: An Introduction to the Study of Rocks in Thin Sections*. 2nd ed. New York: Freeman.

Williams, D. F.

1979 Ceramic Petrology and the Archaeologist. P. 74 in *Pottery and the Archaeologist*, ed. M. Millet. Occasional Publication 4. London: Institute of Archaeology.

Williams, M. H.

1998 *The Jews among the Greeks and Romans: A Diasporan Sourcebook*. London: Duckworth.

Wilson, J., and Saʾd, M.

1984 The Domestic Material Culture of Nabataean to Umayyad Period Buṣrā. *Berytus* 32: 35–147.

Yaalon, D. H., and Ganor, E.

1973 The Influence of Dust on Soils during the Quarternary. *Soil Science* 116: 146–55.

Yeivin, Z., and Edelstein, G.

1970 Excavations at Tirat Yehuda. *ʿAtiqot* 6: 56–67. [Hebrew]

Yellin, J.

1994 Origin of the Lamps from Masada. Pp. 107–28 in *Masada 4: The Yigael Yadin Excavations 1963–1965 Final Reports*, ed. J. Aviram, G. Foerster, and E. Netzer. Jerusalem: Israel Exploration Society and the Hebrew University of Jerusalem.

Žuravlev, D. V.

2007 Lighting Equipment of the Northern Pontic Area in the Roman and Late Roman Periods: Imports and Local Production. Pp. 209–37 in *The Black Sea in Antiquity: Regional and Interregional Economic Exchanges*, ed. V. Gabrielsen and J. Lund. Black Sea Studies 6. Aarhus: Aarhus University Press.

Żurawski, B.

1992 *Magica et Ceramica*: Magic and Ceramics in Christian Nubia. *Archaeologia Polona* 30: 87–107.

Plates

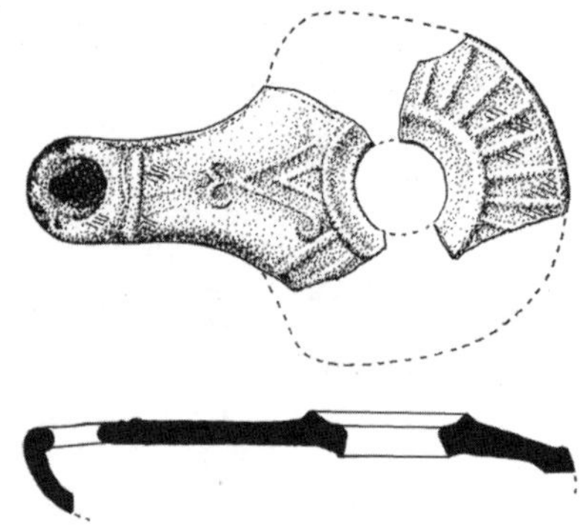

1

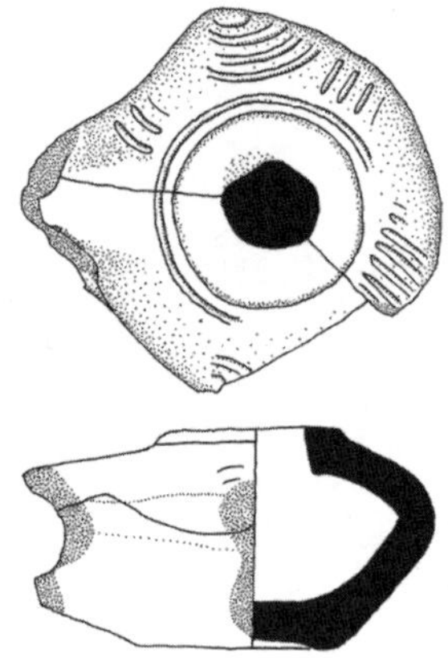

3

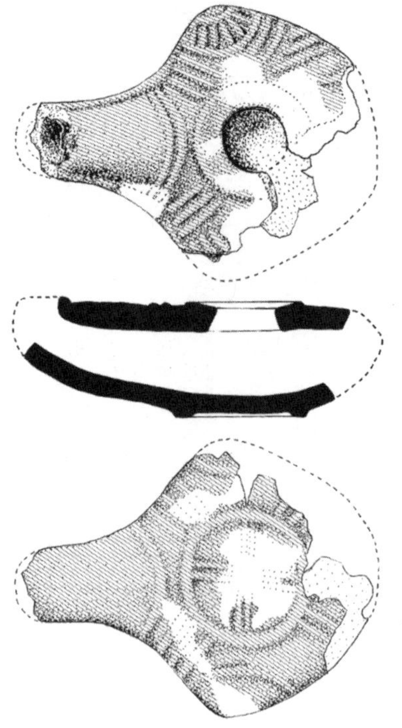

2

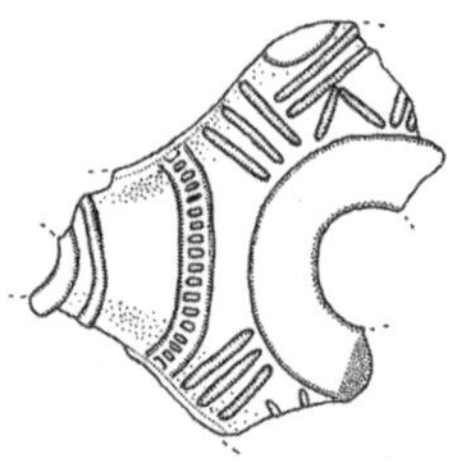

4

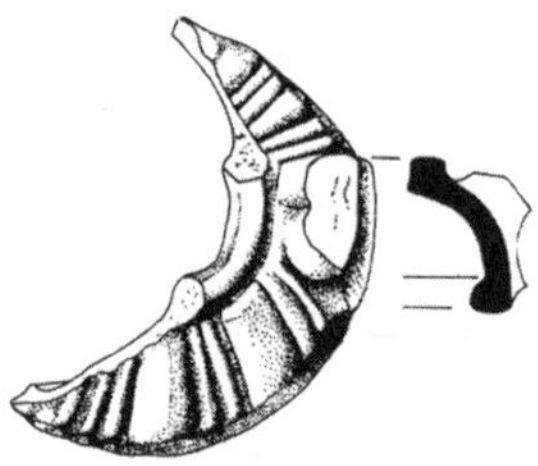

5

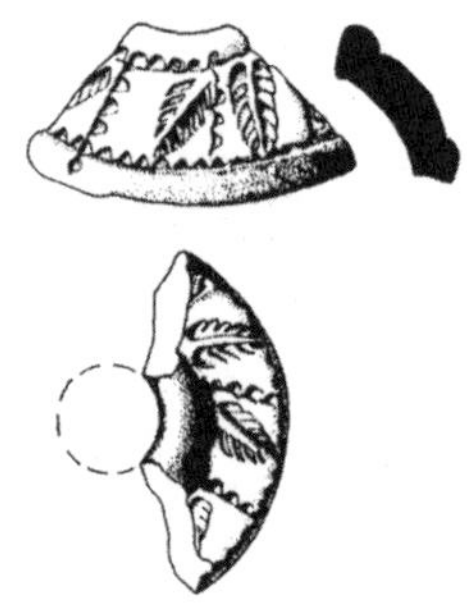

6

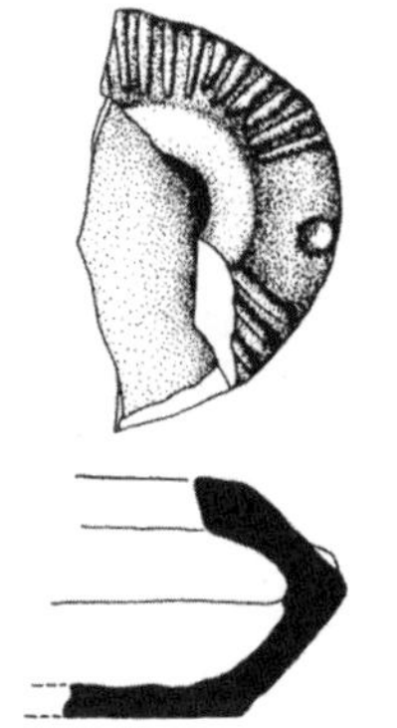

7

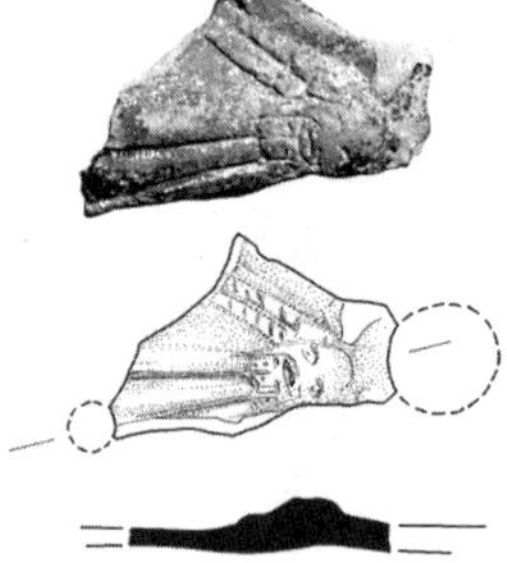

8

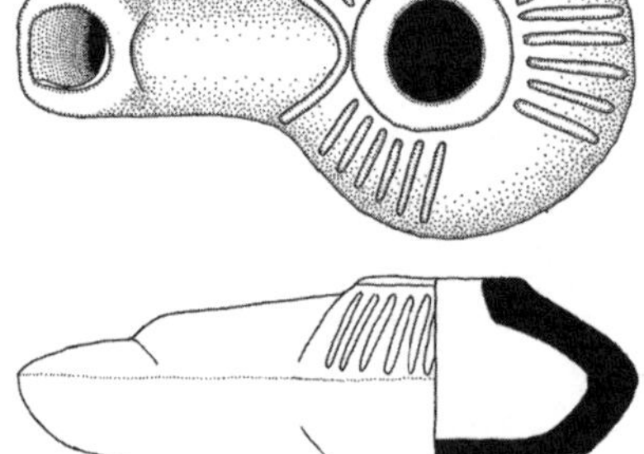

9

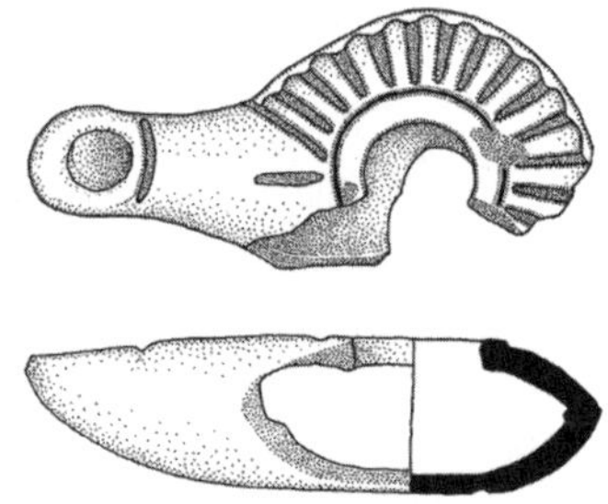

10

0 2cm

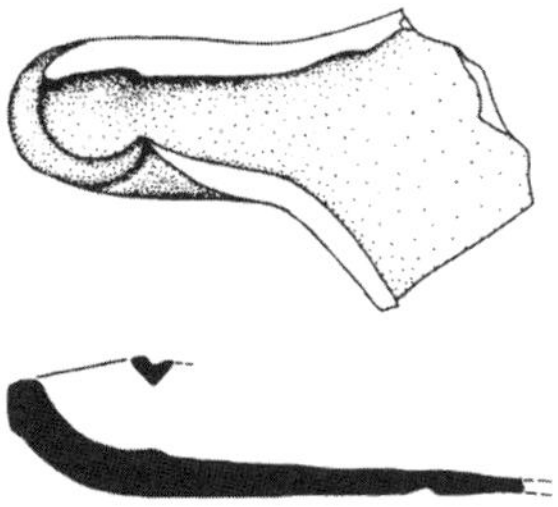

13

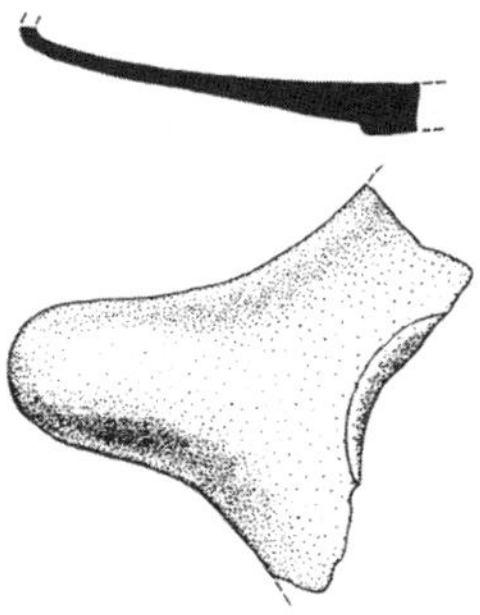

14

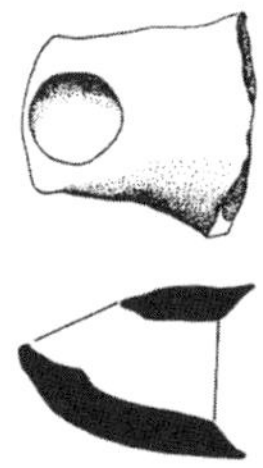

15

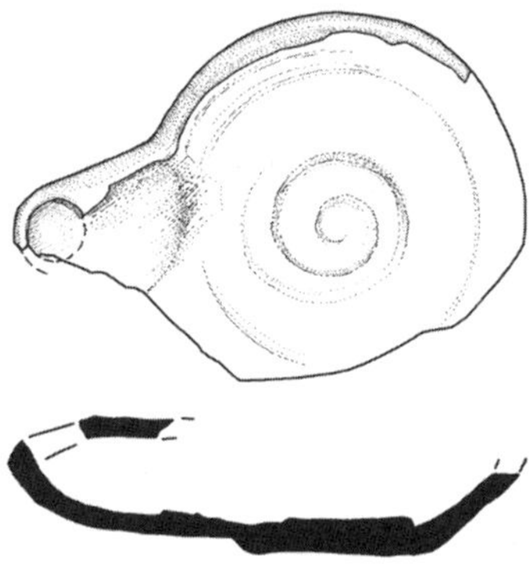

18

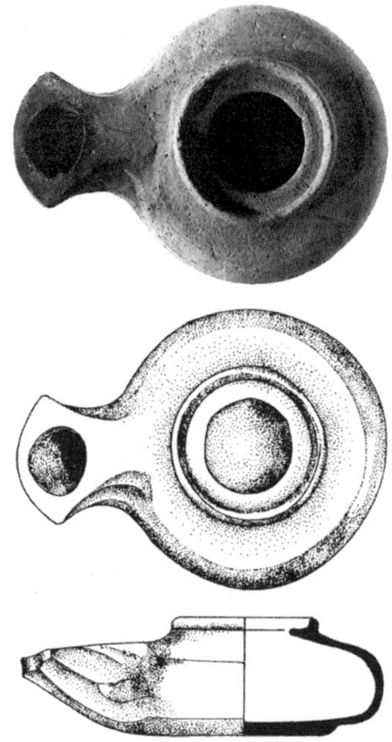

19

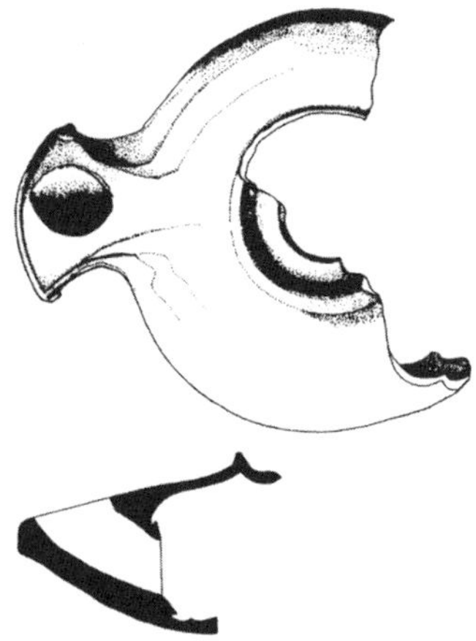

20

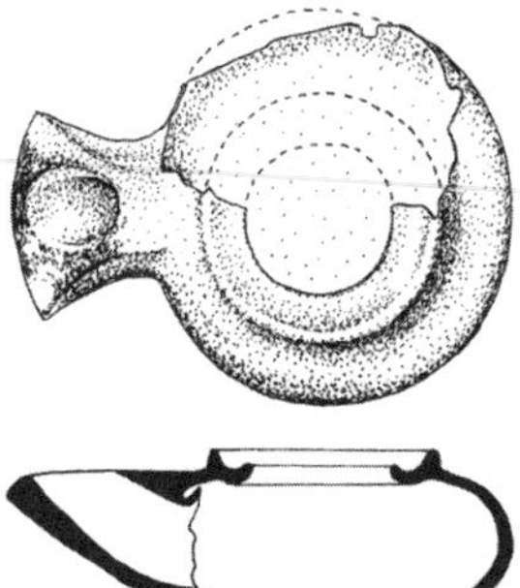

21

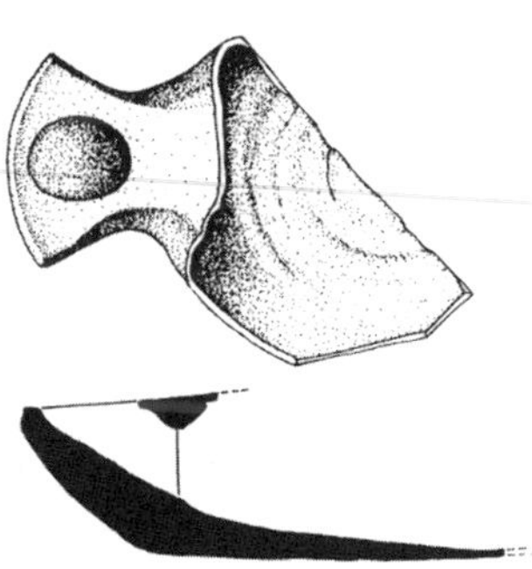

22

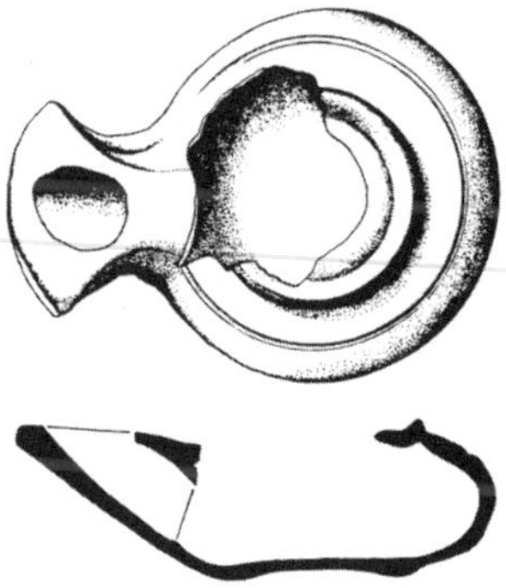

23

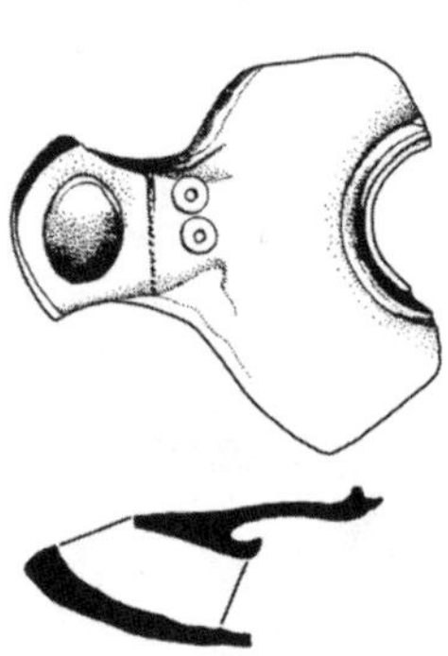

24

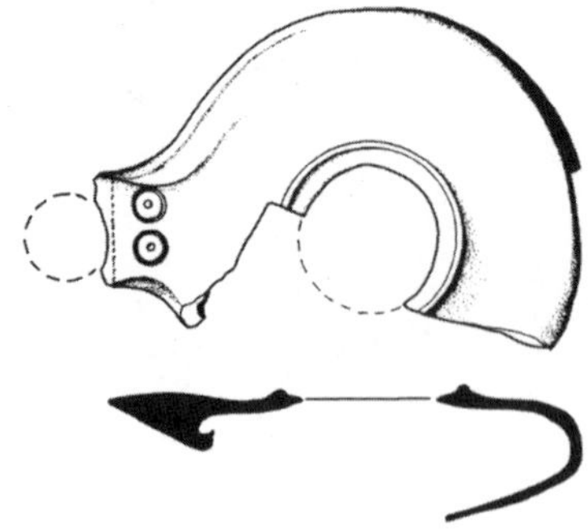

25

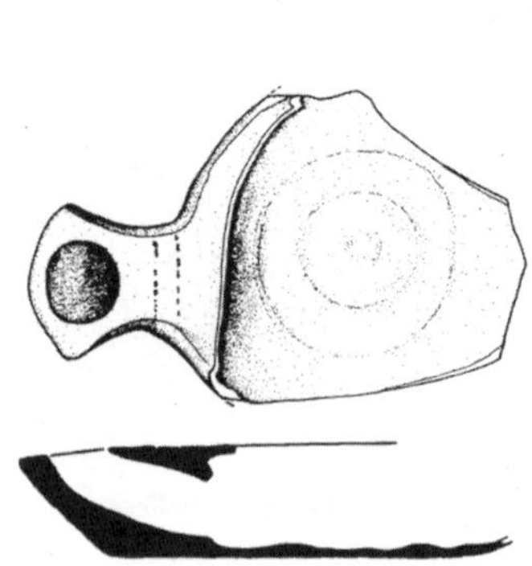

26

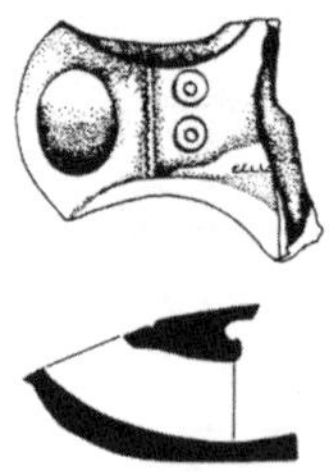

27

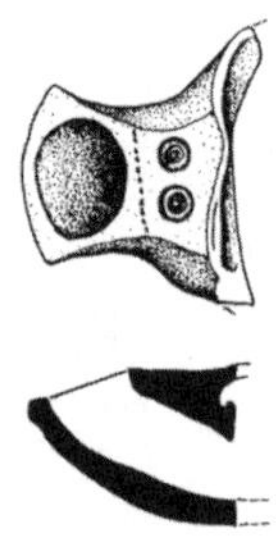

28

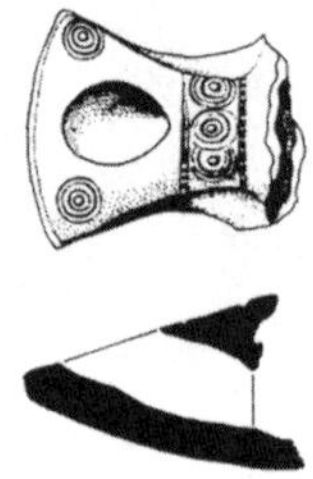

29

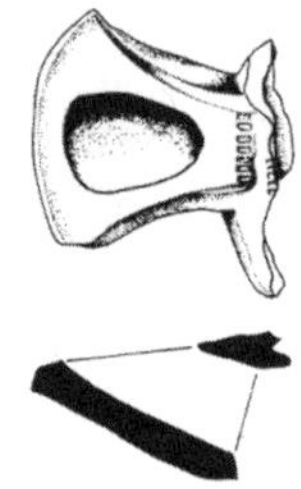

30

31

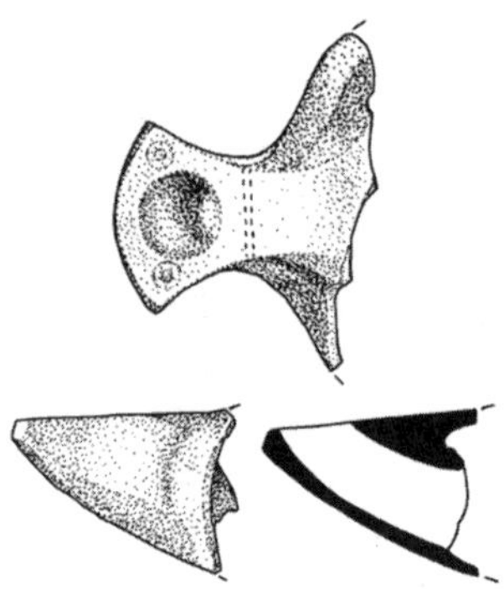

35

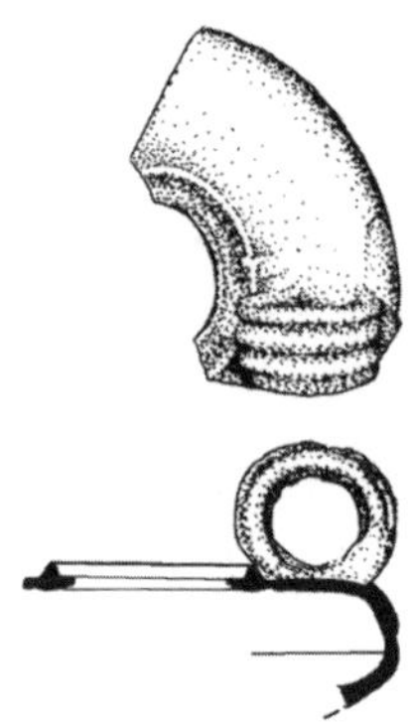

37

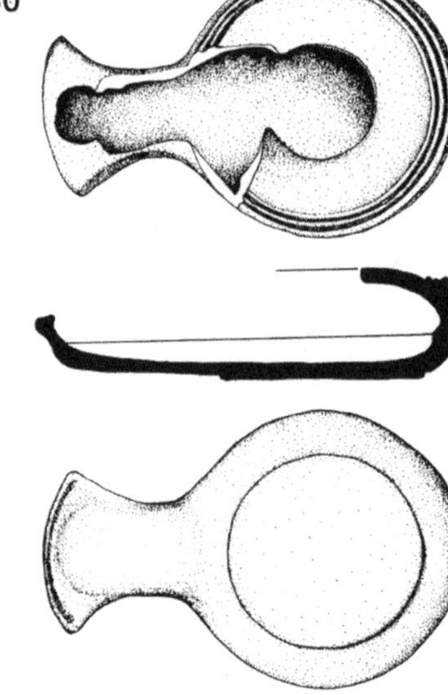

44

45

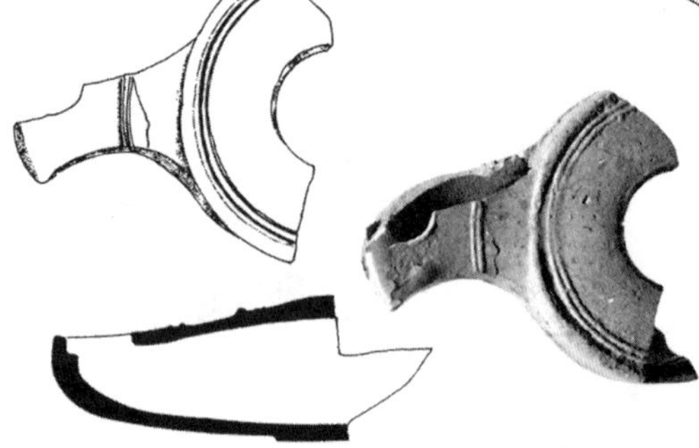

46

0 2cm

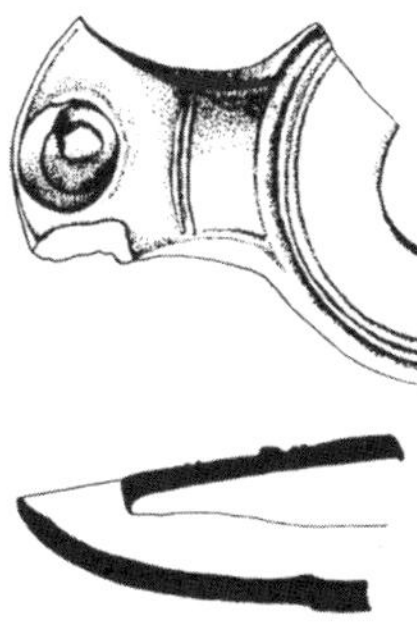

47

48

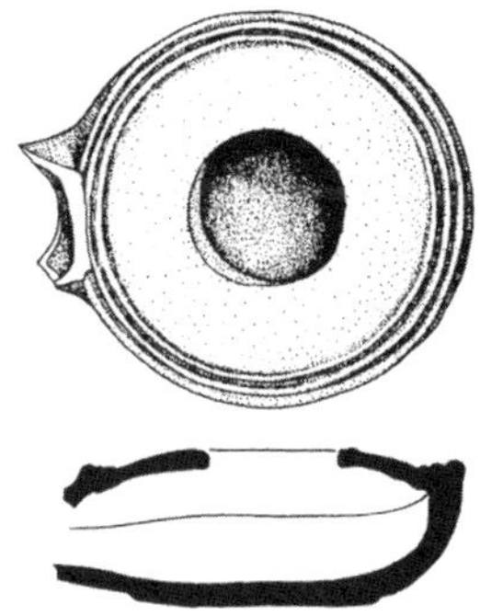

49

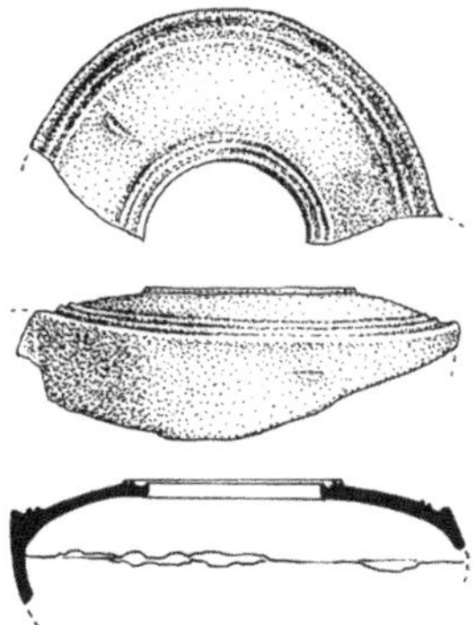

50

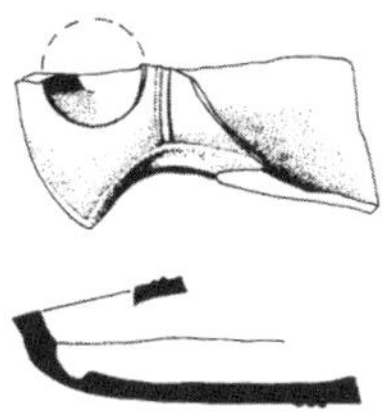

52

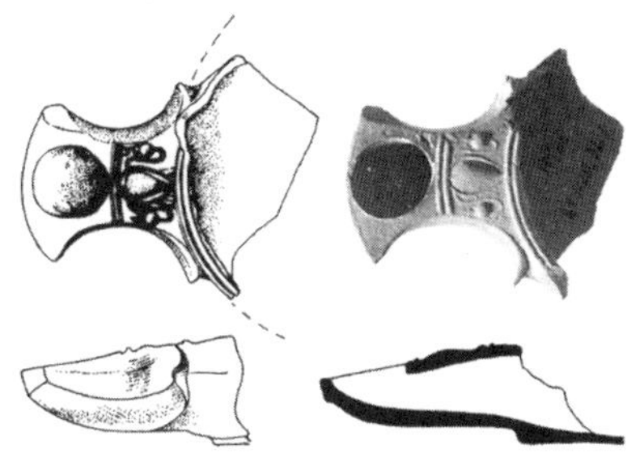

53

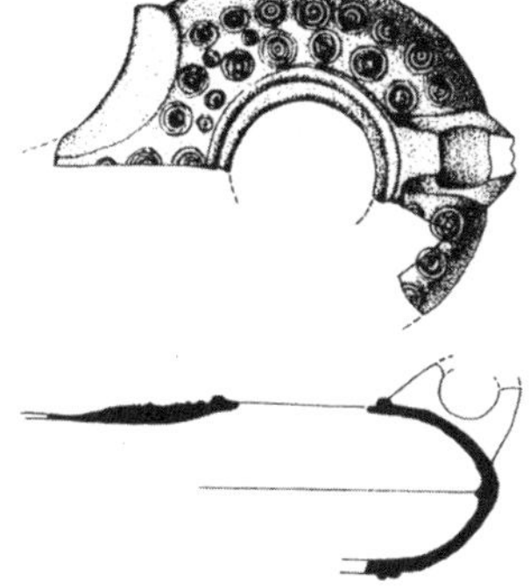

54

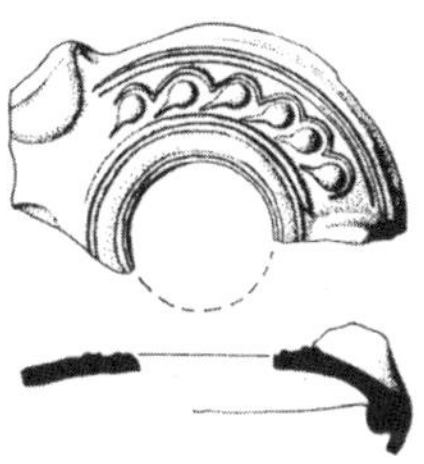

55

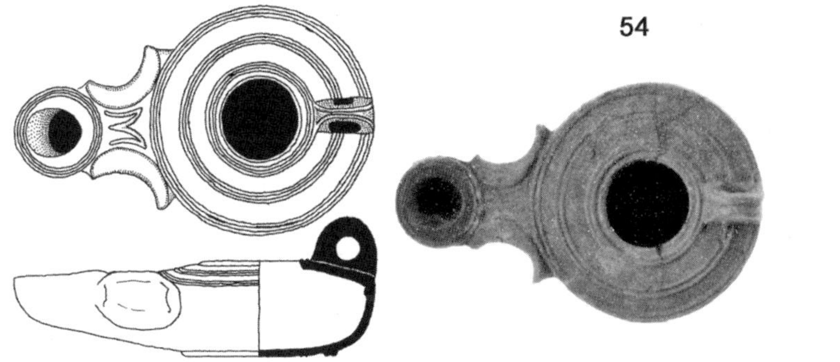

56

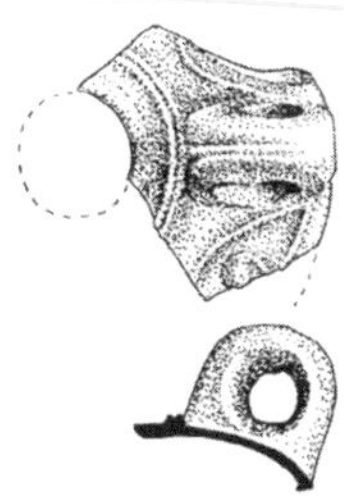

57

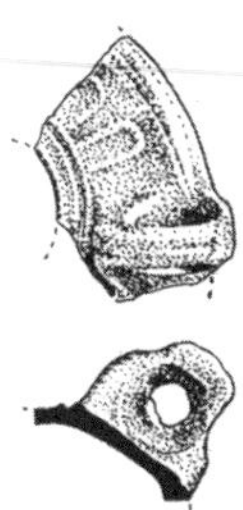

58

59

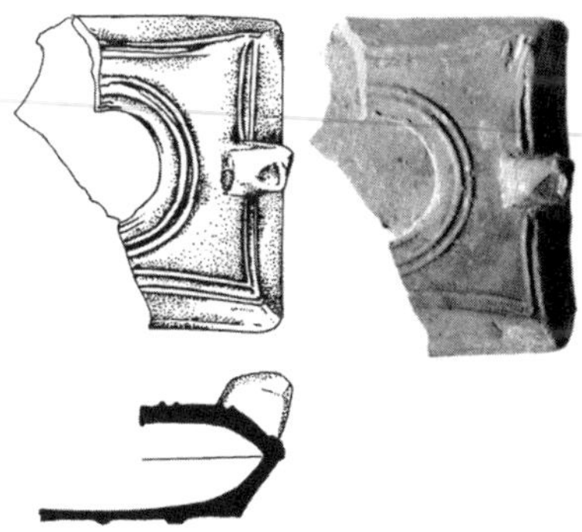

60

0 2cm

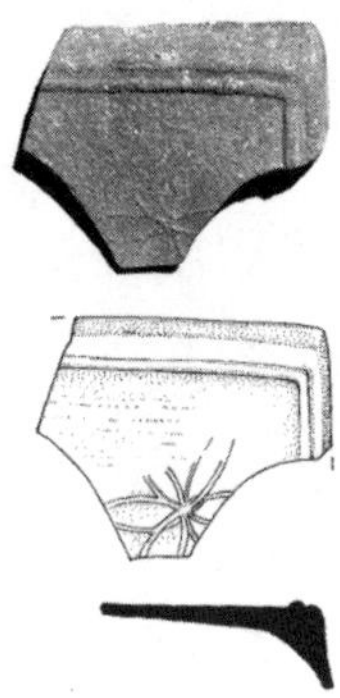
61

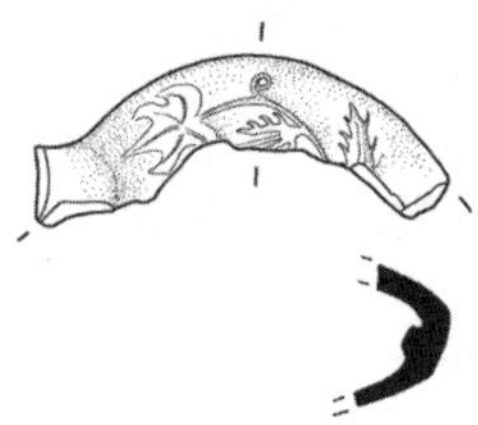
69

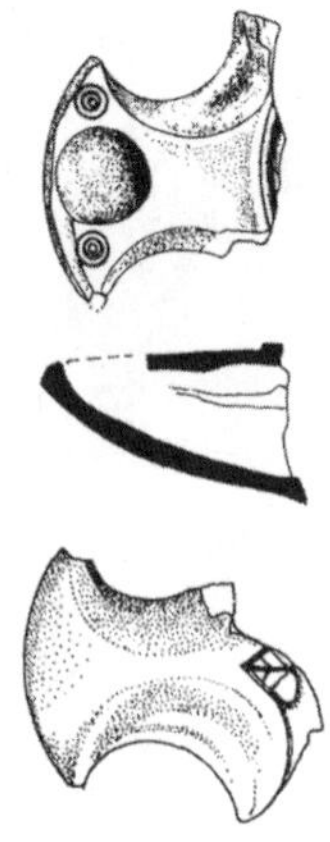
71a

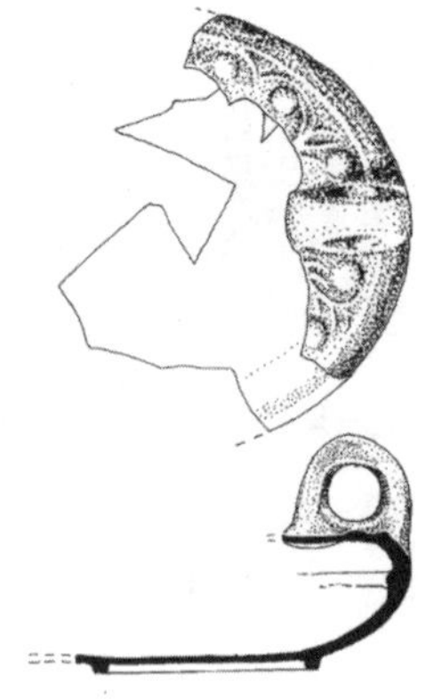
71b

72

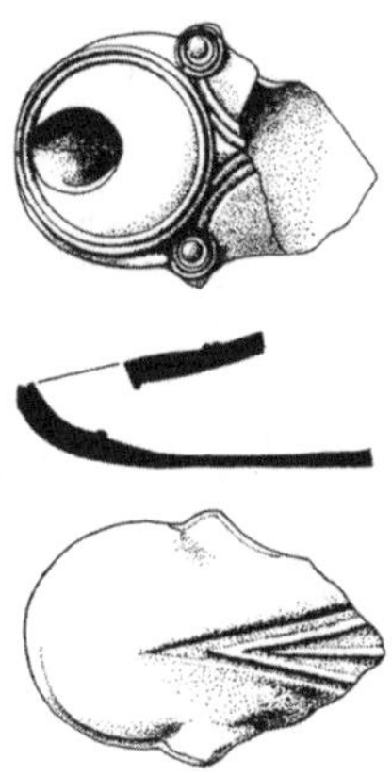
74

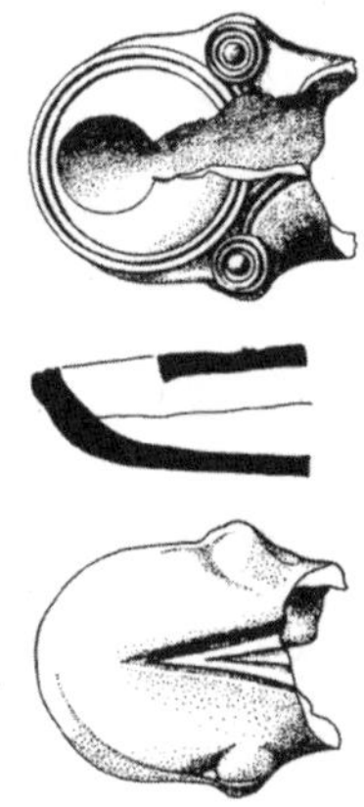
75

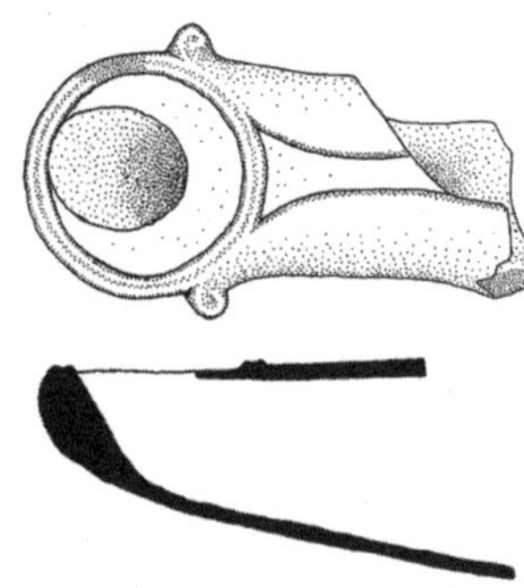
76

77

78

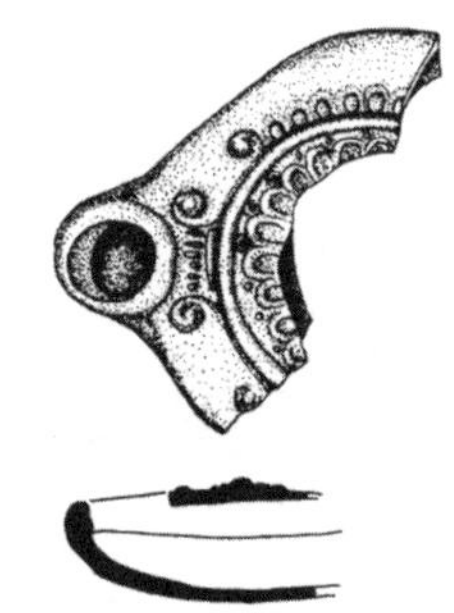
79

0 2cm

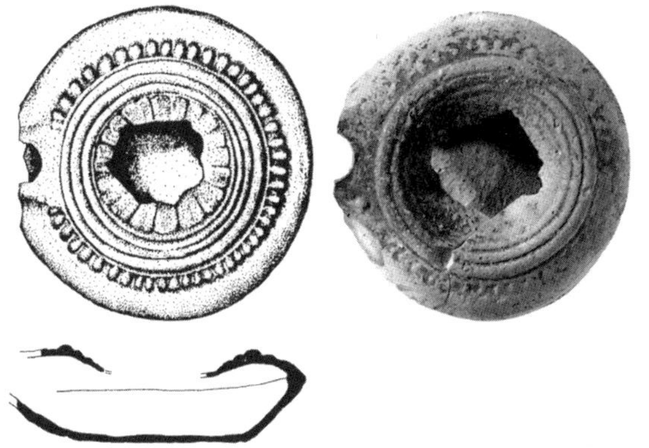

80

81

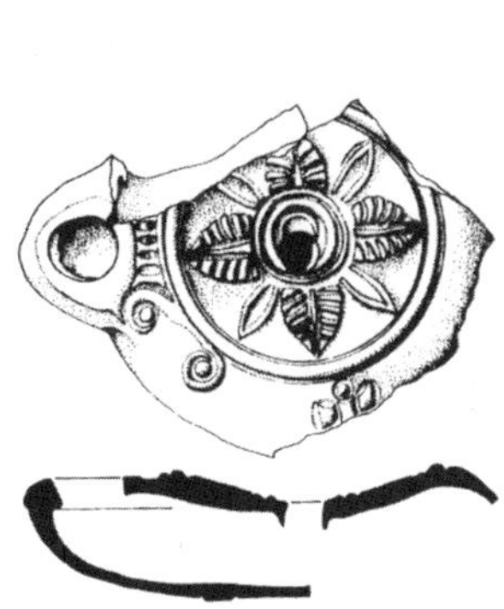

82

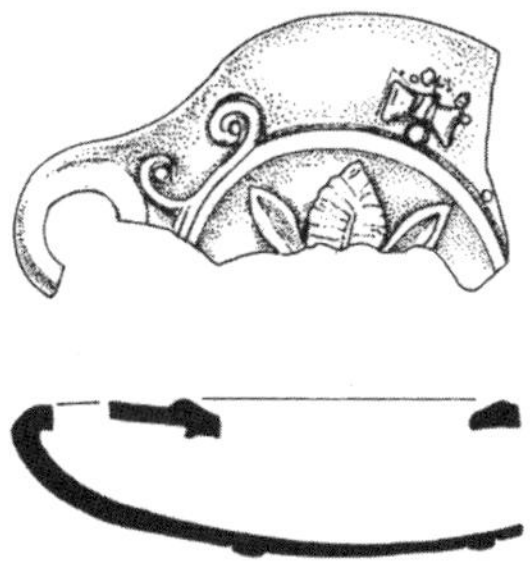

83

85

86

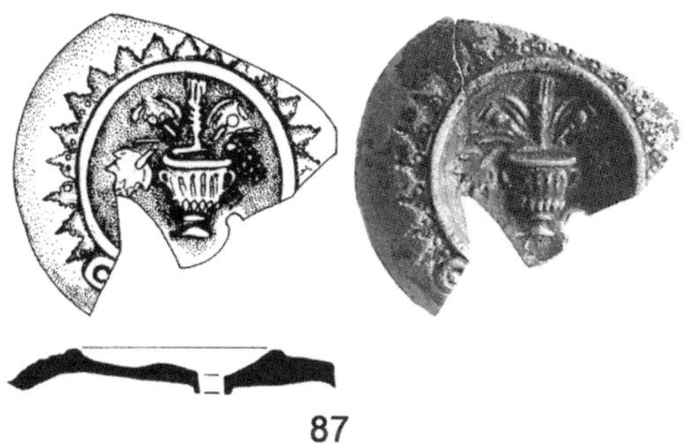

87

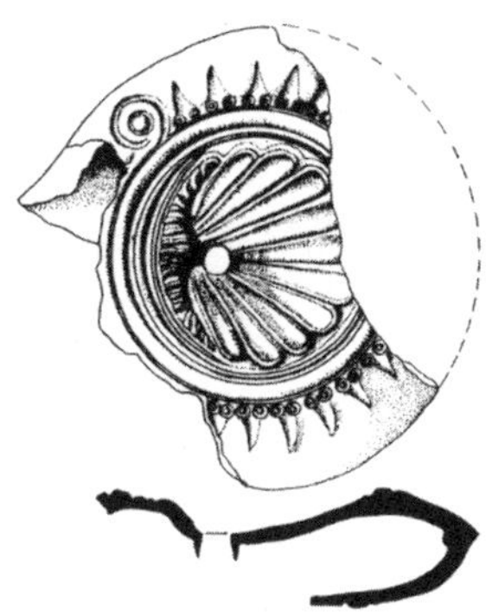

88

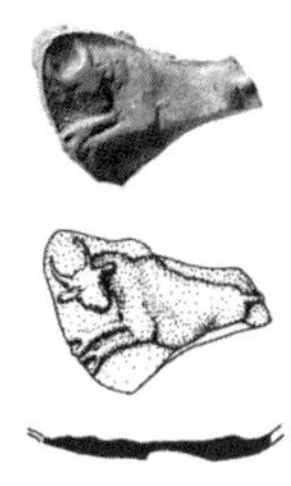

89

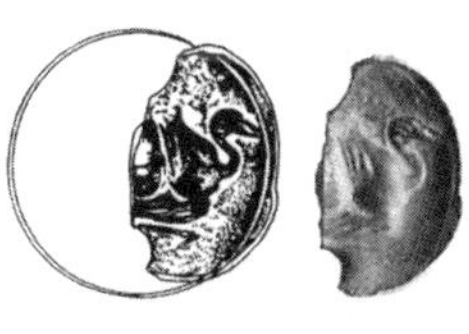

90

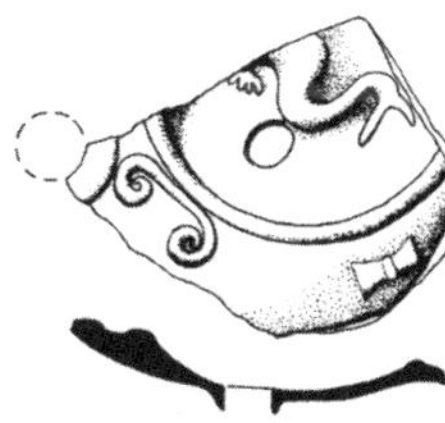

91

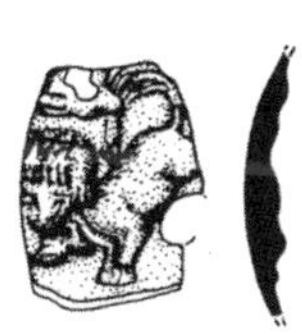

92

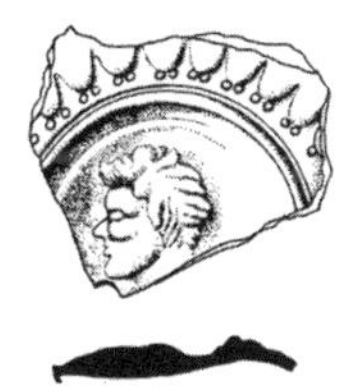

94

96

0 2cm

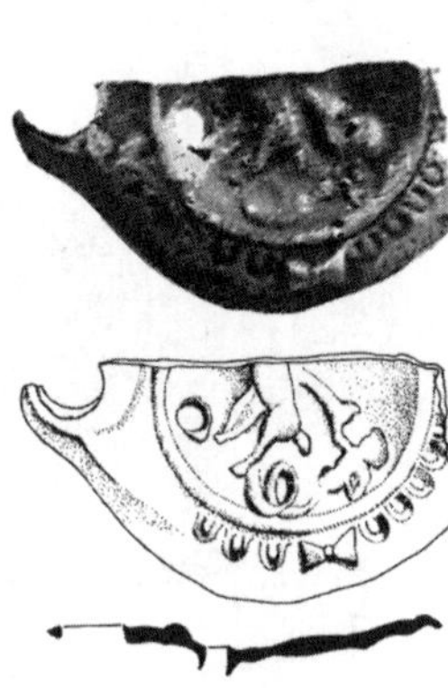

99

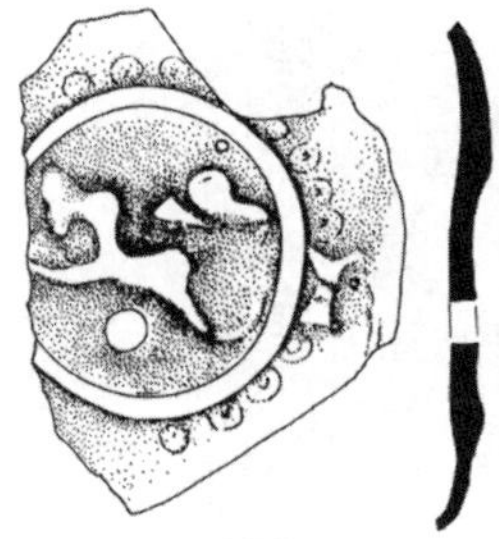

100

101

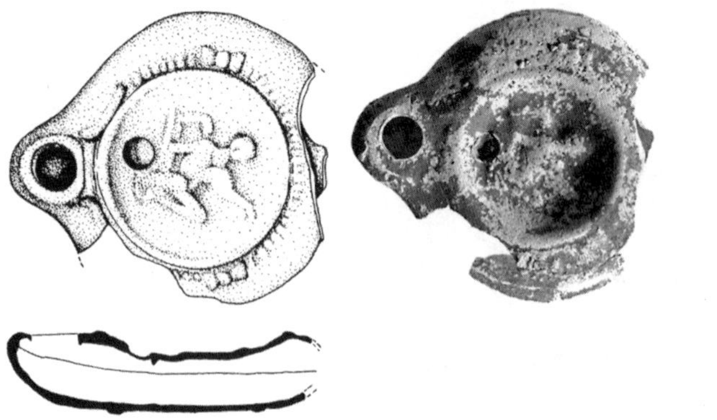

102a

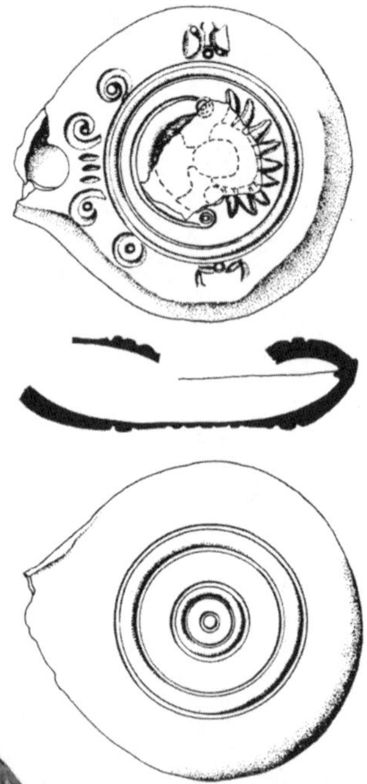

102b

103

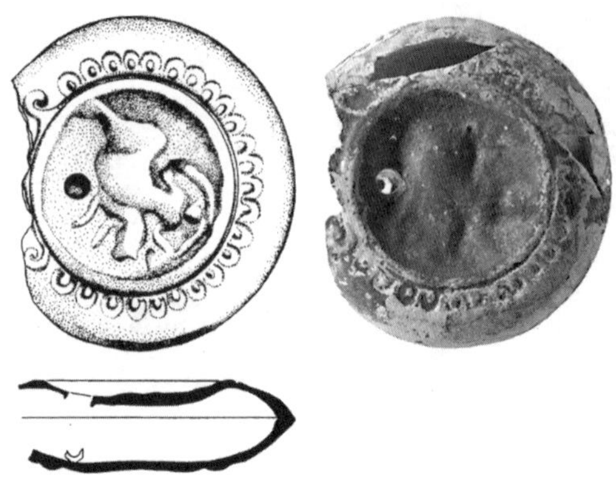

104

105

106

107

108

109

0 2cm

112

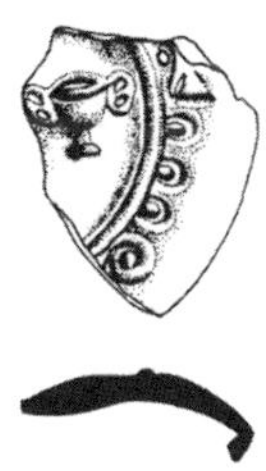

113

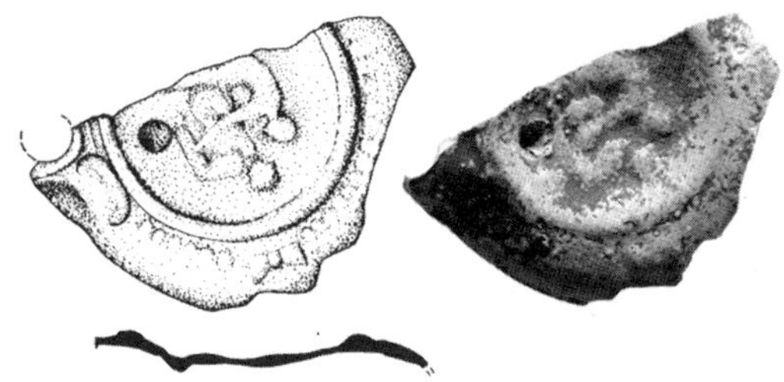

114

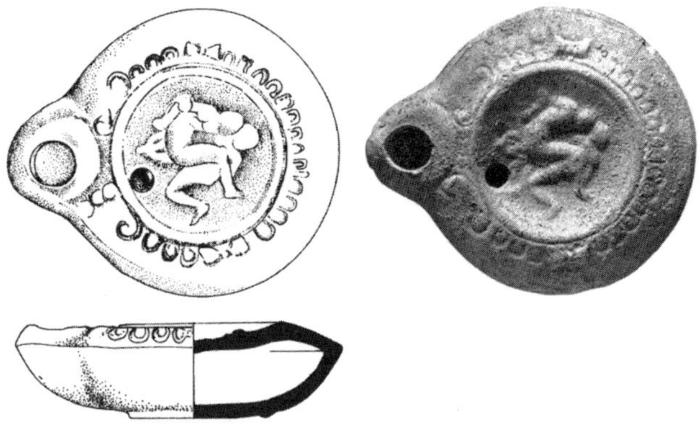

115

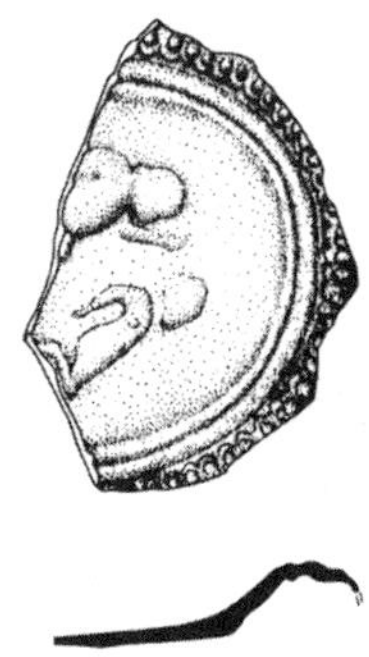

116

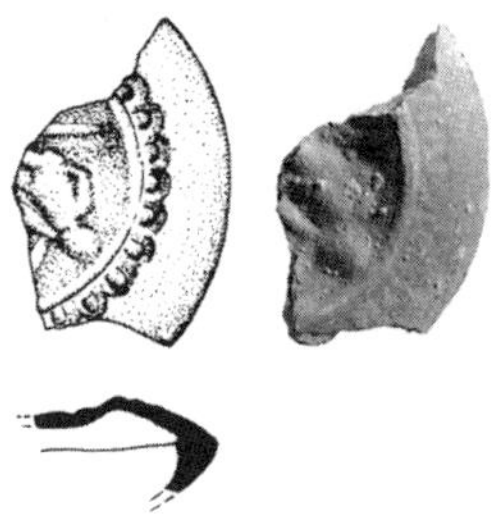

117

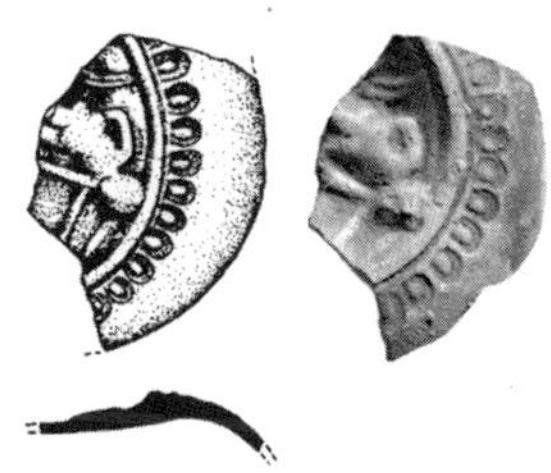

118

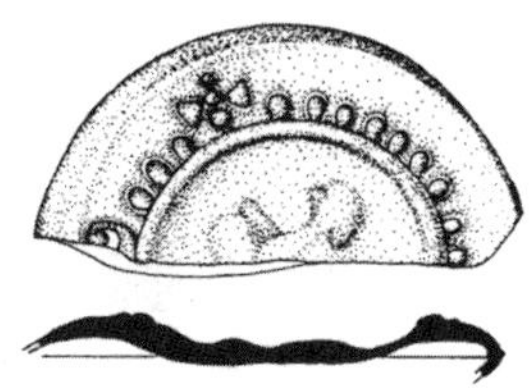

119

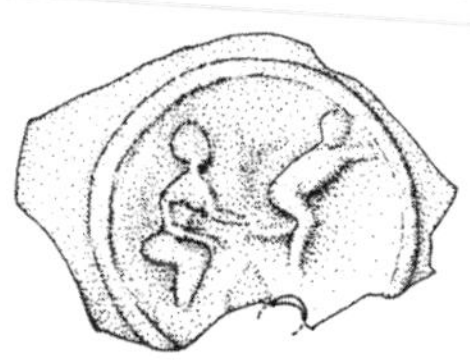

120

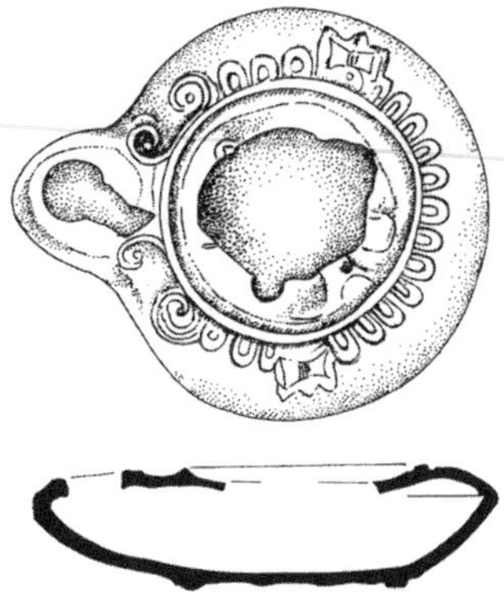

123

0 2cm

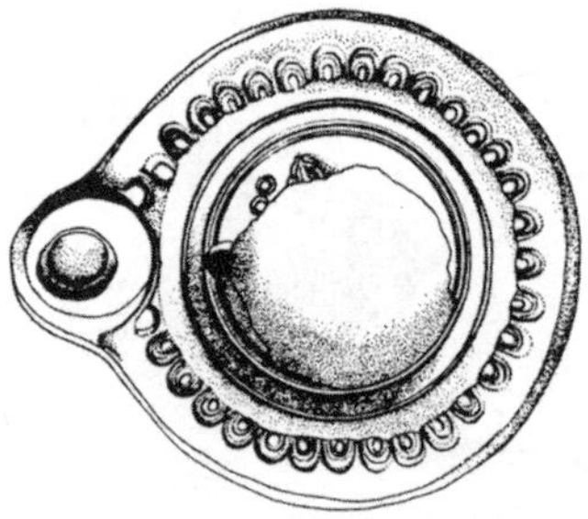

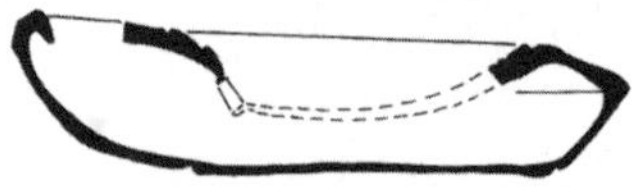

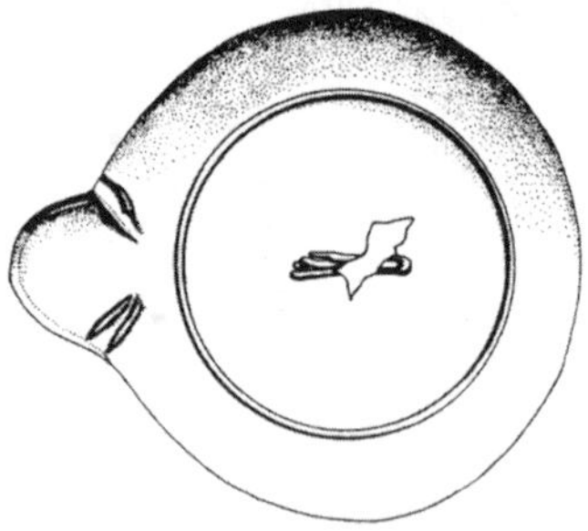

125

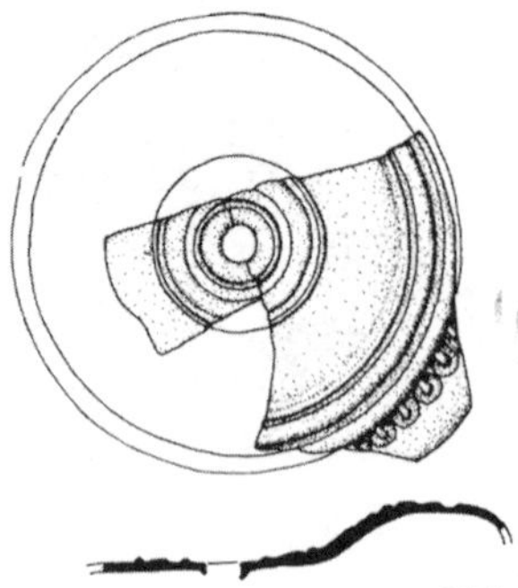

124

126

128

127

129

0 2cm

131

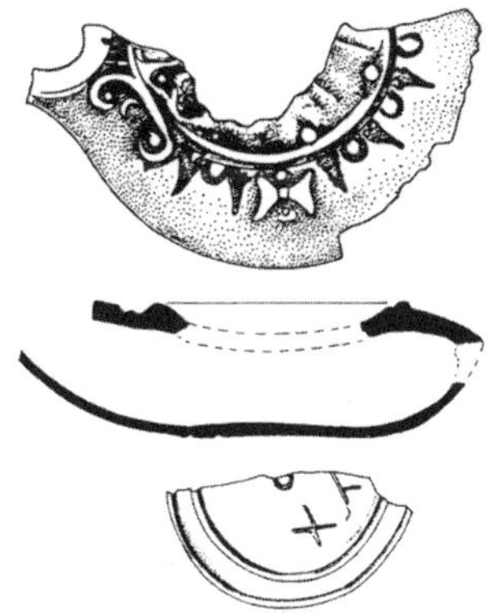

132

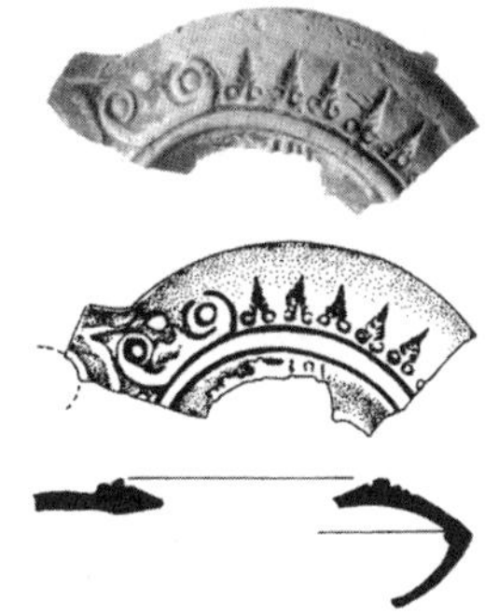

133

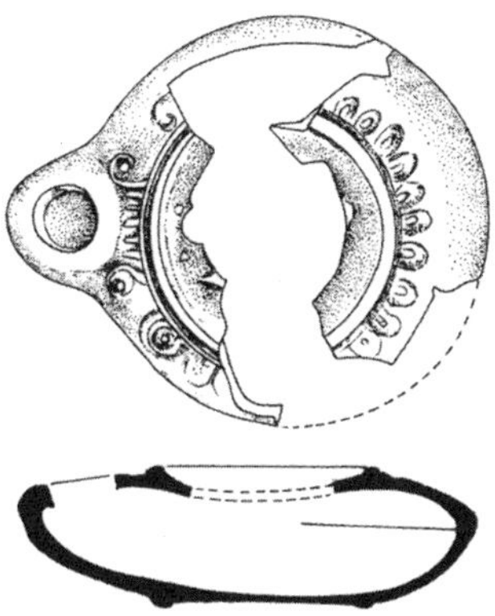

135

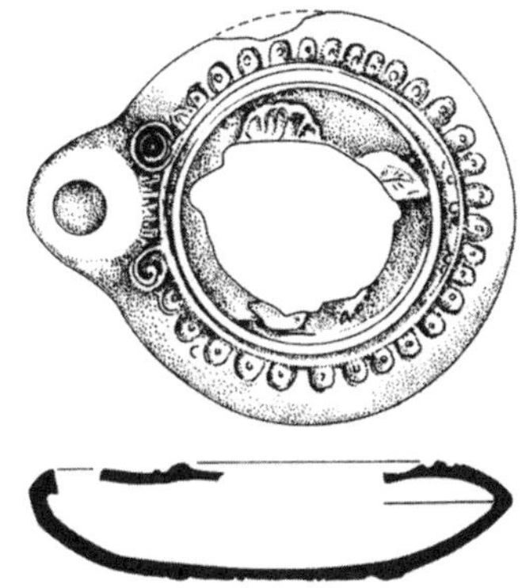

136

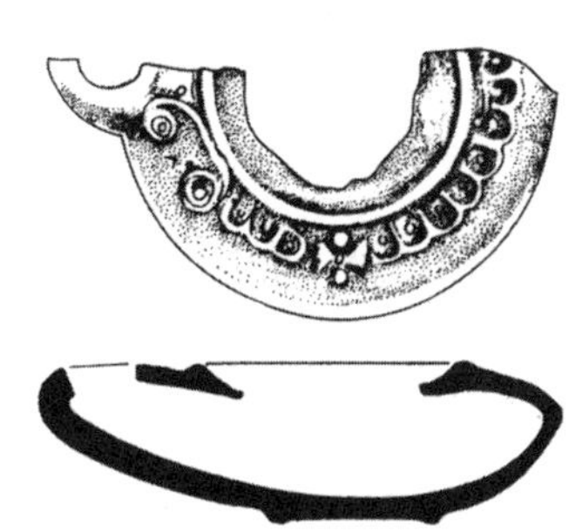

137

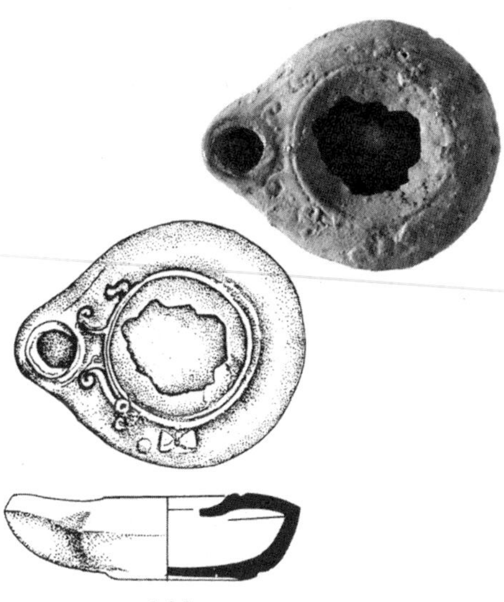

141

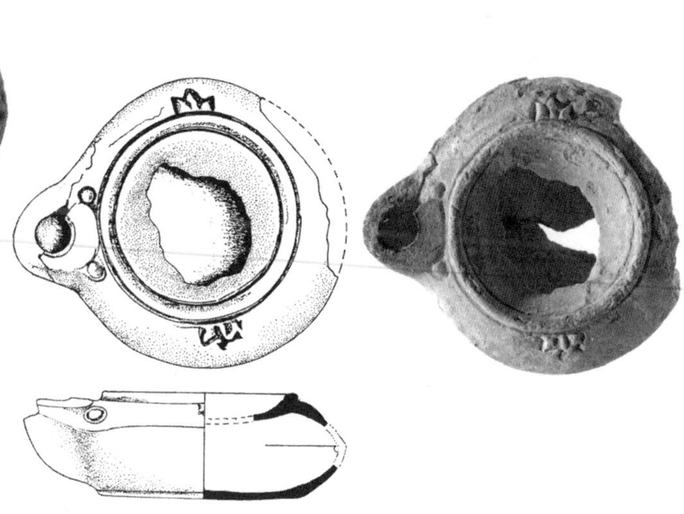

142

0 2cm

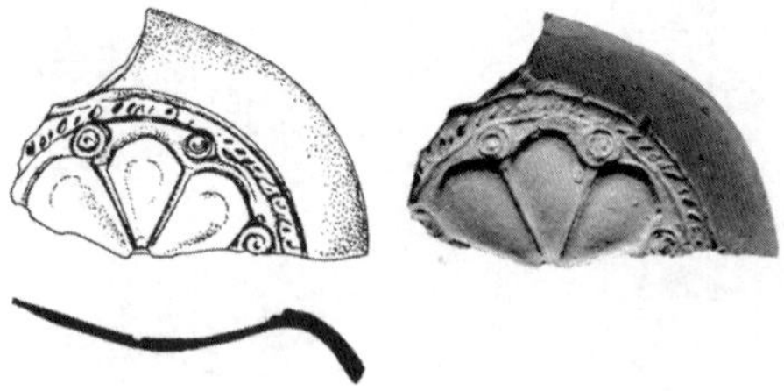

143

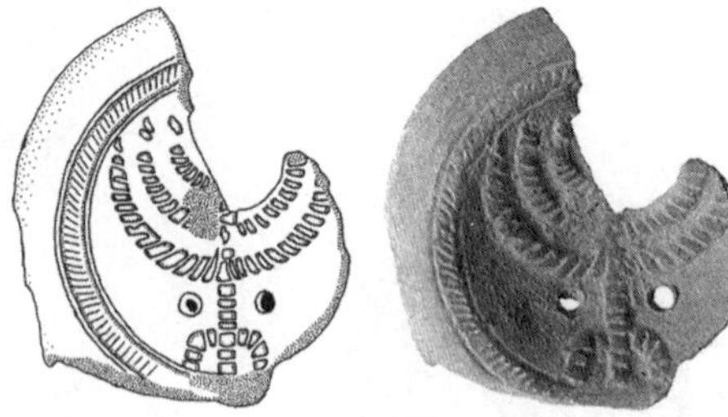

144

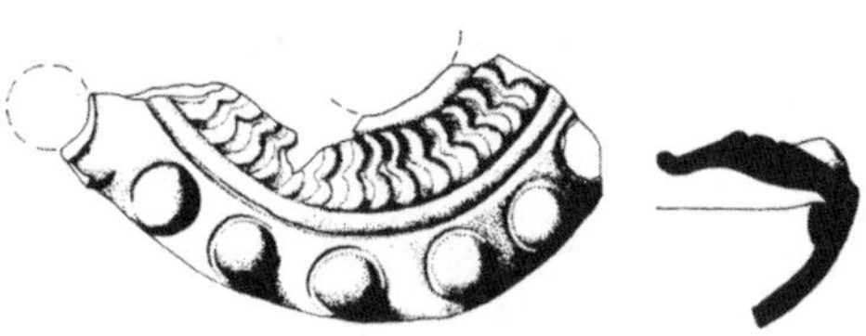

145

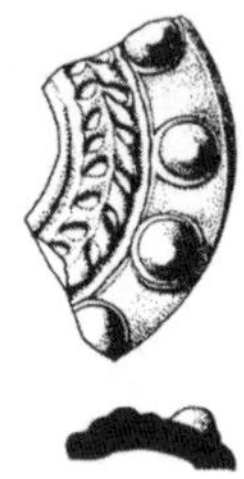

146

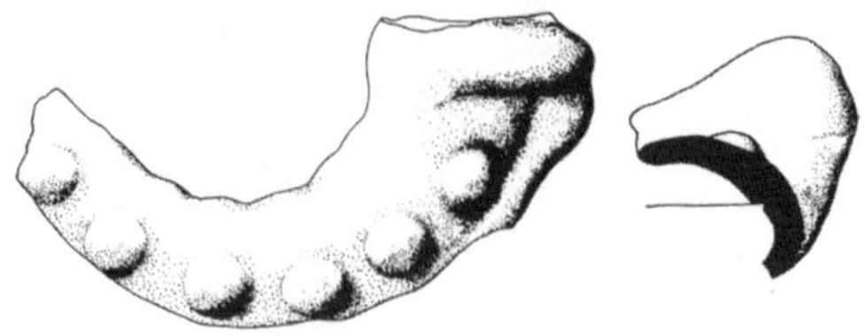

147

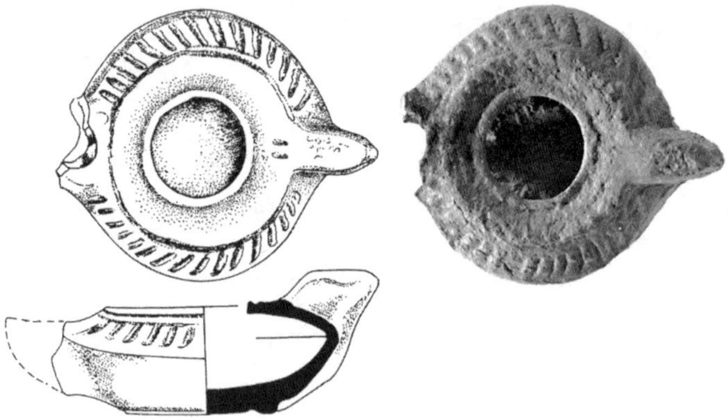

148

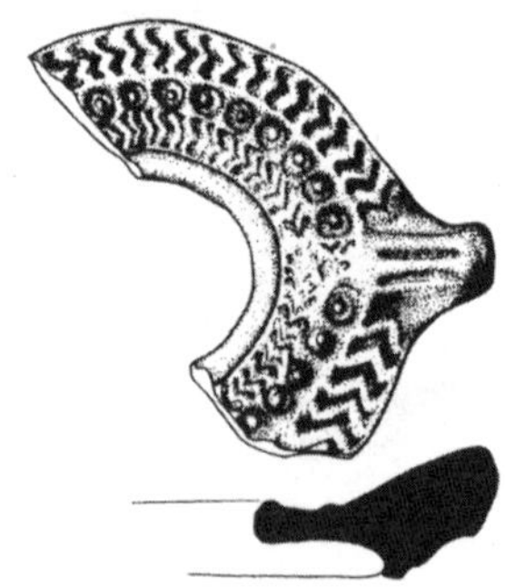

149

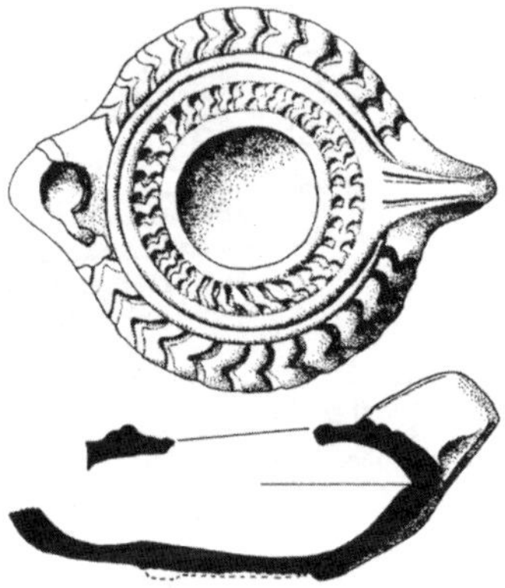

150

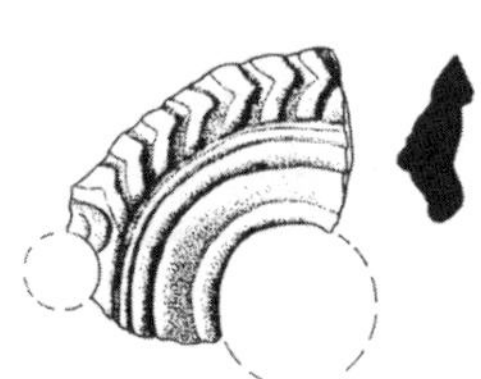

151

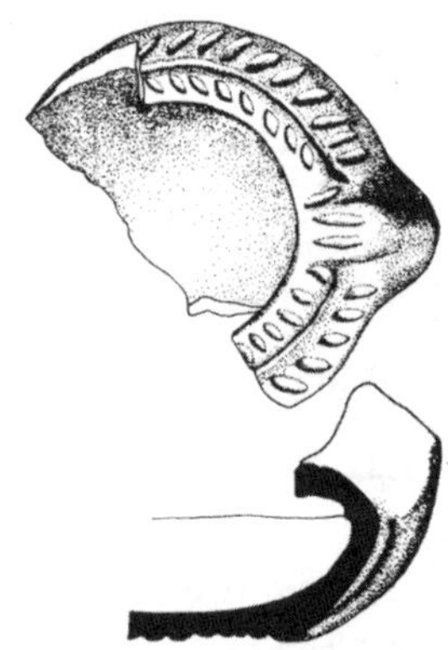

153

0 2cm

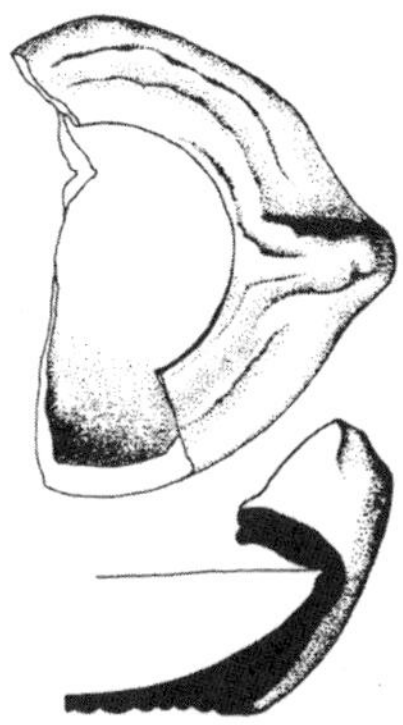

154

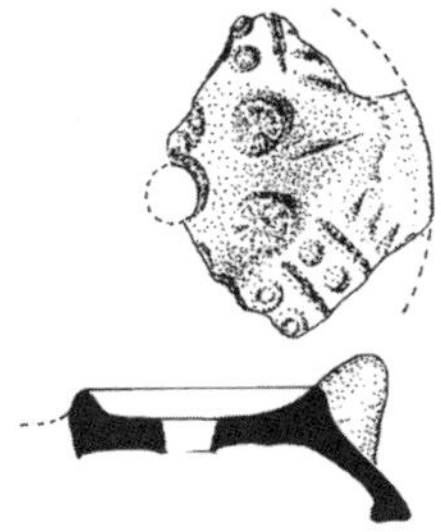

155

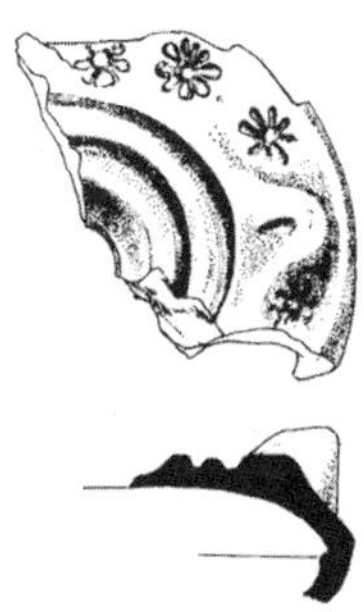

156

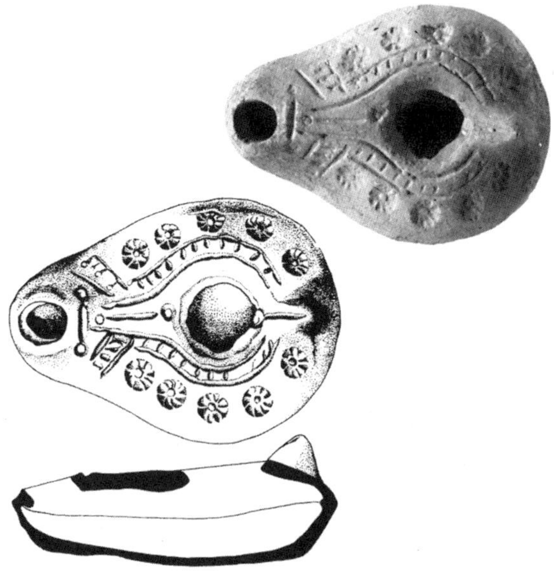

157

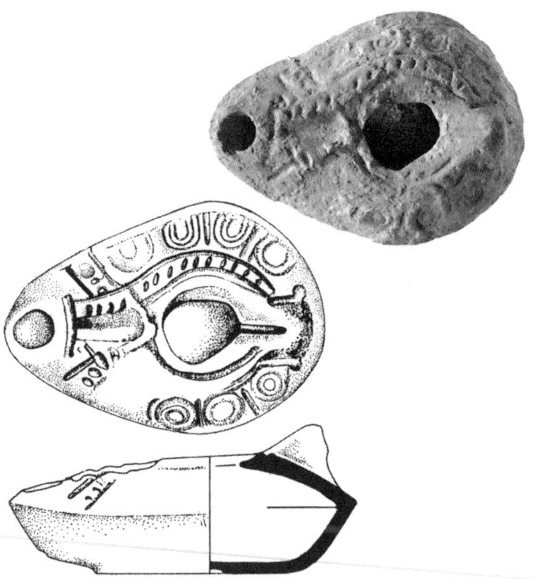

158

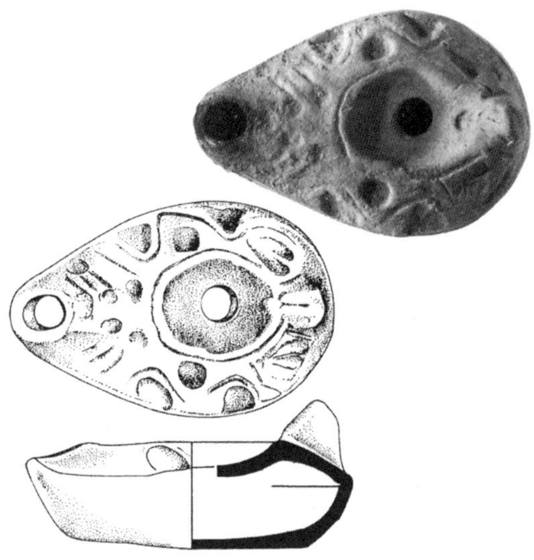

159

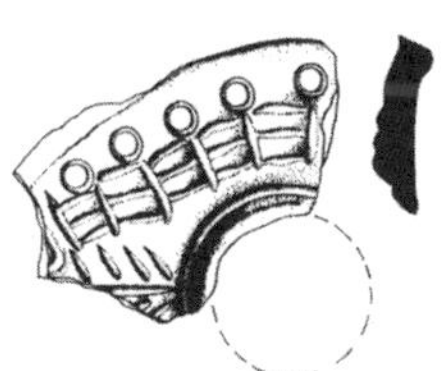

160

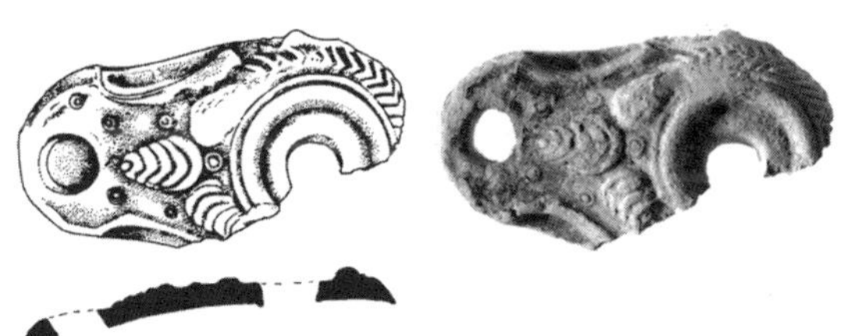

162

0 2cm

163

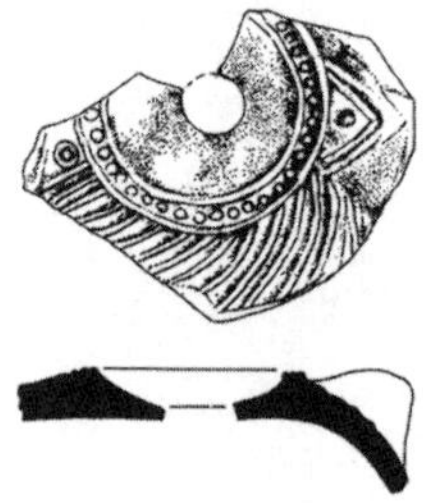

164

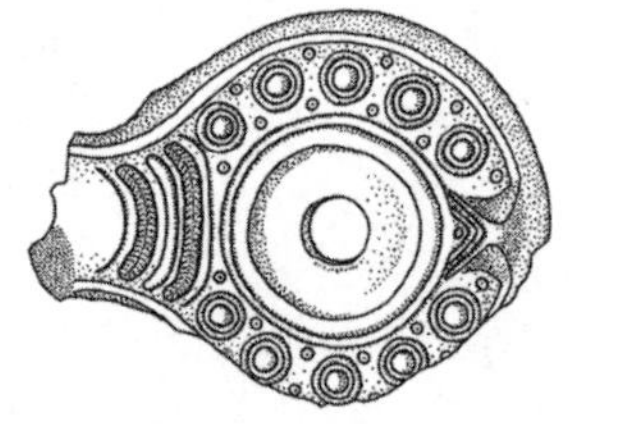

166

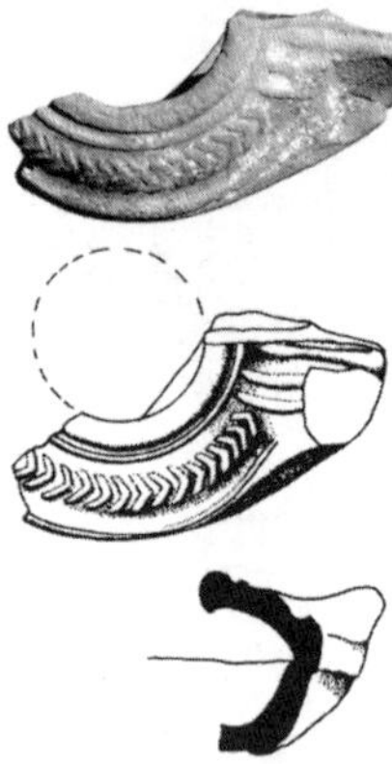

167

168

169

170

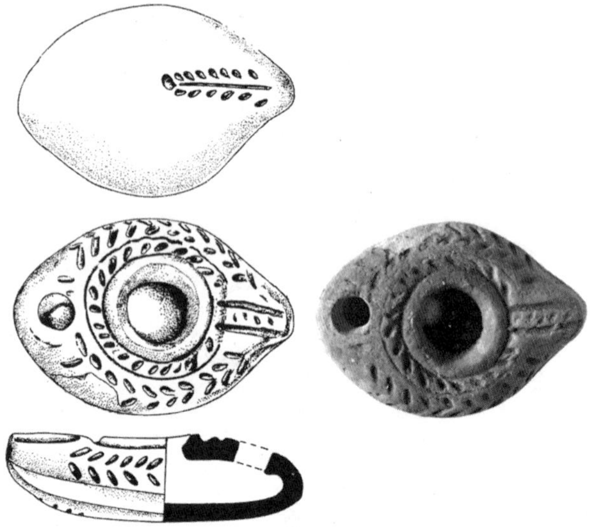

171

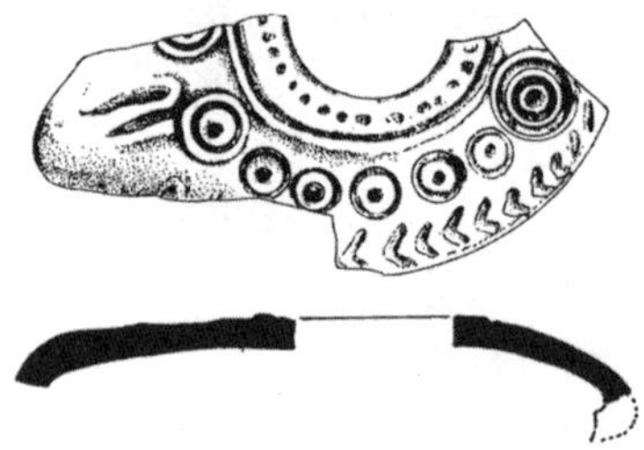

172

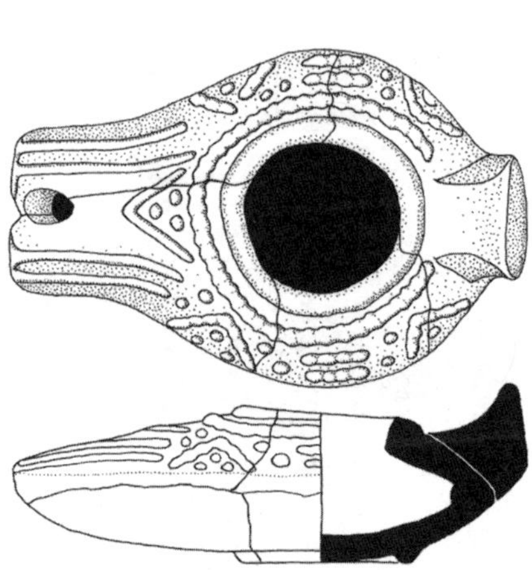

173

0 2cm

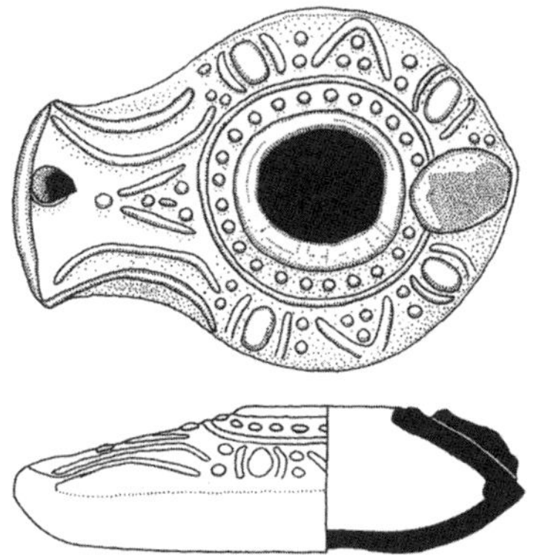

174

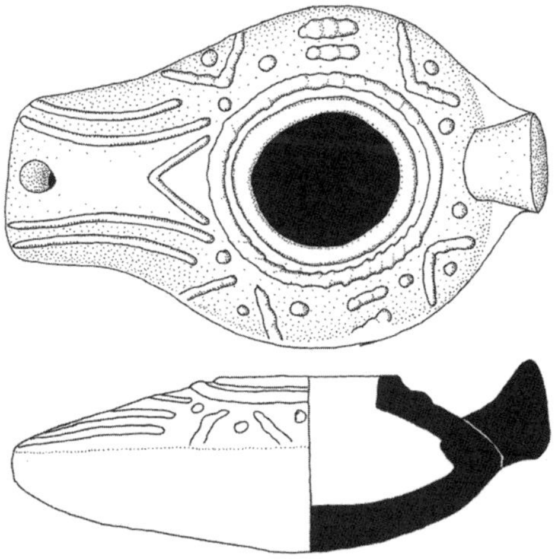

175

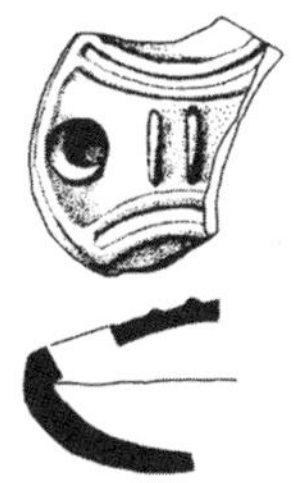

176

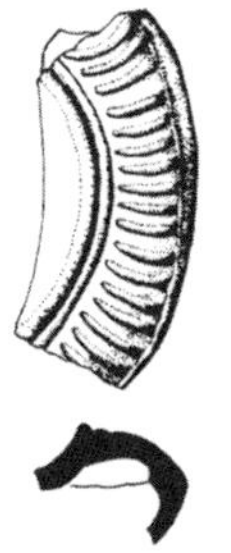

177

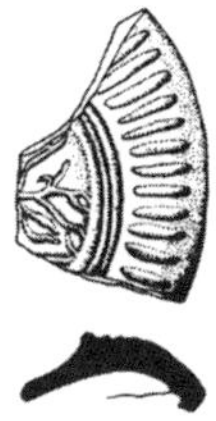

178

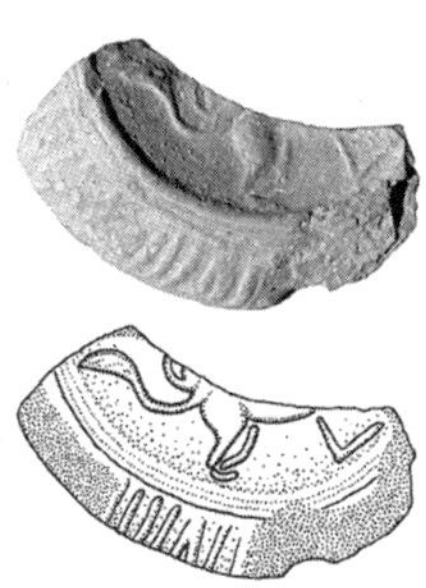

179

180

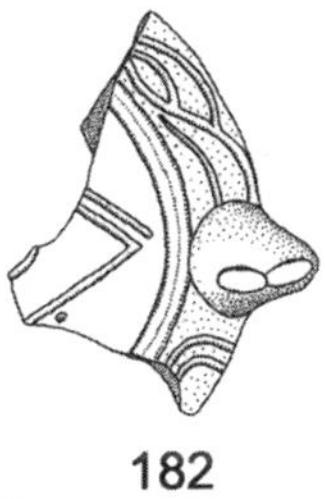

182

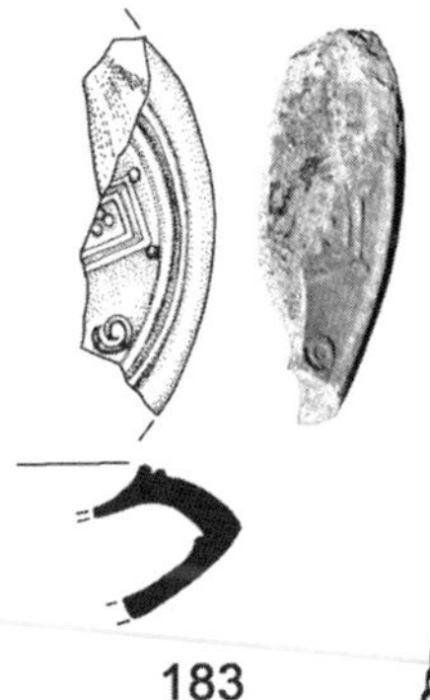

183

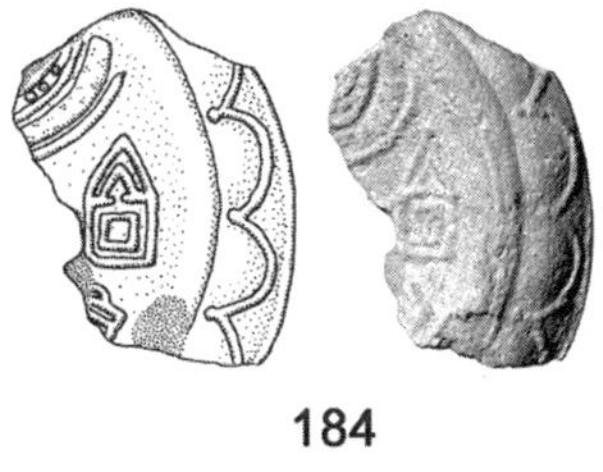

184

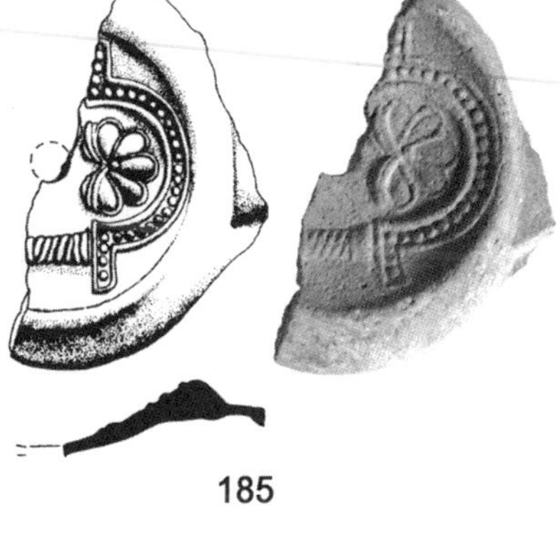

185

0 2cm

186

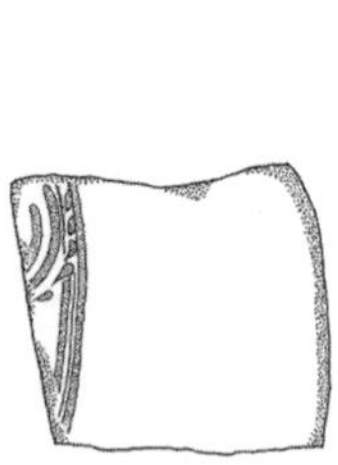
187

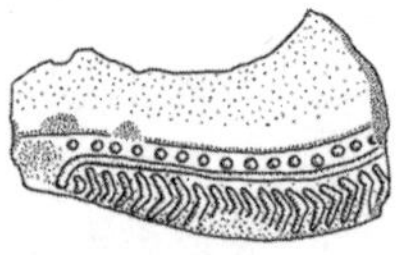
188

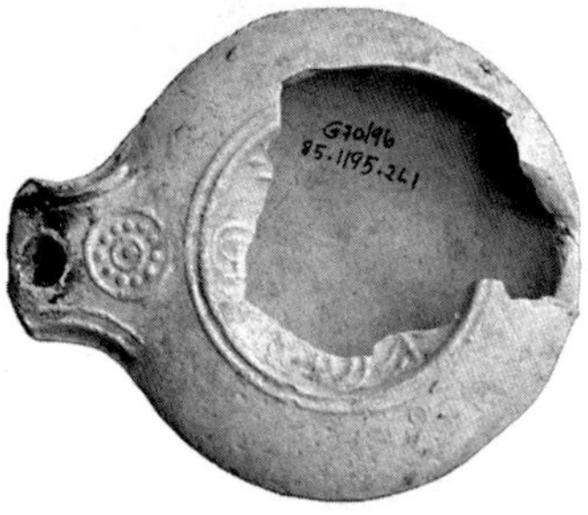
189

190

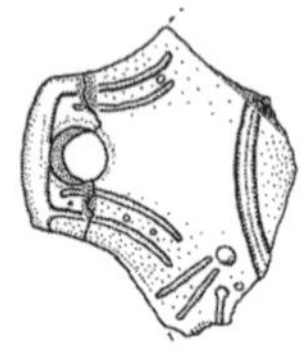
191

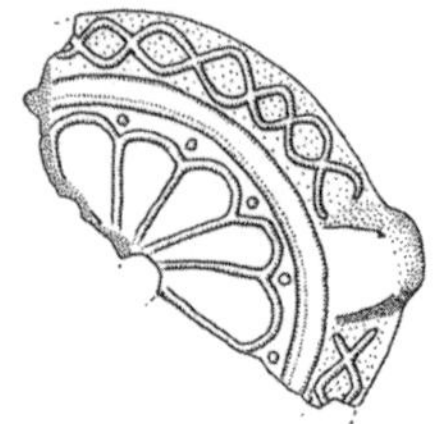
192

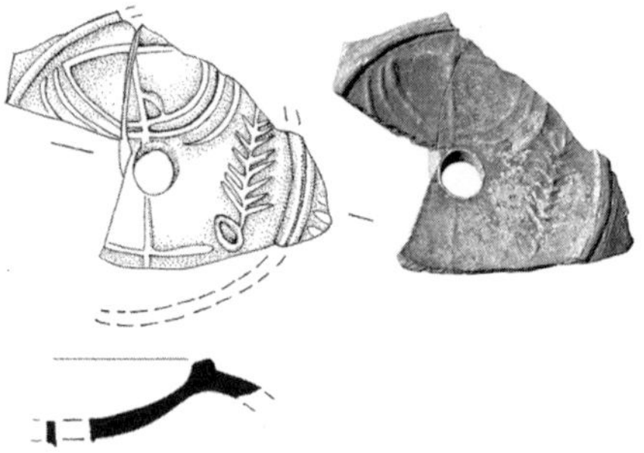
193

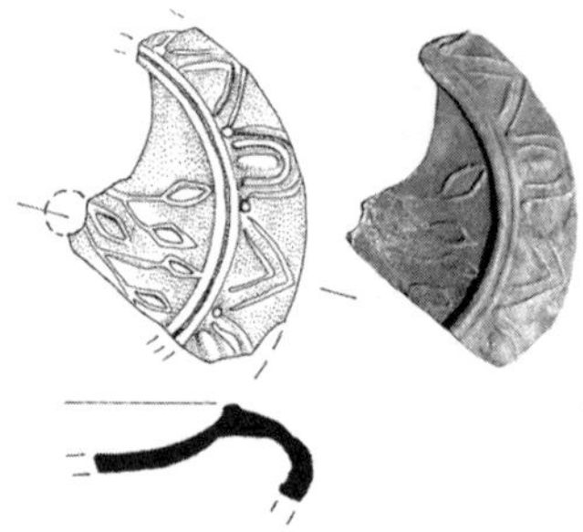
194

195

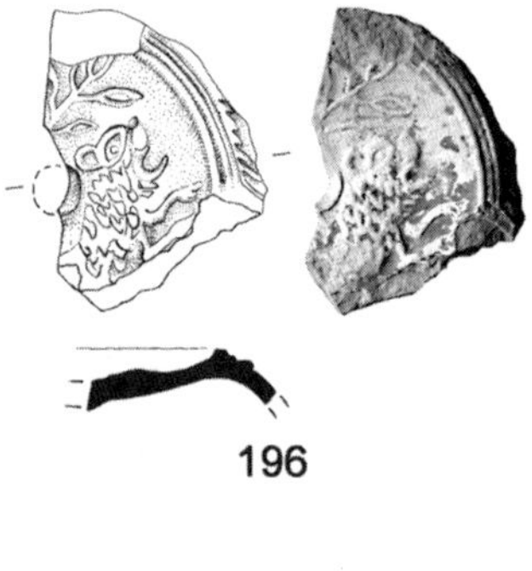
196

209

0 2cm

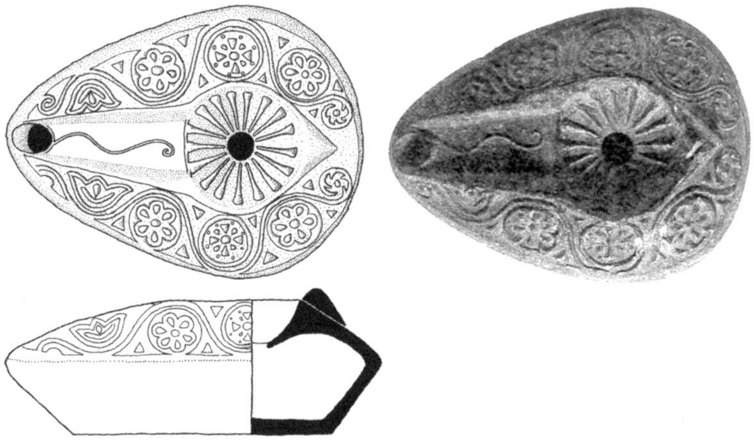

210

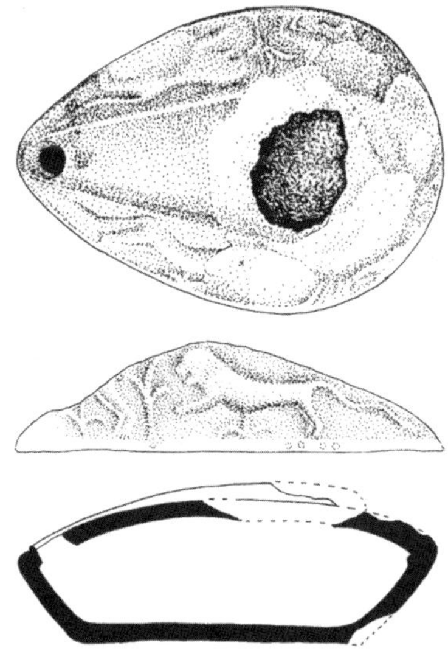

211

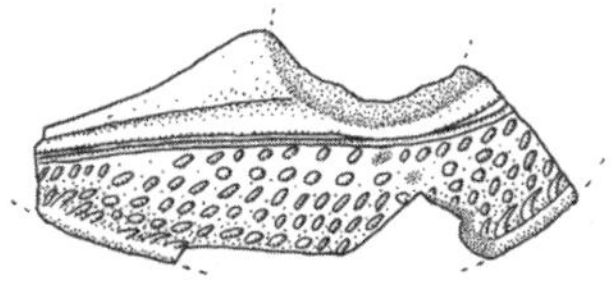

212

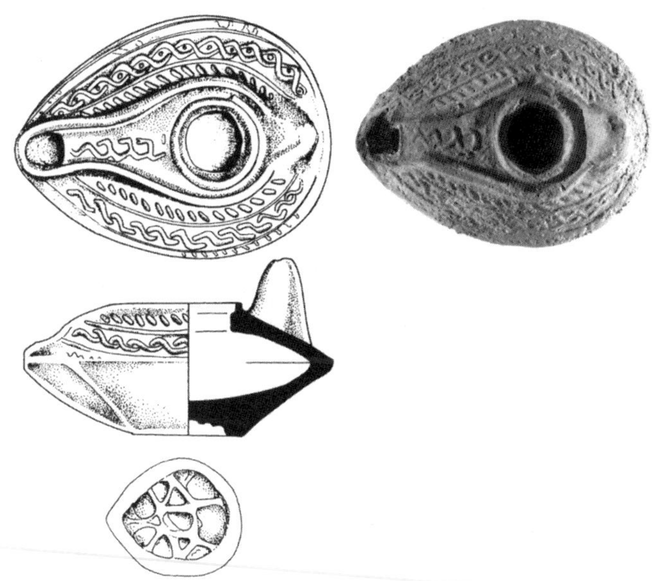

213

0 2cm

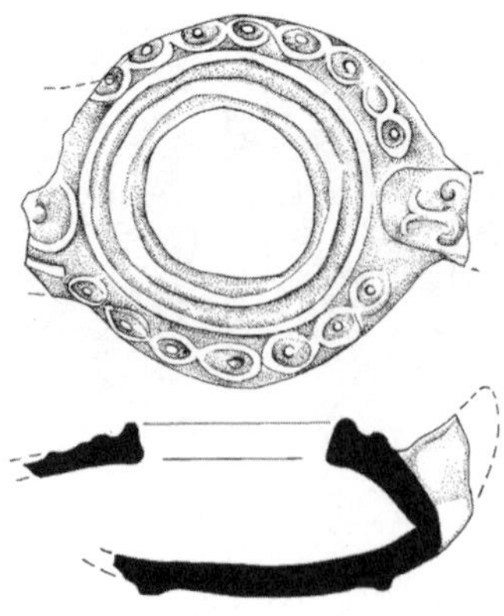

214

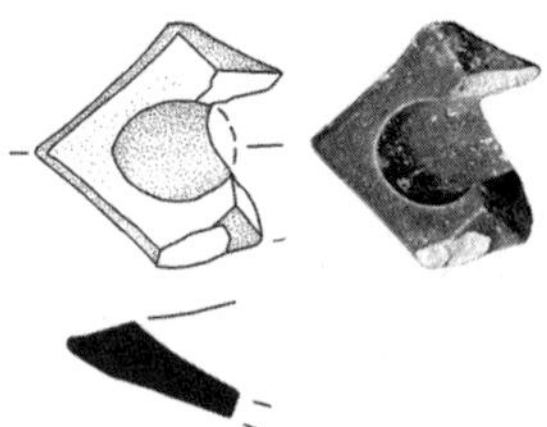

215

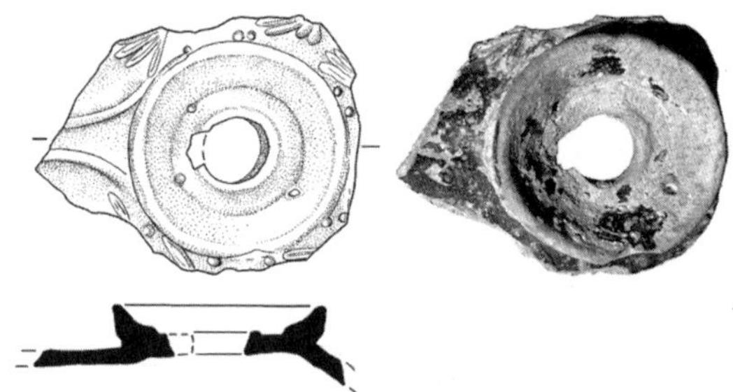

216

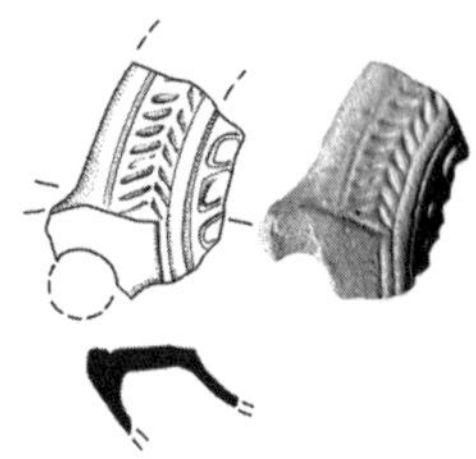

217

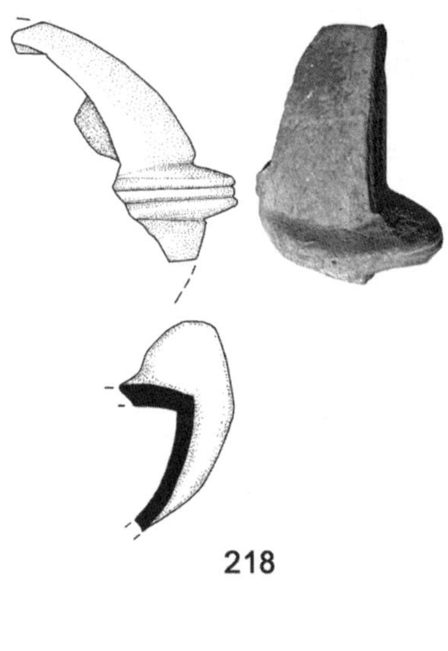

218

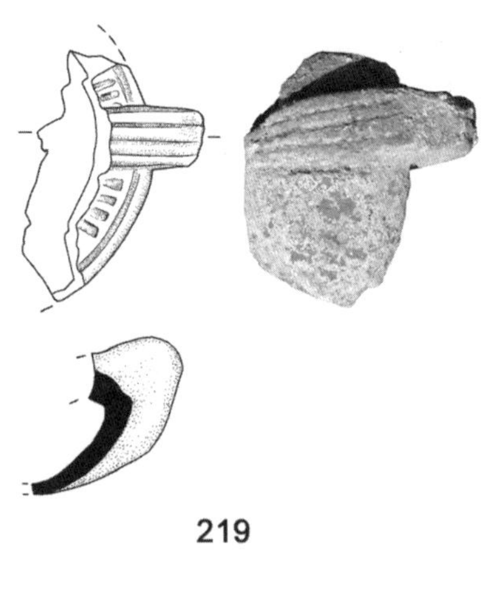

219

0 2cm